Fluvial Dynamics and Cultural Landscape Evolution in the Rio Grande de Nazca Drainage Basin, Southern Peru

Ralf Hesse

BAR International Series 1787
2008

Published in 2016 by
BAR Publishing, Oxford

BAR International Series 1787

Fluvial Dynamics and Cultural Landscape Evolution in the Rio Grande de Nazca Drainage Basin, Southern Peru

ISBN 978 1 4073 0275 1

BAR Publishing is the trading name of British Archaeological Reports (Oxford) Ltd. British Archaeological Reports was first incorporated in 1974 to publish the BAR Series, International and British. In 1992 Hadrian Books Ltd became part of the BAR group. This volume was originally published by Archaeopress in conjunction with British Archaeological Reports (Oxford) Ltd / Hadrian Books Ltd, the Series principal publisher, in 2008. This present volume is published by BAR Publishing, 2016.

Printed in England

BAR
PUBLISHING

BAR titles are available from:

BAR Publishing
122 Banbury Rd, Oxford, OX2 7BP, UK
EMAIL info@barpublishing.com
PHONE +44 (0)1865 310431
FAX +44 (0)1865 316916
www.barpublishing.com

Contents

Appendix

Page 37 is also available as an A3 download from: http://www.barpublishing.com/additional-downloads.html

List of figures

List of tables

Technical notes

This thesis consists of a main part and an extensive appendix containing sketches and detailed interpretations of the investigated sediment profiles as well as graphs showing the results of the laboratory analyses. The appendix is included in print rather than on a CD to enhance accessibility of the data underlying the argumentations made in the main part. A separate detailed table of contents as well as separate lists of figures and tables for the appendix are given on pages A-ii to A-viii. References cited in the appendix are included in the list of references for the main part of the thesis (pages 89–98).

Calibration of radiocarbon dates was performed using OxCal Version 3.10 (Bronk Ramsey 1995, 2001) and southern hemisphere correction (shcal04.14c, McCormac et al. 2004). All dates within the range of southern hemisphere radiocarbon calibration are reported as cal BP (referring to 1950 as present). Throughout this thesis, calibrated age ranges are given as the 2σ-range. For comparability, AD/BC dates from the literature have been converted to cal BP. Calibrated dates from the literature are used as they are given. Uncalibrated dates taken from the literature are calibrated identically to those aquired in this project.

Geographic coordinates are given in UTM (WGS-84, southern hemisphere, zone 18).

Remarks on terminology

The term *Palpa Valley* as it is used in this thesis refers to the common valley of Rio Palpa and Rio Vizcas where the present-day town of Palpa is situated. To refer to the separate valleys upstream, the terms Rio Palpa valley and Rio Vizcas valley are used. For place names, Spanish or local terms are preferred over English ones (e.g. Cordillera de la Costa, Cordillera Occidental, Pampa San Ignacio). Regarding the name Nasca/Nazca, I follow the spelling proposed by Aveni (1990: 3), i.e. referring to *Nazca* as the place and *Nasca* as the culture.

When referring to the archaeological sequence and pre-Columbian cultures I follow the terminology in the literature used, in particular Unkel (2006), Reindel et al. (2001) as well as Silverman (2002) and Silverman & Proulx (2002). While Nasca ceramic chronology has long been well established (Silverman & Proulx 2002: 25–37), as of 2002, absolute chronology of the Nasca phases had been rather poorly defined (ibid.: 37ff.). The dissertation by Unkel (2006) greatly improves this situation in particular for the Palpa Valley. In this thesis I consistently use the chronology of pre-Columbian cultures presented by Unkel (2006: 111). For the purpose of comparability, all data are converted to calendar years before present (cal BP) and reproduced in Table 2 (page 22).

It has to be noted that while the term *Andes* in the context of physical geography is applied to the mountain belt constituted by the Cordillera Occidental and Cordillera Oriental, the term *Andean* in the archaeological literature is commonly used to refer to cultures or societies in the coastal desert as well as in the Andes. In this context, *Andean* principally serves to distinguish both coastal and highland societies from the eastern lowland (rainforest) societies. As stated by Stanish (2001: 55), "[t]he Andean cultural area [is] defined conservatively as the limits of the Inca state in 1532." On the other hand, the term *coastal* as it is used in the archaeological literature on Peru generally refers to cultures of the desert on the western slope of the Andes irrespective of their actual distance from or connection with the coast as such. Therefore, despite its location at only ca. 300 m asl and at a distance of more than 50 km from the sea, pre-Columbian cultures in the Palpa Valley are *costal* and *Andean* cultures in the sense of the archaeological literature.

The terms *clay, silt, sand, gravel* and *boulders* are used to refer to grain-size classes only. Where shape parameters (angular/subrounded/rounded) were recorded, they are explicitly given. Grain size designation for *clay* (<2 μm), *silt* (2–63 μm) and *sand* (63–2,000 μm) follows AG Boden (2005). *Gravel* (2–200 mm) is distinguished from *boulders* (>200 mm) (cf. Selley 1988: 40).

Abbreviations

BP – before present (refers to the year 1950)
cal BP – calendar years before present (calibrated ^{14}C dates, refers to the year 1950)
CSR – clay: silt ratio
DTM – digital terrain model
ENSO – El Niño–Southern Oscillation
IRSL – Infrared stimulated luminescence
ITCZ – Intertropical Convergence Zone
ka – thousands of years
LGM – Last Glacial Maximum
LoI – Loss on ignition
Ma – millions of years
m asl – metres above sea level
mm/a – millimetres per year
OSH – Outer Shelf High
OSL – Optically stimulated luminescence
SASM – South American Summer Monsoon
SEPA – South-East Pacific Anticyclone
SOI – Southern Oscillation Index
SSTA – Sea Surface Temperature Anomaly

Acknowledgments

I wish to thank Dr. Jussi Baade for involving me in a truly interesting research project, for ideas, suggestions and discussions, for the freedom to explore and for the hard but fascinating fieldwork we shared. The project "Jungholozäne fluviale Dynamik und Kulturlandschaftsentwicklung am Rio Grande de Nazca, Süd-Peru" was made possible by a grant (BA1377/6-1,6-2) from the German Research Foundation (DFG). Stephanie Kastner and Susanne Angelstein took part in the fieldwork; Stephanie Kastner wrote a diploma thesis on the investigations at La Muña.

I am grateful to Don Oscar Tijero, Johny Isla Cuadrado as well as Carmen, Alfonso and Miguel Tijero for their hospitality in Palpa. I thank Johny Isla Cuadrado, Dr. Markus Reindel, José Palomino and Judith Astuhuaman Blanco for logistical support during the fieldwork and assistance regarding archaeological questions. Special thanks go to Johny, Martha and Chucho for making Hanaq Pacha feel like home and to Judith for teaching me spanish. The many helpers during the fieldwork are thanked for the hard and committed work in the heat.
Dr. Markus Reindel provided aerial photographs of the Palpa Valley; Dr. Martin Sauerbier and Prof. Dr. Armin Grün of the Institute of Geodesy and Photogrammetry at the ETH Zurich made available high-resolution DTM data.

Laboratory work was carried out in the geo-ecological laboratory (Institute of Geography, Friedrich Schiller University Jena) led by Dr. Gerhard Daut. AAS measurements reported in this thesis were performed by Brunhilde Dressler. For the sometimes tedious work of grain size, LoI and carbonate analysis I wish to thank Stefanie Kastner, Julia Baumert, Anja Thielemann, Nadine Binder and Stephan Sonntag. Radiocarbon measurements were performed by the Poznan Radiocarbon Laboratory led by Dr. Tomasz Goslar. Rosemarie Mendler is thanked for producing the final drawings of the sediment profiles.

I am very thankful to Manja Lohse for proofreading. Many thanks also go to Adriana and Oliver Fügen for corrections of the spanish version of the synopsis and to Judith Astuhuaman Blanco whose corrections reached me mere minutes before the deadline. Kathrin Sander, Brunhilde Dressler, Manja Lohse, Annette Kluge, Marie Koch, Jana and Gabriel Welsche, Judith Astuhuaman Blanco, Anja Wittich and David Wartschinski are thanked for their encouragement through all the highs and lows of writing this thesis.

Last but not least my thanks go to my parents Waltraud and Josef Hesse for their patience and support through all my long years of studies.

Writing this thesis would not have been possible without the help of all these and many other people who in one way or another contributed to this work.

1. Introduction

Relationships between environmental parameters and cultural development have been postulated by a number of authors (e.g. YOFFEE 1988; ROBERTS 1998; DEMENOCAL 2001; HODELL et al. 2001; HAUG et al. 2003; DILLEHAY et al. 2004; BROOKS 2004; BROOKS 2006). These relationships are thought to be particularly relevant in arid regions like coastal southern Peru, where irrigation agriculture based on allogenic river discharge has been the subsistence base for human societies for at least 3,000 years (SILVERMAN 2002; REINDEL et al. 2002). This was the background for the research project "Late Holocene fluvial dynamics and cultural landscape evolution in the Rio Grande de Nazca drainage basin, southern Peru" (BAADE 2001), of which this thesis is an outcome.

A large number of publications deals with human-environment interactions in coastal Peru during pre-Columbian times. Observations of environmental conditions and apparent temporal coincidences of environmental and cultural changes have stimulated interpretations in terms of natural causality (LANNING 1967: 175; PAULSEN 1976: 126), circumscribed (CARNEIRO 1977: 9ff.), constrained or contextual development (WELLS & NOLLER 1999: 784; HABERLE & CHEPSTOW-LUSTY 2000: 350), adaptation and mitigation (DILLEHAY et al. 2004: 279; MOSELEY 1992; SILVERMAN 2002: 24), "punctuated equilibrium" (WELLS & NOLLER 1999: 782, after Moseley 1987), "agrarian collapse" (MOSELEY 1983) and "convergent catastrophes" (MOSELEY 1999). Impacts of long-term changes and short-term extreme events on pre-Columbian cultures have been suggested for the Rio Grande de Nazca drainage by GRODZICKI (1990), SILVERMAN (2002: 24f.) and REINDEL et al. (2003: 127).

The lower Rio Grande de Nazca drainage basin is situated in the hyperarid southern Peruvian coastal desert. It comprises several interconnected river oases constituted by the valleys of the Rio Grande and its major tributaries which have been the location of long-term human settlement and a succession of pre-Columbian cultures, most notably the Nasca culture. Of particular interest under hygrically limited conditions are irrigation systems which in the case of oases watered by allogenic rivers may comprise the entire agricultural landscape. Agricultural landscapes as such "are often very fragile and ephemeral remains in the archaeological record because field technologies and planting surfaces are constructed, cultivated, and abandoned again, often resulting in partial or total destruction of evidence for earlier production stragtegies" (ZARO & ALVAREZ 2005: 717). This aspect of environmental and cultural reconstruction has so far received very limited attention in the Rio Grande de Nazca region.

The aim of this research is to reconstruct the landscape evolution in the lower Rio Grande drainage basin during the Late Holocene and to detect interrelations between landscape evolution, cultural development, climatic changes and extreme events. Central to this is to identify and, if possible, quantify factors of landscape change. In doing so, one has to face the challenge of separating natural from anthropogenic factors, i.e. to determine both the natural and the human impacts on the landscape. An important question is whether climatic changes and extreme events have had an influence on past societies. To answer these questions, this thesis goes beyond physical geography approaches to paleoenvironmental reconstruction and includes the wealth of archaeological evidence and interpretations available for the research area. On the other hand, it adds to the classic archaeological approaches of site excavations and settlement pattern studies by widening the scope to include the entire landscapes used and modified by past societies (cf. ERICKSON 1999: 641 for the Lago Titicaca region).

Field work focused mainly on the Palpa Valley, i.e. the common valley of the Rio Grande tributaries Rio Palpa and Rio Vizcas within approximately ten kilometres upstream of their confluences with the Rio Grande. Research methods from the field of geography include sedimentology, geomorphology, hydrology, climatology, radiocarbon dating and interpretation of remote sensing data. Discussion and interpretation of the field and laboratory data build on HESSE (2004) and KASTNER (2005).

Figure 1 visualises the structure of this thesis. The introduction (chapter 1) is followed by an overview of the current state of knowledge regarding the geology, climate, ecology and archaeology of the research area (chapter 2). In chapter 3, research methods and data used in this thesis are presented. The research results regarding late historic landscape changes, in particular present-day fluvial dynamics and anthropogenic impacts, are discussed in chapter 4. In chapter 5, the results of the sedimentological studies are presented by deriving a facies classification specific to the research area (5.1), describing the investigated sediment profiles in short individual interpretations (5.2) and transects (5.3) and arriving at a generalised stratigraphy of the Palpa Valley (5.4). Building upon all previous chapters, chapters 6 and 7 discuss the Late Pleistocene to mid-Holocene landscape evolution and the Late Holocene cultural landscape evolution, respectively. For the Late Pleistocene to mid-Holocene (chapter 6), the focus is on the incision history of the allogenic rivers and the environmental conditions before the inception of agriculture. The analysis of Late Holocene developments (chapter 7) concentrates on the interconnections between fluvial dynamics, irrigation agriculture and landscape and cultural development. A final synopsis (chapter 8) brings together the main results of this thesis.

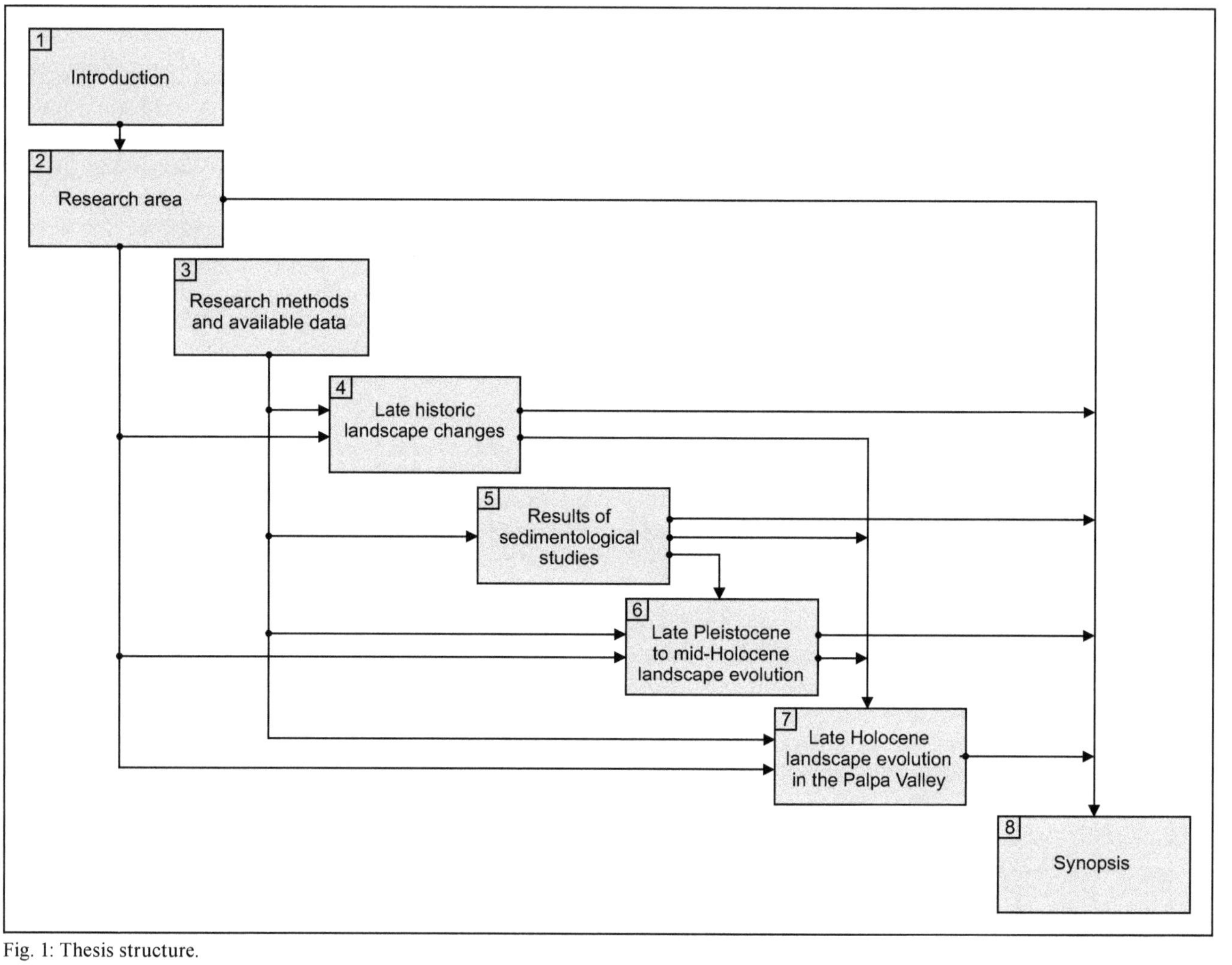

Fig. 1: Thesis structure.

2. Research area

2.1 *Overview*

The research area is defined as the drainage basin of the Rio Grande de Nazca in southern Peru, extending from 13°44'S to 15°12'S and from 74°15'W to 75°30'W (Figure 2). It comprises elevations from sea level to approximately 4,600 m asl and has an area of 11,162 km^2 of which 6,500 km^2 (58%) belong to the arid zone below 2,500 m asl (deduced from SRTM data using elevation zones from ONERN 1971a: 44). This lower part of the catchment lies in the centre of the Peruvian-Chilean coastal desert which extends 3,200 km from Talara in northern Peru to Santiago de Chile (Gay 1962: 37). Within the Rio Grande drainage basin, the focus of this thesis is on the Palpa Valley, i.e. the broad common valley bottom of Rio Palpa and Rio Vizcas within approximately ten kilometres upstream of their confluences with the Rio Grande. While fieldwork concentrated on the Palpa Valley, the scope of this thesis extends beyond this valley segment and even beyond the confines of the Rio Grande drainage basin for the data and literature used as well as for the applicability of the results. Likewise, while the temporal focus is on the Late Holocene, an understanding of geomorphic processes during this period requires a consideration of a much longer time span.

Morphology and environmental processes in the area are characterised by steep gradients. The land surface rises from south-west to north-east, perpendicular to the direction of the main geological structures. From the 5,300 m deep deep-sea trench approximately 100 km southwest of the coastline, the seafloor rises steeply to the only 25 km wide shelf (less than 200 m deep) (Global Seafloor Topography, Smith & Sandwell 1997). The Cordillera de la Costa extends approximately 30 km inland from the coast with average elevations of 500 to 700 m asl and a maximum elevation of 1,791 m asl (Cerro Tunga) (topographic map 1:100,000; SRTM). Along the 200 km stretch of coast from Paracas to Lomas where the Cordillera de la Costa is present, only Rio Ica and Rio Grande reach the sea. To the north-east of the Cordillera de la Costa lies the pre-Andean basin, a 25 km wide sedimentary basin with elevations of 200 to 700 m asl. Into this basin, mainly coarse clastic sediments have been deposited from the Andes (Dornbusch 1997: 9). These sediments form extensive fans and plains (bajadas) which are locally termed pampas. The bajada surfaces lie at elevations of 400 to 700 m asl. North-east of the pre-Andean basin, the Cordillera Occidental (the westernmost chain of the Andes) rises steeply and reaches elevations of up to 4,600 m asl within 150 km from the coast.

The drainage basin of the Rio Grande de Nazca comprises seven main sub-basins (from north-west to south-east, Figure 3): Rio Santa Cruz, Rio Grande, Rio Palpa, Rio Vizcas, Rio Ingenio, Rio Nazca (Rio Aja + Rio Tierra Blanca) and Rio Las Trancas. These rivers are fed by precipitation in the semiarid to humid parts of the Cordillera Occidental above 2,500 m asl and cross the pre-Andean basin as allogenic rivers. After their confluence, the Rio Grande cuts through the Cordillera de la Costa. Due to the seasonality of highland precipitation, discharge is limited to a few months per year. The Rio Grande and its tributaries are deeply incised into the Cordillera Occidental and the bajadas. Incision into the bajadas varies and is greatest for Rio Grande, Rio Palpa, Rio Vizcas (160 m) and Rio Ingenio (120 m) and much less for Rio Santa Cruz (60 m), Rio Nazca (40 m) and Rio Las Trancas (30 m) (topographic map 1:100,000; SRTM). The differences in incision are attributable to differences in discharge and the distance (and hence slope) to the confluences with the Rio Grande.

2.2 ***Geology***

The crucial process for the creation of the cordilleras and for tectonically influenced landscape changes is the subduction of the Nazca Plate under the South America Plate. The height of the Cordilleras is controlled by the balance between boundary forces and gravity forces (isostatic compensation of uplift) (Mercier et al. 1992: 11,973). Subduction of the Nazca Plate began during the Mesozoic (Goudie 2002: 67). With incipient uplift, erosion and sedimentation of continental material at the mountain front and in the adjacent oceanic trench and shelf areas began (Paskoff & Naranjo 1983: 743). The structural trends of the Andes were created during the Late Triassic. Subsidence during the Upper Jurassic and maximum accumulation during the Lower Cretaceous were followed by an Upper Cretaceous uplift period which was accompanied by batholitic intrusions. Tectonic quiescence during the Oligocene was followed by a period of strong uplift, deformation, volcanism and plutonism during the Miocene and Pliocene (Thornburg & Kulm 1981: 394ff.). While Zeil (1979: 160) and Thornburg & Kulm (1981: 394ff., 415) describe discrete orogenic phases based e.g. on the interruption of sedimentation in the marine East Pisco basin at the end of the Eocene, Miocene and Pliocene, Fischer (1985: 252) emphasises that phases of uplift can not be satisfactorily shown. Most of the uplift occurred during the past 40–45 Ma; during the late Miocene the present height was almost reached. The topography of the high cordillera was created between 26 and 6 Ma BP (Mercier et al. 1992: 11,945, after Sébrier et al. 1988). The height of the Andes has not changed significantly during the past 5 Ma (Mercier et al. 1992: 11,976; Fischer 1985: 248).

Between 13°30'S (Peninsula Paracas) and 15°30'S, a Cordillera de la Costa is present (Figure 2). While it is not exposed as a land surface feature from 6°S to 13°30'S and from 15°30'S to 17°S, it can be traced along the seafloor as a structural ridge termed Outer Shelf High (OSH). The presence of the Cordillera de la Costa in the Rio Grande de Nazca region is due to uplift caused by the approximately 1,500 m high Nazca Ridge which exerts a strong influence on the overriding South America Plate (HSU 1992: 93f.; Global Seafloor Topography, SMITH & SANDWELL 1997). Due to the East-West convergence of the oceanic Nazca Plate and the continental South America Plate, the Nazca Ridge is moving in a south-eastern direction relative to the coastline, causing uplift over the southern and subsidence over the northern flank of the Nazca Ridge (HSU 1992: 92ff.). HSU (1992: 95) found a good correlation between the height of Pleistocene marine terraces and the expected pattern of uplift and subsidence due to the movement of the Nazca Ridge.

The creation of the OSH as well as the uplift-related deformation of the Cordillera de la Costa are largely due to block tectonics (HSU 1992: 88; THORNBURG & KULM 1981: 413f.). A number of tectonic faults recorded in the geological map of the area are clearly recognisable in the SRTM DTM and in satellite images, in particular the prominent Falla Tunga (approximately 15°10'S, 75°02'W), a 100 to 300 m high tectonic escarpment. Towards the south-east, the geological maps record an increase in the number of faults corresponding to the area of strongest deformation and highest uplift (up to 0.5 m/ka) over the leading flank of the Nazca Ridge (cf. HSU 1992: 91). Towards the north-west, uplift and deformation are weaker or there is slight subsidence (e.g. in the Paracas area), therefore existing faults are less active and are masked by erosion or sedimentation. Recent tectonic activity is documented by the occurrence of strong earthquakes in the years 1942, 1996 and 2001 in the area influenced by the Nazca Ridge (HAMPEL et al. 2002: 291; SPENCE et al. 1999: 753; SWENSON & BECK 1999: 747ff.). Significant uplift occurs in conjunction with these earthquakes, e.g. the coastline of the Bahia de San Juan was uplifted by approximately one metre during the earthquake of 1942 (BROGGI 1946: 30).

Based on a plate convergence of 100 mm/a, HSU (1992: 92f.) assumes a velocity of 71 mm/a of the Nazca Plate relative to the coastline. However, according to a more recent study by NORABUENDA et al. (1998: 358ff.) the plate convergence velocity is only 68 mm/a of which only 30-40 mm/a are explained by plate margin processes and 10-15 mm/a by shortening of the continental crust at the sub-Andean foreland fold and thrust belt. HAMPEL (2002: 665ff.) and HAMPEL el al. (2002: 293) likewise assume a migration velocity of the Nazca Ridge relative to the coast of 43 mm/a during the past 4.9 Ma. Therefore, a lower migration velocity has to be taken into consideration when applying the findings of HSU (1992): the mouth of the Rio Grande de Nazca has been situated over the south-east edge of the Nazca Ridge for approximately 1 Ma based on the velocity assumed by HSU (1992) or for almost 2 Ma based on the velocity determined by NORABUENDA et al. (1998: 358ff.). This lower velocity also helps explain the remaining discrepancies between reconstructed (approximately 0.5 m/ka) and modelled (approximately 0.7–0.9 m/ka) uplift rates (cf. HSU 1992: 96; ORTLIEB & MACHARÉ 1990).

2.3 Climate

2.3.1 Controlling factors and present climatic conditions

The research area is situated in the north-western part of the South American "arid diagonal" which crosses the Andes at approximately 30°S from north-west to south-east. Due to the prevailing circulation patterns (influence of the ITCZ from the north-east and west wind circulation from the south-west), atmospheric moisture is higher on the eastern side north of the diagonal and on the western side south of the diagonal (GARLEFF et al. 1991: 366). North-east of the diagonal, the precipitation maximum occurs in summer, south-west of the diagonal in winter. Under current climatic conditions, the northern limit of cyclonal winter precipitation lies at approximately 32°S (GARLEFF et al. 1991: 366) or 30°S (MESSERLI et al. 1993: 119). For the research area, both local precipitation in the coastal desert and the hygric conditions in the Andean highlands which control discharge in the allogenic rivers are relevant.

The coastal desert

The main factors controlling the climatic conditions in coastal Peru are the Andean mountain chains and the South-East Pacific Anticyclone (SEPA) (MESSERLI et al. 1992: 257ff.; MESSERLI et al. 1993: 117, 119; GOUDIE 2002: 64). The Andes block moisture transport from the Amazon Basin and reinforce and stabilise the strength and meridional position of the SEPA (WEISCHET 1966: 1; GOUDIE 2002: 64). The south-east trade winds cause the upwelling of cold deep water in the Peru or Humboldt Current along the west coast of South America (e.g. SCHRADER & SORKNES 1991: 233). The Humboldt Current has been described as a number of upwelling centres rather than a continuous ocean current (KOEPCKE 1961: 57, after Schweigger 1947); its width varies spatially as well as seasonally between 150 and 400 km (KOEPCKE 1961: 57). Temperatures in the Humboldt Current are approximately 10 K (near the coast of northern Chile at 18°30'S) to 12 K (near the coast of Peru at 12°S) colder than in other parts of the ocean at similar latitudes (KOEPCKE 1961: 56, after Carranza 1946 and Mann n.d.). The temperature difference between sea and adjacent land is approximately 5 K (DE ABREU & BANNON 1993: 2,952). Descending air in the SEPA combined with the thermal effect of upwelling causes a very stable inversion in the lower troposphere (e.g. GAY 1962: 37, 39; WEISCHET 1966: 1; GOUDIE 2002: 64) at an altitude of approximately 1,000 m (base: 300–1,400 m, top: 650–2,250 m), which has a strength of 10 to 15 K (PROHASKA 1973: 97; DE ABREU & BANNON 1993: 2952). This inversion as well as the low evaporation rates over the cold ocean and the strong thermal sea-land gradient effectively

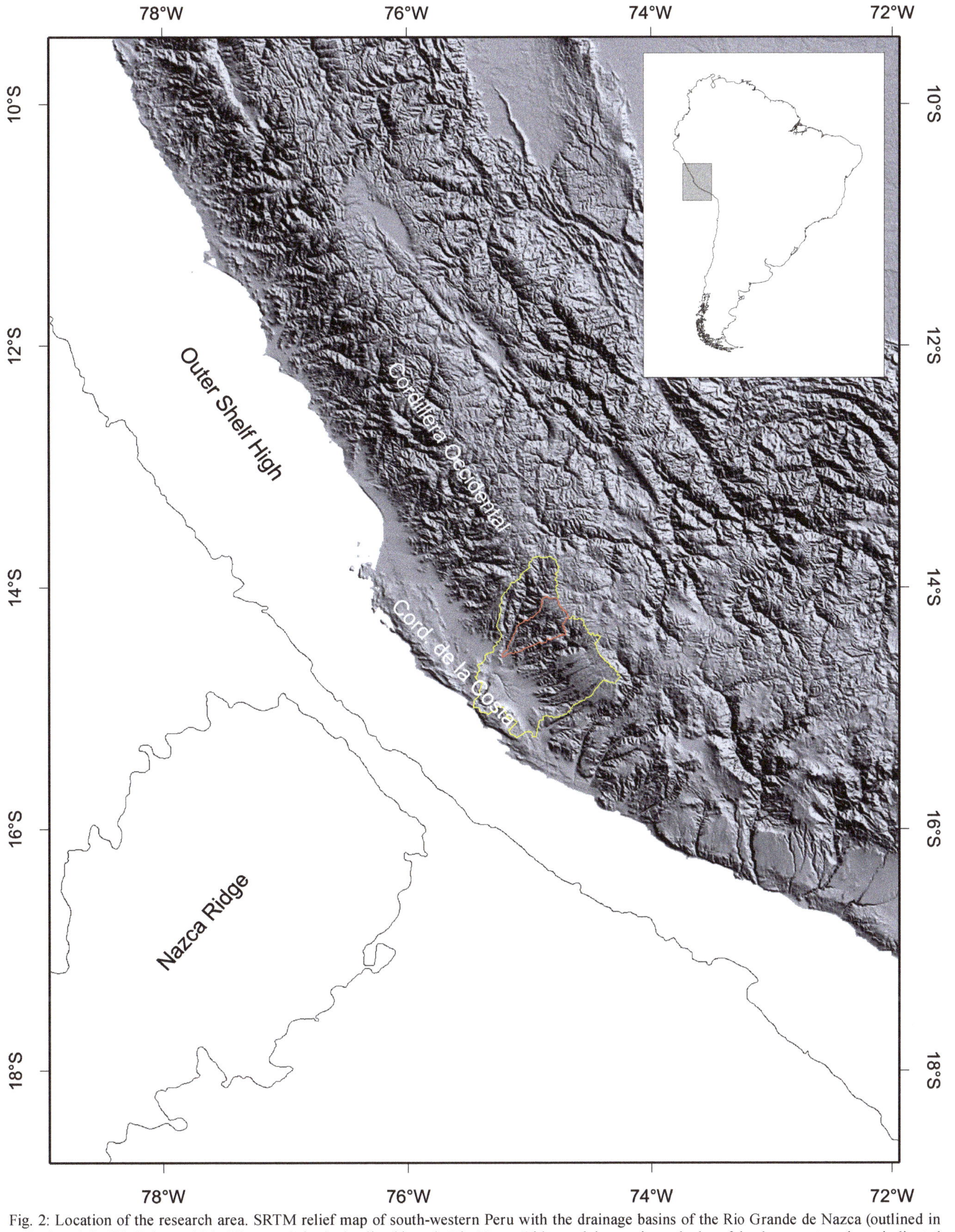

Fig. 2: Location of the research area. SRTM relief map of south-western Peru with the drainage basins of the Rio Grande de Nazca (outlined in yellow) and the Rio Palpa and Rio Vizcas sub-basin (outlined in red). The Nazca Ridge and the continental edge of the deep see trench are indicated by the 4,000 m bathymetric contour (from Global Seafloor Topography, SMITH AND SANDWELL 1997).

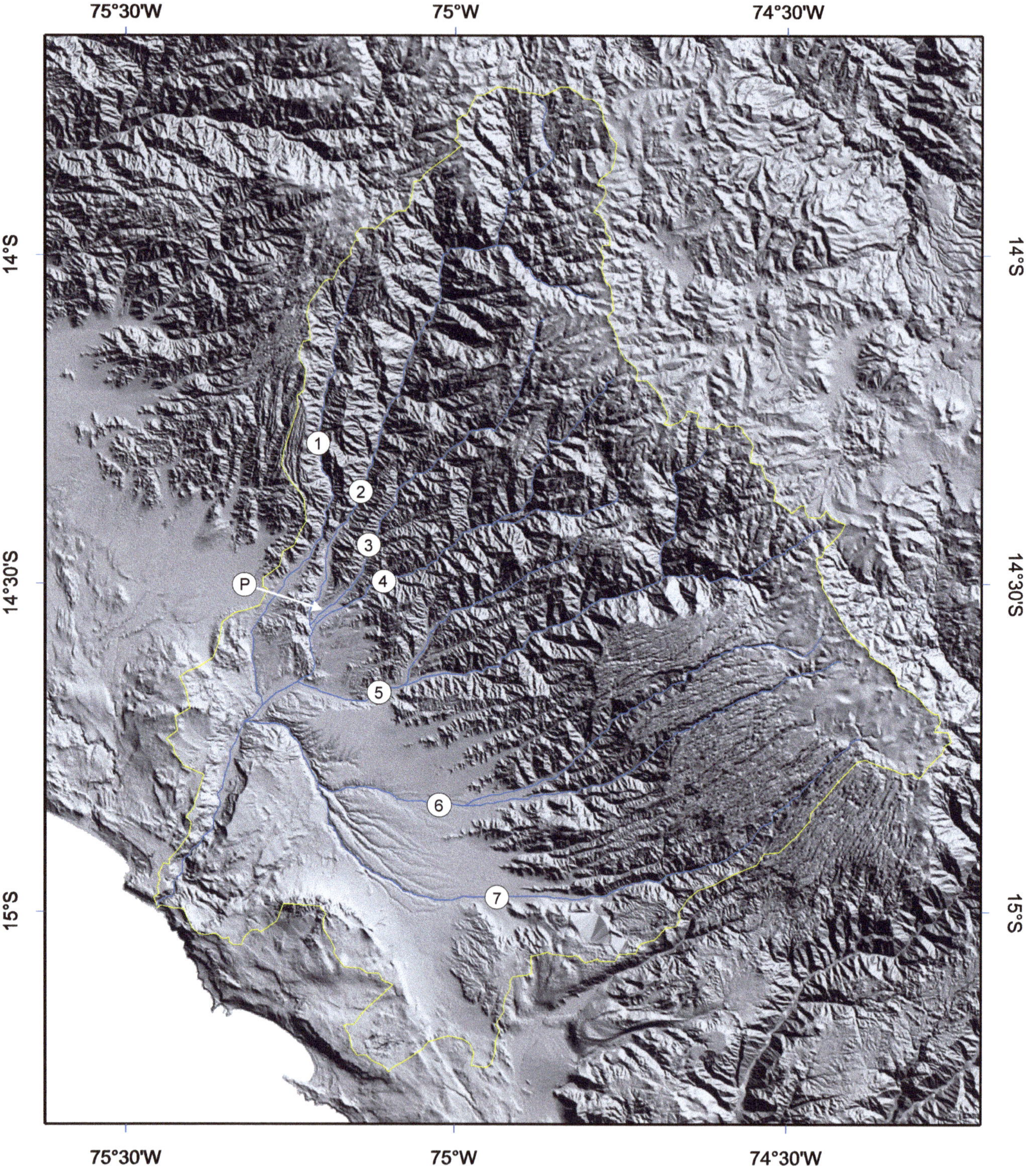

Fig. 3: SRTM relief map of the Rio Grande de Nazca drainage basin (outlined in yellow) with the main tributaries (blue): **1** – Rio Santa Cruz, **2** – Rio Grande, **3** – Rio Palpa, **4** – Rio Vizcas, **5** – Rio Ingenio, **6** – Rio Nazca, **7** – Rio Las Trancas; **P** – Palpa Valley.

inhibit convection and convective precipitation over the subtropical south-east Pacific and adjacent land areas (e.g. Caviedes 1973: 119; Weischet 1996: 394). The SEPA and sea-land temperature gradients further cause a diversion of winds: near-surface air movement over the coast is dominated by sea winds (de Abreu & Bannon 1993: 2952 for Lima; Gay 1962, 1999 and 2005 for southern Peru). These phenomena cause the hyperarid coastal climate and create the world's largest west coast desert, extending from northern Peru to Santiago de Chile (Goudie 2002: 61). It is also the world's driest coastal desert, with average annual precipitation in coastal southern Peru below 10 mm (Abele 1992: 7f.; Garleff et al. 1991: 361; ONERN 1971a: 42–44) and measurable precipitation occurring only at intervals of several years (Weischet 1966: 1; Oberlander 1994: 27). Precipitation is therefore not a regular, characteristic phenomenon but rather results from exceptional meteorological conditions which occur at intervals of several years and are not necessarily linked to seasonality. The coastal desert of southern Peru can thus be considered rainless during the normal course of a year (Weischet 1966: 7). Below 2,800 m asl, precipitation is negligible and largely insufficient for the development of vegetation (Dornbusch 1998: 42; cf. ONERN 1971a: 44).

However, immediately along the coast (in particular where it rises steeply), the uplift of cool, moist air frequently leads to the development of fog (termed *garúa* in Peru and *camancháca* in Chile) which in places allows plant growth by surface condensation or light drizzle (Prohaska 1973: 98; Caviedes 1973: 116f.; Gay 1962: 38). The cause for the development of the stratus layer is cooling by long wave radiation loss (Rundel et al. 1991: 9). The coastal fog belt is generally considered to be restricted to distances of up to a few kilometres from the coastline and elevations from 100–200 to 700–800 m asl (Prohaska 1973: 98) or up to 1,000–1,200 m (Caviedes 1973: 118; Berninger 1925: 383). However, the fog layer may penetrate inland as far as 60 km from the coast as ascertained by Koepcke (1961: 220, after Ferreyra or. commun.) and as evident in austral winter satellite images (e.g. GOES-12 data of July and August 2006). *Garúa* precipitation is strongest in southern winter and spring (Prohaska 1973: 99; Caviedes 1973: 117). With increasing distance from the sea, the air is warmed over the intensely insolated ground surface. This leads to a strong reduction of relative humidity. Under normal conditions, the dew point of rising air would not be reached at altitudes below 3,000 m asl (Koepcke 1961: 61) or even 4,000 to 5,000 m asl (Gay 1962: 39); however, the atmospheric inversion over the coastal desert inhibits convection to these altitudes. Taking into account the existing thermal and humidity gradients in the northern Chilean desert, Weischet (1966: 1ff.) and Caviedes (1973: 116ff.) differentiate between a very narrow (a few kilometres wide) coastal desert proper with high low-altitude cloudiness, low diurnal and annual temperature amplitudes, high relative humidity and very low precipitation on one hand and an interior desert with very little cloud cover, extreme temperature variability, low relative humidity and even lower precipitation on the other hand. The limit between the two desert types is the topographically controlled occurrence of coastal fog (Weischet 1966: 1ff.). Due to the topographic differences between the coasts of southern Peru and northern Chile, this transition is more gradual in southern Peru.

The climate of the interior desert can be described thermally as a diurnal climate (Koepcke 1961: 60, after Troll 1952; cf. Troll 1968: 19). At the station Palpa, the annual temperature amplitude is significantly lower than the diurnal temperature amplitude: the difference between the mean monthly temperatures of the warmest and the coldest month is 9.7 K while the average difference between mean maximum monthly temperature and mean minimum monthly temperature is 20.2 K. Mean annual temperature is 21.6°C, mean maximum monthly temperature and mean minimum monthly temperature is 34.1 and 5.0°C, respectively (ONERN 1971a: 42, based on data from 1964–1969). Potential evaporation in the interior desert is very high with 1,648 mm/a for the station Palpa (ONERN 1971a: 42).

Occasionally, the coastal desert is impacted by El Niño events. These events are characterised by a weakening of the SEPA and the trade winds and a southward displacement of the ITCZ along the west coast of South America in conjunction with increased sea surface temperatures, a weakening of the upwelling and a covering of the cold Peru Current by warm equatorial waters. The weakened trade winds lead to a lowering of the thermocline and a slight increase in sea level along the coast of Peru (Caviedes 1984a: 428; Aceituno 1988: 520; Barber & Chavez 1983: 1,203–1,209). These anomalies allow the development of tropical convection which leads to intense precipitation along the coast of southern Ecuador and northern Peru (Koepcke 1961: 59; Aceituno 1988: 519; Rundel et al. 1991: 2; Murphy 1926 for El Niño 1925; Caviedes 1975 for El Niño 1972–73; Caviedes 1984b for El Niño 1982–83). The periodicity of El Niño is between two and seven years in northern Peru (cf. Murphy 1926: 53; Cane et al. 2000: 3; Gu & Philander 1997: 805, after Philander 1990; Quinn & Neal 1992: 624). With increasing distance from the equator, El Niño conditions occur later and with decreasing intensity (Barber & Chavez 1983: 1,207); their southern limit varies considerably (Koepcke 1961: 72; Cane 1983). At 8°S, strong events occur on average every 15 to 16 years (Moseley 1983: 783, after Nials et al. 1979). Only rarely do El Niño events cause strong anomalies in the Palpa area. However, both the very strong 1998 and the moderate 2007 events caused intense precipitation in Palpa (Beaumont 1998; Silverman & Proulx 2002: 49f.; Anonymous 2007a, 2007b, 2007c). The strength of an event in terms of SOI (Southern Oscillation Index) or SSTA (Sea Surface Temperature Anomaly) appears to be a poor indicator for its southward extent.

The Andes

The Andes act as an effective barrier for moisture transport from the tropical Amazon Basin related to the South American Summer Monsoon (SASM) (Zhou & Lau 1998). Maximum elevations in the Andes are more than 6,000 m asl, and in Peru and northern Chile no pass across the Cordillera Occidental is lower than 4,000 m asl. For tropical precipitation from the Amazon basin, the Andes are therefore an insurmountable obstacle (Koepcke 1961: 39). The coastal desert can thus also be termed a rain-shadow desert (Macharé et al. 1990: 159, after Bowman 1916, Fenner 1948 and Rauh 1979). Tropical convective summer precipitation is therefore largely restricted to the eastern slope of the Andes (Messerli et al. 1993: 119) and increases towards the north (Goudie 2002: 64). Amazonian spillover precipitation reaching the coastal desert is slight (Pro-

HASKA 1973: 104) and affects inland locations rather than the coastal desert (RUNDEL et al. 1991: 10).

Seasonal precipitation in the high Cordillera Occidental feeds the allogenic rivers descending the western slopes which are the almost exclusive source of water in the coastal desert. In the Cordillera Occidental of southern Peru, precipitation exceeds 400 mm/a at altitudes above 4,000 m asl (ONERN 1971a: 42ff.; ONERN 1971b: 22ff.; cf. ABELE 1992: 7f.; GARLEFF et al. 1991: 361). Highland precipitation is controlled by continental warming in mid-summer and the related seasonal shift of the ITCZ (CAVIEDES 1982: 62). On the western slope of the Cordillera Occidental, precipitation steeply decreases with decreasing altitude (Figure 4). Thus the size of the high-altitude portions of drainage basins largely controls the amount and reliability of water flow in the lower courses of the rivers. Interannual variability of highland precipitation is very high as shown by the interannual variability in river discharge (Figure 5).

2.3.2 *Climate history*

Antiquity of aridity along the coast

The initiation of arid conditions along the west coast of South America due to the uplift of the Andes as a moisture barrier is widely accepted (MACHARÉ et al. 1990: 167). Generally arid conditions have therefore been in place for millions of years (e.g. OBERLANDER 1994: 27, after Ericksen 1981 and 1983; CLARKE 2006). Uplift of the Andes during the Tertiary led to a stabilisation of the SEPA and the creation of rain-shadow conditions. With the development of cold Antarctic bottom waters and the cold Humboldt Current approximately 15 to 13 Ma BP as a result of the glaciation of Antarctica, conditions similar to today were created. Arid conditions in the Atacama have persisted at least since the Late Eocene and hyperarid conditions since the Middle to Late Miocene (GOUDIE 2002: 71; ALPERS & BRIMHALL 1988; ARANCIBIA et al. 2005; RECH et al. 2006; cf. FISCHER 1985: 251). It is disputed whether the changes have occurred gradually or in discrete phases (cf. ABELE 1992: 5). During the Quaternary, the height of the Peruvian Andes did not change significantly; thus fundamental climatic changes in the coastal desert should not be expected (MACHARÉ et al. 1990: 159, after Sébrier et al. 1982 and 1985). The well-preserved marine terraces southeast of the mouth of the Rio Grande de Nazca (up to 27 terraces between sea level and 800 m asl) which possibly cover the entire Quaternary indicate conditions of extremely reduced erosion (BROGGI 1946: 21ff.; HSU 1992: 91; cf. MACHARÉ et al. 1990: 163 for northern Peru); commonly excellent preservation of marine shells on these terraces also indicates long-term aridity (HSU et al. 1989: 256). The high levels of endemism in lomas formations are a further argument for the long-term persistence of hyperaridity in southern Peru and northern Chile, while lower levels of endemism and closer floristic links to the Andes point to a more recent origin of lomas communities and hence pronounced climatic changes during the Pleistocene in northern Peru (RUNDEL et al. 1991: 1f., 39ff.).

However, despite generally hyperarid conditions at least since the Miocene, paleo-shorelines document the existence of paleolakes in northern Chile and hence indicate either wetter local conditions or increased runoff from the Andes into the coastal desert (OBERLANDER 1994: 27). Substantial precipitation must have occurred to deposit and dissect the ubiquitous large alluvial fans in the coastal desert (cf. EVENSTAR et al. 2005; MACHARÉ et al. 1990: 163, after Dollfus 1965, Laharie 1970, Teves 1976, Sébrier & Macharé 1980 and Macharé 1981). These alluvial deposits are at least partially attributable to local precipitation in the desert; however, most of the sediment originates from intermediate to high elevations in the Andes and thus may have to be attributed to climatic changes outside the coastal area (MACHARÉ et a. 1990: 163). The presence of autochthonous terraces throughout the coastal desert of central and southern Peru indicates the occurrence of periods with pronounced local rainfall despite the general permanence of arid conditions (SÉBRIER & MACHARÉ 1980: 20).

Late Pleistocene climate changes

During the last Glacial, precipitation at mid-latitudes in South America was higher than today, however, the position of the westerly precipitation belt remained relatively stable and did not reach further north than 30°S (MESSERLI et al. 1993: 121; GARLEFF et al. 1991: 371). Pleistocene soils in the lowlands west of the Andes do not occur north of 27°S (VEIT 2000: 8). North of the arid diagonal, the southern limit of tropical summer precipitation in high altitudes shifted southward by 9 to 12° latitude, reaching 27°S (MESSERLI et al. 1993: 124, 126). While the position of the arid diagonal remained largely unchanged under Glacial conditions, the temperature and moisture gradients between tropics and mid-latitudes were amplified (MESSERLI et al. 1993: 121; GARLEFF et al. 1991: 370, 373; VEIT 2000: 8). The relative importance of temperature and moisture for changes in glacier extent changes with latitude (cf. MACHARÉ et al. 1990: 171; DORNBUSCH 1997: 43). Precipitation was the limiting factor for more extensive glaciation between 14–18 and 25°S while temperature was the limiting factor south of 29°S (AMMAN et al. 2001: 323f.; KULL et al. 2003: 3ff.).

Maximum Late Pleistocene glacier extent was asynchronous in different parts of the Andes (cf. KULL et al. 2002: 359) and was regionally not contemporaneous with the global maximum land-based ice volume as deduced from sea-level minimum at 22–19 ka cal BP (YOKOYAMA et al. 2000). According to HEINE (2000: 10ff.), maximum glacier extent during the last Glacial was reached between 30 and 25 ^{14}C ka BP in Ecuador and between 25 and 15 ^{14}C ka BP in Bolivia. In parts of the northern Andes, gla-

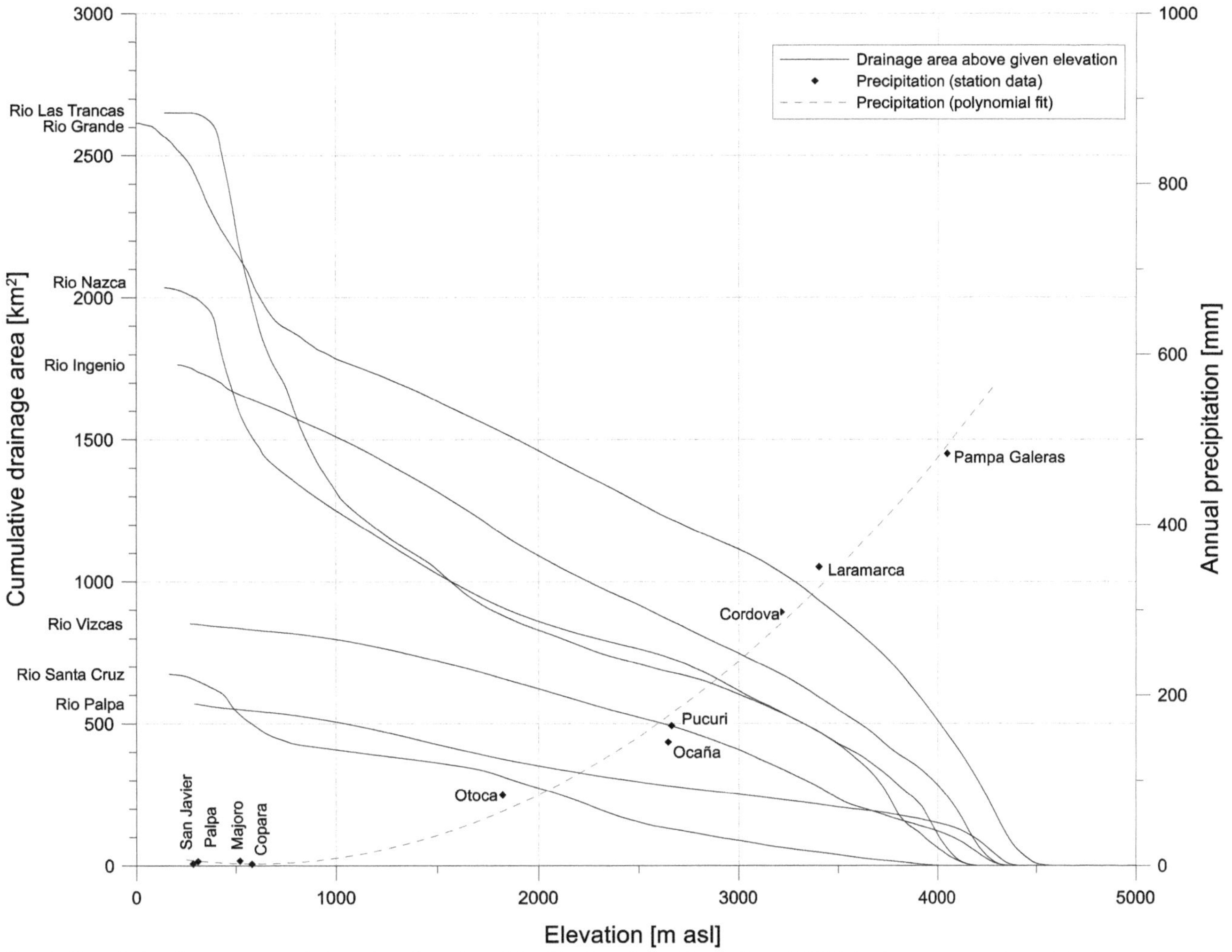

Fig. 4: Cumulative drainage area of the main tributaries of the Rio Grande de Nazca (from SRTM data), mean annual precipitation station data (ONERN 1971a: 41–43) and polynomial fit for the relationship between elevation and precipitation.

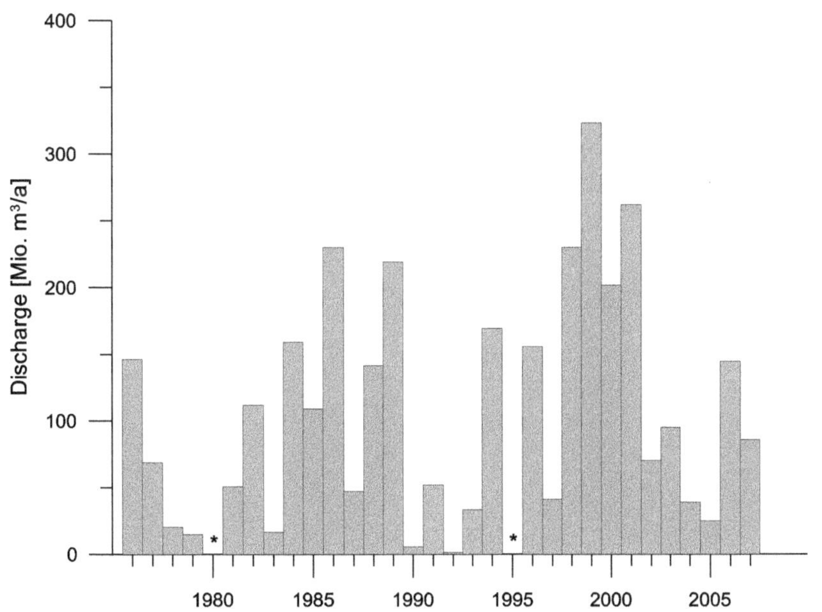

Fig. 5: Annual discharges if the Rio Grande de Nazca from 1976/76 to 2006/07 (data provided by Junta de Usarios Palpa) (* = no data).

cier advance culminated around 27 ^{14}C ka BP (Clapperton 1993: 190, after Palacios 1977, Van der Hammen et al. 1981 and Clapperton 1986, 1990). According to Smith et al. (2005: 678f.), greatest glacier extent in the Andes of Peru and Bolivia was reached around 34 ka BP (^{10}Be) and glaciers were retreating by ca. 21 ka BP (^{10}Be). In the Lago Titicaca (16°S) and Lago Junin (11°S) watersheds, glacial maximum conditions occurred between ca. 30 ka and 20 cal ka BP and deglaciation from glacial maximum positions took place from 22 ka to 19.5 cal ka BP (Seltzer et al. 2002: 1686). However, a late glacial maximum in Peru and Bolivia at 14 to 12 ^{14}C ka BP resulting from a precipitation increase has been deduced by Seltzer (1993: 136). Glacial conditions were rapidly terminated between 14–12 and 10 ka BP (Macharé et al. 1990: 161, after Mercer 1984, Seltzer & Wright 1989, Seltzer 1990 and Rodbell & Hansen 1990).

Holocene climate changes

Significant Holocene climate changes in the high Andes have been described by various authors, mainly on the basis of glacial/periglacial and lacustrine evidence. Both the general progression of warm/cold and wet/dry phases and chronological issues appear at least partially contradictory and are still debated (e.g. Goudie 2002: 71ff.; Grosjean 2001; Grosjean et al. 2003). Figures 6 and 7 compare the results of several paleoclimatic studies of South America with a focus on the occurrence of dryer and wetter phases on the Altiplano of Peru, Bolivia and Chile. A number of obvious discrepancies strongly suggest that climatic conditions during the Holocene do not follow the same trends over the entire Andes. Furthermore, the results of some studies are not unambiguous. For example, increased dust accumulation in ice cores has been interpreted as a result of both dryer (Thompson et al. 1998: 1,863 for the Holocene until 3,400 BP) and wetter (Thompson et al. 1988: 763; cf. Ortloff & Kolata 1993: 199) conditions. Reconstructions of lake level changes, in particular of Lake Titicaca, play a major role in the study of Holocene climate in the Andes. Lake Titicaca is considered a reliable indicator for precipitation changes in wide areas of South America (Baker et al. 2001: 640). Given that lake levels of Lake Titicaca are used as indicator for annual changes in precipitation (e.g. Kessler 1990: 93), the time lag of 2,500 years between reduced ice accumulation and lower lake levels on the Altiplano (from 9,000 BP to 3,000 BP) as described by Thompson et al. (1998: 1,863) appears illogical.

Most publications are in general agreement regarding relatively high humidity on the Altiplano during the late Pleistocene and early Holocene; however, the mid-Holocene climate has long been a matter of debate. Grosjean et al. (2003) and Grosjean et al. (2005: 168) reconcile a number of apparently contradictory paleoenvironmental reconstructions by stressing the space-time-transgressive nature of climatic changes and the fact that different proxies record different aspects of climatic changes on different time scales. Grosjean et al. (2003: 256) summarise that in the high Andes the "mid-Holocene was on average very dry (low lake levels), but the variability of humid and dry years was very large (cycles in lakes), and humid years were frequent enough to keep the plant communities alive." The extremely dry mid-Holocene was interrupted by "a century-scale more humid interval around ca. 5.5–6 kyr B.P. that appears systematically in lake sediment archives" (Grosjean et al. 2003: 247). The mid-Holocene dry phase on the Altiplano is considered to be underlying the *silencio arqueológico* between 8,500 and 4,800 BP (Messerli et al. 1993: 125, after Nuñez unpubl.). By ca. 4,000 cal BP, modern climatic conditions were established (Grosjean et al. 2005: 163).

For the coastal desert, only limited data exist regarding possible Holocene climate changes. The most significant phenomena discussed in the literature are the variability of winter fog and of the frequency and strength of El Niño events. Both are linked with the intensity of the SEPA and the related strength of coastal upwelling and the position of the ITCZ. The variability of winter fog has been discussed with regard to declining lomas extent (see 2.4); however, no definite conclusions have yet been made. As for the occurrence of El Niño events, the assertion of a "birth of El Niño" around 5,000 BP (Rollins et al. 1986) has been extensively discussed (e.g. DeVries & Wells 1990: 11ff.; Wells & Noller 1997; Sandweiss et al. 1997: 966; Sandweiss et al. 2001: 603). A long-term hiatus of El Niño events can not be maintained. Rather, changes in El Niño frequency and intensity occurred with a mid-Holocene phase of less frequent and/or weaker events (e.g. Rodbell et al. 1999; Riedinger et al. 2002; Wells 1987; Keefer et al. 1998; Keefer et al. 2003; Sandweiss 2003; Koutavas et al. 2006).

A major mechanism underlying the Holocene climate changes both in the highlands and in the coastal desert is the latitudinal displacement of the ITCZ controlled by orbitally (precession) forced changes in insolation (Rowe et al. 2002: 179, 194f.; Garreaud et al. 2003: 5; Seltzer et al. 2000: 38; Abbott et al. 2000: 1,817). A long-term southward shift in the mean latitude of the ITCZ since the mid-Holocene was deduced by Haug et al. (2001) from a Cariaco Basin Ti record. Besides the general trend towards higher moisture conditions on the Altiplano since the mid-Holocene, the wet interval interrupting the mid-Holocene dry phase may be related to a short-term southward displacement of the ITCZ due to a northern hemisphere ice rafting event (cf. Brooks 2006: 32; Bond et al. 1997; Koutavas & Lynch-Stieglitz 2004; Wang et al. 2005); however, it does not appear in the Cariaco Basin record (cf. Haug et al. 2001). A southward displacement of the ITCZ creates moister conditions in the high Andes by intensifying precipitation and lengthening the wet season (Rowe et al. 2002: 194f.; Koutavas & Lynch-Stieglitz 2004). It is also compatible with a weakening of the SEPA, with more frequent and stronger El Niño events (e.g. PULWARTY et al. 1992: 285; CAVIEDES 1982: 68, after Caviedes 1973 and Pittock 1980) and with reduced winter fog occurrence on the Peruvian coast due to shorter and weakly developed winter seasons.

2.4 *Ecology*

The extreme altitudinal and climatic gradients from the coast to the heights of the Cordillera Occidental result in the development of altitudinal vegetation belts, the lower limit of vegetation being controlled by available moisture and the upper limit controlled by temperature (GROSJEAN et al. 1991: 103). Figure 8 shows the correlation of vegetation cover with altitude for the geographic latitudes 12, 15 and 18°S on the western slope of the Andes (total vegetation cover calculated as sum of herbal and tree cover percentages from MODIS Vegetation Continuous Fields data of November 2000 to November 2001 in ±0.5° latitude segments centred on each degree latitude; elevation data used: SRTM). Vegetation below ca. 2,000 m asl is located mainly in the river valleys and lomas areas. The altitudinal control of vegetation zones on the western slopes of the Andes (mainly through moisture changes) has been described by several authors (e.g. KOEPCKE 1961; TROLL 1968: 33; RICHTER 1981: 24ff.; HOLMGREN et al. 2001: 244; cf. PROHASKA 1973: 104; BAIED & WHEELER 1993: 146). The zonations differ in terminology as well as latitudinal and altitudinal resolution but agree in the general pattern of latitudinal and altitudinal control on vegetation. Table 1 summarises the ecological zonation for the Arequipa area (ca. 16°S, after HOLMGREN et al. 2001: 244) and the equivalent elevations for the Palpa area as deduced from comparison of MODIS Vegetation Continuous Fields data (cf. Figure 8). Note that resolution of vegetation zones based on MODIS vegetation cover data was only possible for the transition from absolute desert to Pajonal. The absolute limit for plant growth is around 4,800 m asl (RICHTER 1981: 26). Snowline in central and southern Peru is between 4,800 and 5,800 m asl, its east-west and north-south gradients showing a greater influence of temperature in central Peru and precipitation in southern Peru (SELTZER 1990: 139f.).

The coastal desert up to an elevation of approximately 2,000 m asl is practically devoid of vegetation except in the valleys of allogenic rivers and in areas of strong fog influence. However, even the bajada surfaces can not be considered absolutely lifeless desert as exemplified by a living insect larva found in a desert loess deposit at approximately 700 m in February 2006. Shells of land snails of the genus *Bulimulus* can be found in the cactus belt and even in the hyperarid desert below (e.g. EITEL et al. 2005: 144); however, according to CRAIG (1992: 127ff.) their presence does probably not indicate long-term changes in climatic moisture but short-term excursions from the extreme aridity. *Bulimulus* shells are ubiquitous in some lomas areas (e.g. ENGEL 1991: 70; KOEPCKE 1961: 232) and live specimens have been reported from the Paracas Peninsula by CRAIG & PSUTY (1968: 133). In the large valleys, seasonal river discharge as well as related groundwater recharge allow the development of riverine forests with tall trees (up to 20 m) and a species-rich undergrowth (KOEPCKE 1961: 141). KOEPCKE (1961: 142f.) argues that before the begin of agriculture, large parts of the river valleys in the Peruvian desert were covered by forests of the phreatophyte *Prosopis juliflora* without significant undergrowth during the dry season. These river valley environments of dense riverine forest and sparse valley floor forests were the foci of late Holocene human settlement and have been largely transformed by agriculture with only remnants of the natural vegetation remaining today (cf. BERESFORD-JONES 2004).

A notable exception from the extreme aridity below 2,000 m asl occurs in disconnected localities along the desert coasts of Peru and Chile where fog is common for at least a part of the year. There, the so-called lomas vegetation persists despite an almost complete lack of rain. While the total area of lomas vegetation is small, the species richness is extremely high (over 1,400 species of vascular plants have been recorded). Very high levels of endemism and the existence of two separate centres of endemism north and south of the arid diagonal respectively (less than 10% of all species are present both north and south of 18–20°S) strongly support the long-term isolation due to persisting hyperarid conditions (DILLON 2005: 131, 135, 141f.).

During southern hemisphere winter (May to October, RUNDEL et al. 1991: 6; June to August, FERREYRA 1983: 242), fog conditions almost continuously affect the entire coast of Peru south of ca. 7°S (RUNDEL et al. 1991: 6), the upper limit of the fog reaching ca. 700 to 1,000 m (FERREYRA 1983: 242; KOEPCKE 1961: 220) or 1,200 m (PINTO et al. 2006: 545f.). The resulting fog precipitation is not measurable with normal rain gauges because it depends on the capture by surfaces exposed in the moving moisture-laden air (e.g. ELLENBERG 1959: 51ff.). It may reach 100 to 200 mm/a (WEISCHET 1996: 395; however, cf. SCHEMENAUER & CERECEDA 1993: 184). Capture and accumulation of fog moisture is enhanced by topographic conditions (hillsides) (RUNDEL et al. 1991: 10; RICHTER 1981: 24), air movement (SCHEMENAUER & CERECEDA 1993: 185; ELLENBERG 1959: 61f.) and the presence of vegetation (vegetation creates its own moisture climate by increasing the condensation surface, PROHASKA 1973: 98f.; ELLENBERG 1959: 58, 62). Besides providing moisture, the presence of fog strongly reduces insolation and evaporation loss (PROHASKA 1973: 98f.; WEISCHET 1966: 395). The maximum development of lomas vegetation is therefore not only controlled by the amount of fog precipitation but by the daily number of hours and the annual number of days with fog cover (KOEPCKE 1961: 27). The lower limit of lomas vegetation reaches sea level in southern Peru but lies between 100 and 300 m in central Peru (KOEPCKE 1961: 220).

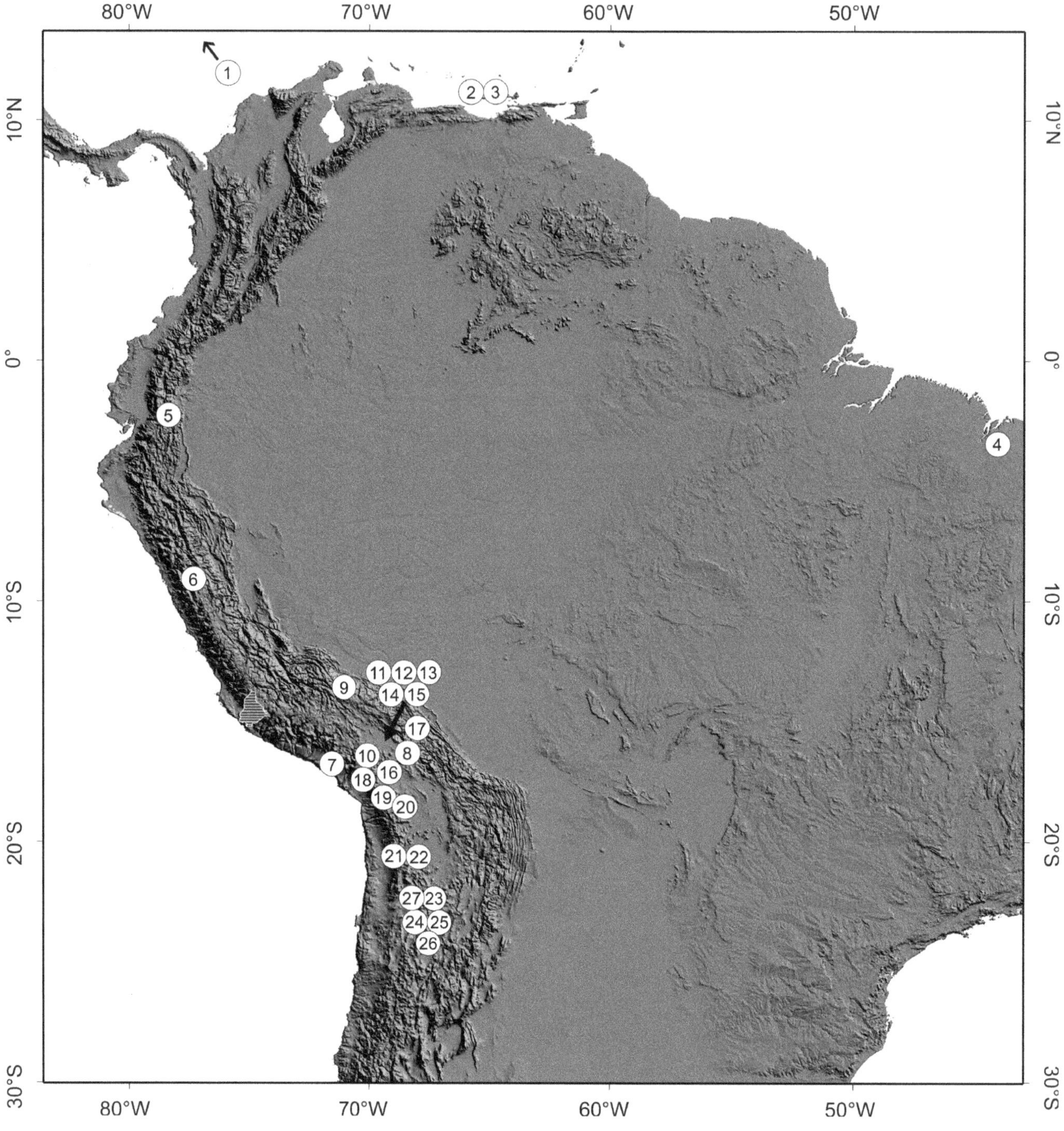

Fig. 6: Map of Holocene paleoclimate records used in Figure 7; Rio Grande de Nazca drainage basin shown hatched. **1** – Poore et al. 2003 (Gulf of Mexico, foraminiferal $\delta^{18}O$); **2** – Haug et al. 2001 (Cariaco Basin, Ti concentration); **3** – Tedesco & Thunell 2003 (Cariaco Basin, foraminiferal oxygen isotopes); **4** – Pessenda et al. 2005 (Lagoa do Caçó, pollen, charcoal and $\delta^{13}C$); **5** – Hansen et al. 2003 (Lagunas Chorreras and Pallcacocha, sediment cores); **6** – Thompson 2001a (Huascarán ice core $\delta^{18}O$); **7** – Holmgren et al. 2001, Latorre et al. 2005 (Arequipa area, rodent middens); **8** – Abbott et al. 2000 (Lago Taypi Chaka Kkota, Cordillera Real, $\delta^{18}O$ and paleolimnology); **9** – Thompson 1992 (Quelccaya Summit and Quelccaya Core 1 accumulation rates); **10** – Rigsby et al. 2003 (Rio Ilave, fluvial terraces); **11** – Abbott et al. 1997 (Titicaca, sediment cores); **12** – Baker et al. 2001 (Titicaca, sediment cores); **13** – Tapia et al. 2003 (Titicaca, diatom stratigraphy); **14** – Cross et al. 2000 (Titicaca, sediment cores and seismic profiles); **15** – Rowe et al. 2002 (Titicaca, sediment cores); **16** – Baucom & Rigsby 1999 (Rio Desaguadero, fluvial sediments); **17** – Servant & Fontes 1984 (Bolivian Andes, fluvial incision and sedimentation); **18** – Placzek, Quade & Betancourt 2001 (Lago Aricota, lake levels); **19** – Thompson 2001b (Sajama ice core $\delta^{18}O$); **20** – Baied & Wheeler 1993 (Laguna Seca, palynology); **21** – Latorre et al. 2006 (Rio Salado basin, rodent middens); **22** – Rech et al. 2002 (Tilomonte Springs, Rio Loa, Rio Salado, wetland deposits); **23** – Betancourt et al. 2000 (central Atacama, rodent middens and wetland deposits); **24** – Bobst et al. 2001 (Salar de Atacama, salt core); **25** – Grosjean et al. 2001 (Laguna Miscanti, sedimentology and palynology); **26** – Grosjean et al. 1995 (Atacama Altiplano, i.a. lake levels); **27** – Rech et al. 2003 (Quebrada Puripica, paleowetlands).

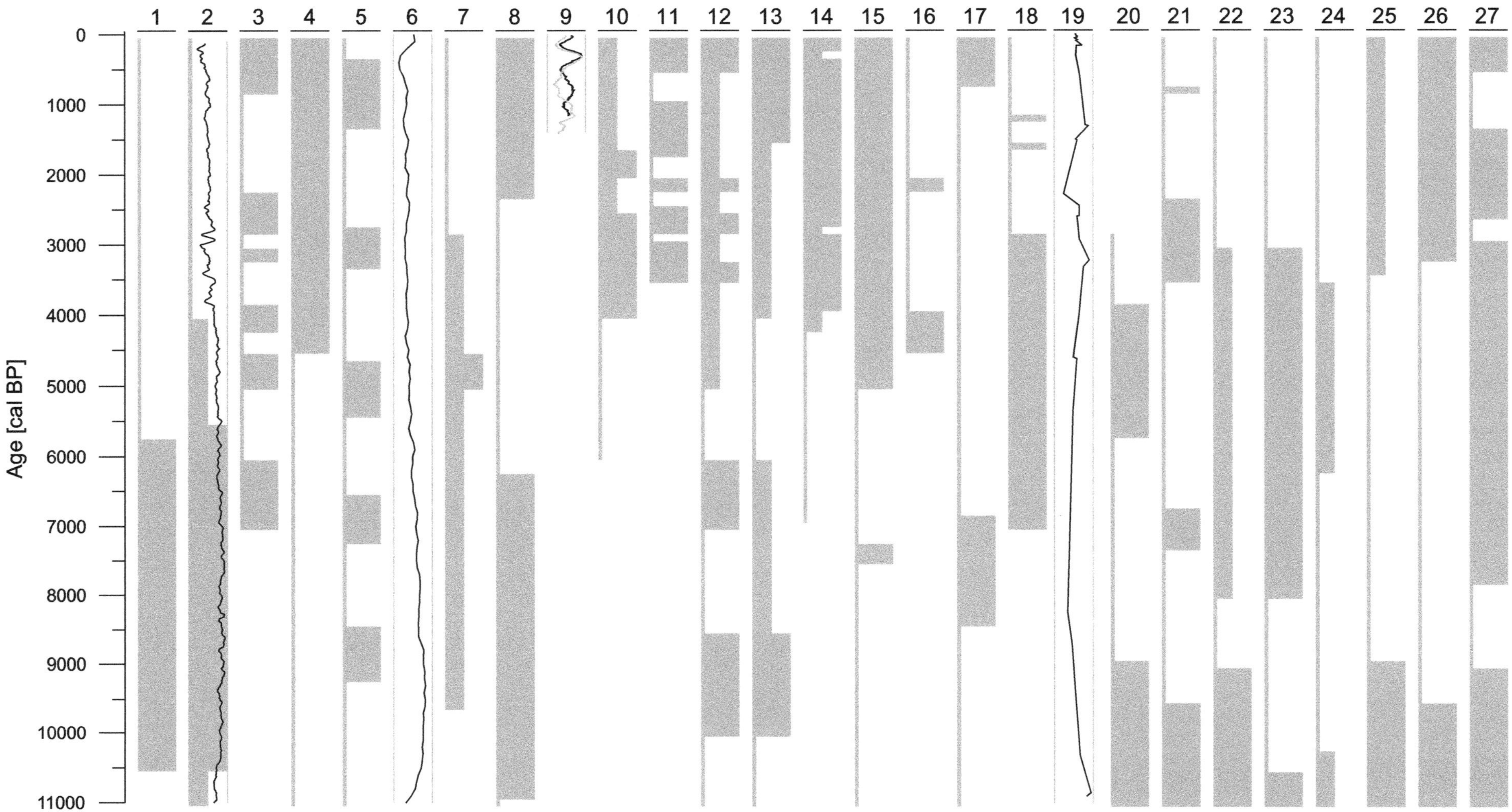

Fig. 7: Holocene moisture variability in South America north of 30°S. Records used are listed and mapped in Figure 7. Wide bars indicate that records have been interpreted as showing relatively moist conditions. Data for records 2, 6, 9 and 19 are shown as line graphs: **2** – 11-year running average for the Cariaco Ti record, scale: 0.0 to 0.4%; **6** – Huascarán ice core $\delta^{18}O$, scale: -20 to -15‰; **9** – 101-year running average for the Quelccaya Summit (black) and Quelccaya Core 1 (grey) accumulation rates, scale: 1.0 to 1.8 m/a (note the discrepancies between the two cores before ca. 500 cal BP); **19** – Sajama ice core $\delta^{18}O$, scale: -8 to 8‰.

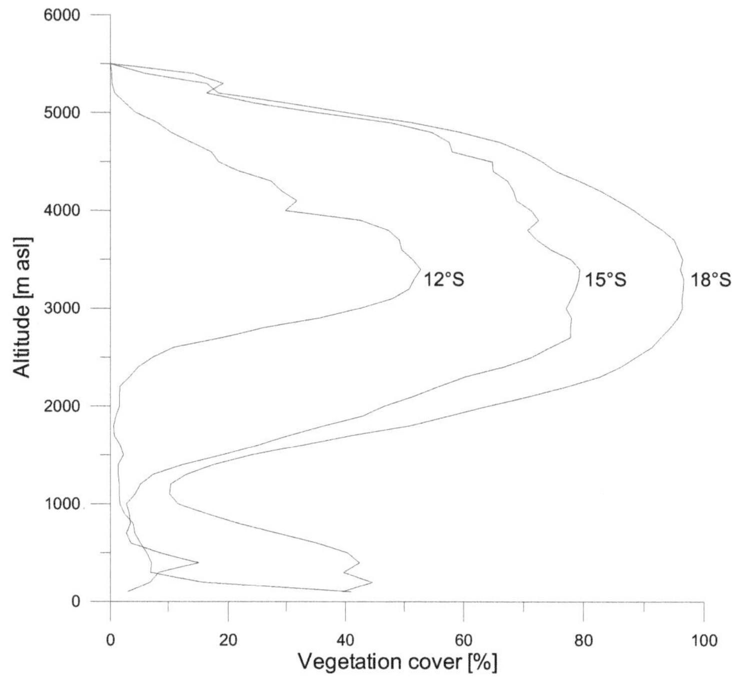

Fig. 8: Relationship between elevation and total vegetation cover (deduced from comparison of MODIS Vegetation Continuous Fields of November 2000 to November 2001 in ±0.5° latitude segments centred on each degree latitude; elevation data from SRTM) for the western slope of the Cordillera Occidental including the pre-Andean basin and the Cordillera de la Costa at 12, 15 and 18°S.

Tab. 1: Ecological zonation for the Arequipa area (ca. 16°S; after HOLMGREN et al. 2001: 244) and the equivalent elevations for the Palpa area as deduced from comparison of MODIS Vegetation Continuous Fields of November 2000 to November 2001 in ±0.5° latitude segments centred on each degree latitude; elevation data from SRTM).

Ecological zone	Elevation at 16°S	MODIS total vegetation cover (herb+tree)	Equivalent elevation at 15°S
hyperarid, absolute desert	below ~2,400 m	<15%	below ~2,000 m
Matorral desert: xerophytic shrubs, columnar cacti, small herbs and grasses	2,400 – 2,800 m	15 – 40%	2,000 – 2,400 m
Matorral (shrubland) desert with columnar cacti and Pajonal (grassland)	2,800 – 3,300 m	40 – 65%	2,400 – 3,000 m
Pajonal (grassland)	3,300 – 4,000 m		above ~3,000 m
Pajonal-Pastizal vegetative community	4,000 – 4,300 m		
Tolar-Pajonal	4,300 – 4,500 m		
dispersed cushion plants	above ~4,500 m		

Lomas communities typically consist of ephemeral, hygrophilic herbs (FERREYRA 1983: 242; KOEPCKE 1961: 220; RICHTER 1981: 24). A few lomas areas even support perennial trees or cacti (KOEPCKE 1961: 221; RICHTER 1981: 24). The presence of almost pure Tillandsia lomas is common at the upper limit of fog influence in southern Peru and northern Chile (RUNDEL et al. 1991: 27ff.; PINTO et al. 2006: 550; KOEPCKE 1961: 230f., after Ferreyra 1953 and Rauh 1958). Their dependence on fog has been described as early as 1925 (BERNINGER 1925: 383f.) and their importance as bioindicators of climatic change has been noted (PINTO et al. 2006: 551, after Rundel et al. 1997).

Where topographic conditions allow the fog to extend inland (e.g. near Tacna, ca. 17°55'S), lomas vegetation (in particular Tillandsia lomas) can occur 40 to 50 km (RUNDEL et al. 1991: 27; RICHTER 1981: 24) or even 60 km (KOEPCKE 1961: 220, after Ferreyra or. commun.) from the coast in southern Peru and up to 45 km from the coast in northern Chile (PINTO et al. 2006: 545f.). Beside the furthermost inland extent of lomas formations their strongest development also occurs in southern Peru. According to KOEPCKE (1961: 221), the lomas of Atiquipa near Chala represent the maxmium lomas development, with evergreen woody vegetation and sufficient *garúa* precipitation to feed a seasonal stream. According to RUNDEL et al. (1991: 30), lomas vegetation is not well developed around Nazca; however, this statement may be due to the severely restricted accessibility of the coast because the Panamericana runs ca. 40 to 50 km inland between the town of Lomas and the Peninsula Paracas. In the desert between Palpa and Ica the characteristic vegetation stripes of Tillandsia lomas (cf. CRAIG & PSUTY 1968: 123ff.; RUNDEL et al. 1991: 30, after Velande 1949 and Masuzuwa 1982) can be discerned on high resolution satellite images (Quickbird) at a distance of 55 km from the coast and elevations of 800 to 900 m, indicating that fog precipitation and lomas resources may have been of some importance to pre-Columbian settlements as far inland as Palpa.

ELLENBERG (1959: 66ff.) poses the question whether anthropogenic changes have led to an irreversible degradation of lomas communities – existing lomas forests may maintain themselves due to the amount of fog moisture caught, but once destroyed they may not be able to re-establish themselves (RUNDEL et al. 1991: 40). The question of Holocene changes in lomas extent and composition and the role of climatic vs. anthropogenic factors in these changes is of great importance for the interpretation of climatic and human history in coastal Peru. However, while impacts due to human utilisation (in particular grazing) have been identified, until now no firm answers have been presented (cf. ELLENBERG 1959: 66ff.; RUNDEL et al. 1991: 40; KOEPCKE 1961: 241; cf. 2.5.1).

2.5 Archaeology and cultural landscape

2.5.1 The archaeological sequence of the coastal desert

Humans have been living in the coastal desert of southern Peru since the terminal Pleistocene (Table 2). Most of the earliest occupations (13–11 cal ka BP: SANDWEISS et al. 1998: 1,830; ca. 13–9 cal ka BP: DEFRANCE & ÁLVAREZ 2004: 267 and KEEFER et al. 1998: 1833; see LLAGOSTERA 1992: 87ff., REITZ 2001 and SANDWEISS 2003 for overviews of Late Pleistocene to mid-Holocene coastal sites) have been found close to the coast where people depended on marine resources. Early occupation of mid- and high altitudes also occurred at 13.46–11.70 cal ka BP (GROSJEAN et al. 2005: 157ff.). Many of the earliest coastal sites may have been submerged by rising sea level (SANTORO et al. 2005: 250; cf. SANDWEISS et al. 1998: 1,830). Inland settlements along the valleys of the allogenic rivers appear after the seashore-based earliest occupations, in the Rio Grande de Nazca area by ca. 3,000 cal BP (REINDEL et al. 2001). This has led Moseley to the formulation of his hypothesis of the maritime foundations of Andean civilizations (MOSELEY 1992; cf. MANN 2005: 34f.; cf. STOTHERT 1992: 43ff. for Ecuador). However, while coastal sites above the present sea level are well preserved in the desert climate, a possible early use of the inland valleys may be masked by fluvial erosion and sedimentation and by intense landscape modification by later cultures (cf. FISH 1999: 205). The paucity of early inland sites can therefore not be taken as a proof of exclusive use of coastal resources, in particular since links between the coast and the Andes are evident e.g. in the presence of highland obsidian on the coast (cf. BONAVIA & CHAUCHAT 1990: 402; REINDEL & ISLA CUADRADO 2004: 149; ISLA et al. 2003: 232; SANTORO et al. 2005: 251). Given the extreme aridity, questions arise regarding the provision of potable water on the coast. Several authors (LANNING 1963: 360ff.; ENGEL 1973: 271ff.; COHEN 1975; cf. QUILTER 1992: 118) have suggested a more vigorous development of lomas formations and the existence of springs or wells during the early to mid-Holocene and that overuse or climatic change may have led to the degradation of lomas and subsequently to a depletion of local water resources. For a discussion regarding LANNING's (1963) fossil lomas hypothesis, see PARSONS (1970: 300ff.) and COHEN (1975: 101ff.).

People were living at the mouth of the Rio Ica between 7,520 ±300 ^{14}C BP (9,050–7,650 cal BP) and 870 ±80 ^{14}C BP (930–650 cal BP), with indications for the use of both marine (molluscs and sea mammals) and terrestric resources (land snails) present and cultigens (yams and beans) appearing in a layer dated at 5,430 ±140 ^{14}C BP (6,450–5,750 cal BP) (ENGEL 1981: 20). Small settlements also existed at the mouth of the Rio Grande de Nazca, perhaps as early as 7,000 to 6,000 years ago and certainly by 4,720 ±120 ^{14}C BP (5,650–4,950 cal BP) (ENGEL 1981: 21). As on the Atacama Altiplano (cf. GROSJEAN et al. 2005: 168), the *silencio arqueológico* on the coast of southern Peru (cf. SANDWEISS 2003) can therefore not be seen as an absolute hiatus in human occupation.

The earliest agriculture in the Andes dates back to ca. 10 ka BP (FELDMAN 1992: 69, after Kaplan et al. 1973 and Lynch et al. 1985) and irrigation agriculture in the Peruvian desert can be traced back to at least 5,400 ^{14}C BP (ca. 6,100 cal BP) and possibly 6,700 ^{14}C BP (ca. 7,500 cal BP) (DILLEHAY et al. 2005: 1,7241ff.). The earliest complex societies with monumental architecture were based on an economy featuring intense links between coastal and inland populations and were established by 4,600–3,900 cal BP (SOLIS et al. 2001: 723ff.). However, QUILTER (1992: 115) states that the "complete range of Andean domesticates was widespread and available in the Central Andes only during the Early Horizon and perhaps not until later during that period for some areas." Coast–inland links existed in the Rio Grande de Nazca area by the time of the Early Horizon: marine mollusc shells and highland obsidian are found in Paracas culture contexts in the Palpa Valley (REINDEL & ISLA CUADRADO 2004: 149; ISLA et al.

2003: 268), in the Nazca Valley (Silverman 2002: 157) and in the Paracas culture coastal settlement of Carhua (Engel 1991: 137).

Engel (1963: 14) places the appearance of cotton around 4,500 ^{14}C BP (ca. 5,000 cal BP) and considers it a marker important enough to warrant the separation of an early (without cotton) and a later (with cotton) preceramic phase. He hypothesizes that the introduction of maize or the arrival of a more dynamic population by ca. 3,200 ^{14}C BP (ca. 3,300 cal BP) (Engel 1963: 20; cf. Wilson 1981: 103; cf. Feldman 1992: 72, after Fung 1972; cf. Silverman 1996: 122f., after Kaulicke 1994) created a disequilibrium of food supply and demand which led to the abandonment of coastal sites and to the development of irrigation agriculture in the allogenic river valleys (Engel 1963: 65ff.). Ceramic appears in southern coastal Peru during the Initial Period by ca. 3,000 ^{14}C BP (ca. 3,200 cal BP) (Engel 1987: 89ff.), almost five millennia after the earliest American pottery in Brazil at ca. 8,000 cal BP (Roosevelt et al. 1991). However, sites with and without pottery coexisted for considerable time (Pozorski & Pozorski 1990: 488). Contrary to the development of complex societies with urban centres and monumental architecture during the late Preceramic through Initial Periods on the northern coast of Peru, the southern coast was characterised by a small population living in scattered villages (Silverman 1996: 107ff.).

In the Rio Grande de Nazca drainage basin, the pre-Columbian cultural sequence following the Initial Period can be divided into five main periods (Table 2). The different tributaries have been investigated by Reindel et al. (2001) and Reindel et al. (2002) (Rio Grande, Rio Palpa, Rio Vizcas) and by Silverman (1994, 1996 and 2002) and Schreiber (1999) (Rio Ingenio, Rio Nazca and south of the Rio Nazca). During the Early Horizon, the Paracas culture was "the first regional culture of the south coast" (Silverman 1996: 134). While Silverman (1996: 131) suggests that "[t]he valleys of the Rio Grande de Nazca drainage were essentially outside the Paracas Tradition", survey in the valleys of the more northern tributaries (Reindel et al. 2001: 94; Reindel et al. 2002) resulted in a number of evenly distributed sites from that period (see also Isla et al. 2003; Reindel & Isla Cuadrado 2004).

The Initial Nasca Period was characterised by a strong increase in the number of settlements located along the valley margins. There appears to have been a preference for the upper valley segments and several centres with formal architecture existed (Reindel et al. 2001: 94). The Early Intermediate Period was characterised by the rise and decline of the Nasca culture. In the Early Nasca period, the number of settlements, including those with public architecture, increased (Reindel et al. 2001: 94). The dominant (likely in terms of religious importance) centres were Cahuachi in the lower Rio Nazca Valley (Silverman 2002: 12, after Silverman 1993) and Los Molinos as a secondary centre near Palpa (Reindel & Isla Cuadrado 2001). In the Middle Nasca Period, the number of settlements in the upper valley segments remained constant while it decreased in the lower segments. Cahuachi and Los Molinos were abandoned and La Muña was the dominant settlement in the Palpa Valley in this period (Reindel et al. 2001: 95; Reindel & Isla Cuadrado 2001). In the Late Nasca period, there was a decline in the number of settlements in the lower and middle valley segments, in particular in the Palpa area. These changes did not affect the upper valley segments (Reindel et al. 2001: 95).

During the Middle Horizon, most of the numerous settlements were located in the middle and upper segments of

Tab. 2: Chronology according to Unkel (2006: 111) (AD/BC ages converted to cal BP with reference to AD 1950; note that mean values are assigned to phase boundaries for which begin and end of subsequent phases were not identical.

Period	Cultures / ceramic styles	Begin [cal BP]	End [cal BP]
Historic Period		418	
Late Horizon	Inca	550	418
Late Intermediate Period	Ica, Chincha	950	550
Middle Horizon	Wari, Loro	1,300	950
Early Intermediate Period	Late Nasca (Nasca 6 and 7)	1,520	1,300
	Middle Nasca (Nasca 4 and 5)	1,655	1,520
	Early Nasca (Nasca 2 and 3)	1,870	1,655
Transition	Initial Nasca	2,150	1,880
Early Horizon	Late Paracas (Ocucaje 8 and 9)	2,325	2,150
	Middle Paracas (Ocucaje 5, 6 and 7)	2,485	2,325
	Early Paracas (Ocucaje 3 and 4)	2,780	2,485
Initial Period		3,750	2,780
Archaic Period	no ceramics	13,950	3,750

the Rio Grande valley (Reindel et al. 2001: 95f.). Schreiber (1999: 168) sees a decrease in regional population during this period. In the Late Intermediate Period, there was a strong increase in settlement with a population size comparable with or higher than during the Early and Middle Nasca period. However, this population was concentrated in a few large settlements instead of numerous small, scattered villages. These large settlements were often located in quebradas at some distance from the valley floors, but usually close to the main agricultural areas (Reindel et al. 2001: 96). During the Late Horizon (Inca), most settlements were located in the middle valley near the best agricultural land (Reindel et al. 2001: 96). The early Historic Period (since 1532) was characterised by a dramatic decline of the native population and land use changes from subsistence to commercial farming (e.g. grapes and sugarcane) (Cook 1981: 160ff.). To a varying degree, the observed changes in population size and cultural traits have been interpreted as being influenced by climatic changes (Silverman 2002: 24f.; Reindel et al. 2003: 127; Mächtle et al. 2006) or extreme events (Reindel & Isla Cuadrado 2000: 95).

2.5.2 *Cultural landscape and human-environment interactions*

The term "cultural landscape" as it is used here refers to the entirety of the built and modified environment, including in particular "landscape capital" (e.g. irrigation infrastructure, terraces, fields) (cf. Erickson 2000: 314) and settlements, but also travel routes and the parts of the environment which are used or exploited but not necessarily shaped by a society (cf. Koepcke 1961: 236). While primarily aimed at the physical interactions between societies and their environment, cultural is also meant in the widest sense of bearing meaning for a population (e.g. including the sense of "sacred space", cf. Lambers 2004: 138ff.; Silverman & Proulx 2002: 96ff.; cf. Aveni 1990 and references therein). However, the main focus of this work lies on the physical interactions between natural and anthropogenic processes, i.e. on the impact of natural processes on societies and of anthropogenic processes on the landscape. Particularly, the impact of fluvial processes on agriculture and settlements and the impact of irrigation agriculture on the morphology and the sediments in the Palpa valley are analysed. Settlement patterns as one important aspect of the cultural landscape have received considerable attention in both coastal and highland Peru (e.g. Billman 1999; Browne 1992; Schreiber 1999; Silverman 2002; Silverman & Proulx 2002: 96ff.; Reindel et al. 1999; Reindel et al. 2001; Reindel et al. 2002; Stanish 1999: 125ff.; cf. Oliver-Smith 1999: 77; Wells 2001: 136f.; Fish 1999) and will be taken into account.

Given the spatially extensive remains of pre-Columbian cultures in the research area, an "archaeology of landscapes" as proposed by Erickson (1999: 641) for the Lake Titicaca region provides a conceptual framework for the interpretation of archaeological as well as geomorphological and sedimentological evidence and the reconstruction of the evolution of the cultural landscape. This not only applies to the obvious cases like the spatial character of the geoglyphs (e.g. Lambers 2004) or settlement patterns (e.g. Silverman 2002), but in particular to the agricultural infrastructure. With respect to the Rio Grande de Nazca drainage basin, this aspect has so far received no attention other than a few marginal remarks.

Publications dealing with the question of whether or not climate influences cultural development are often biased towards finding evidence *for* such an influence. This is in large part due to selective use of the available data: only those climatic events which correlate with cultural changes and only those cultural changes which correlate with climatic events are presented, while climatic events as well as cultural changes which do not fit into this picture remain undiscussed. While covering almost an entire climatic belt, even the thoughtful discussion by Brooks (2006) suffers from not asking the questions why other cultures affected by the mid-Holocene climate change did *not* develop into complex societies and why previous climatic changes of comparable magnitude did *not* trigger the described cultural changes. Given the long history of human societies that repeatedly lived through profound environmental changes, the question why complex societies emerged quasi simultaneously throughout the world in the mid-Holocene remains an enigma to be solved.

There are four fundamentally different ways of thinking about the interaction between the environment and culture:

(a) The deterministic view holds that nature *determines* human activities and *causes* certain cultural traits and developments, i.e. that human societies are at the mercy of favourable or detrimental natural forces (e.g. Lanning 1967: 175; Paulsen 1976: 126).

(b) In the contextual model, the natural environment is seen as the context in which human societies actively develop. Individuals and societies are seen as having to *adapt* and *respond* to changing external conditions (e.g. Dillehay et al. 2004: 279; Moseley 1992; Silverman 2002: 24; Dillehay & Kolata 2004: 4,325).

(c) The co-evolutionary model of environment and human societies focuses on the mutual influence and positive as well as negative *feedbacks* between nature and society (Hoffman 1999: 308, after Ingold 1992); however, focus is often only on one side of the interaction (e.g. Wells & Noller 1999: 756; Brooks 2006: 30).

(d) The anthropocentric view contends that humans are and have for a long time been *actively shaping* their environment and are largely independent of environmental fluctuations in the sense that the outcome of their interaction with nature depends on cultural traits rather than on environmental processes (e.g. ERICKSON 1999).

To convincingly show a climatic influence on cultural development requires secure and accurate dating of both climatic and cultural changes, a mechanism by which climate could have exerted an influence on culture and a discussion of alternative explanations (in particular intra- and inter-societal) which shows that the climatic hypothesis is the best explanation. In particular the last requirement is almost impossible to fulfil given the inherent dynamics of human societies. While parsimony is acknowledged as a basic scientific principle, the interlinked, multivariate nature of both environment and human culture and their often non-linear behaviour imply that simplistic models will exclude relevant phenomena and might therefore result in misleading conclusions. The most parsimonious explanation will therefore likely have to be complex. Even if a climatic influence on society can be shown, the question remains why the society did not respond in different ways – the outcome is therefore not determined, rather, climate can at best be seen as a trigger for society-internal changes. Crisis (e.g. reduced food resources) can be dealt with in different ways (e.g. conflict, co-operation, technology), the choice of which depends on culture and on the (cumulative) decisions of individuals.

Since neither natural systems nor cultures are stable, some sort of natural change is likely to be found within the range of dating uncertainty of both climatic and cultural change, however, climatic influences other than direct physical destruction are hard to prove. All this is not to dismiss the possibility of such an influence but to clearly show the deep-seated limitations of most studies published so far on this subject.

3. Research methods and available data

3.1 *Field methods*

Preliminary field investigations for research in the Palpa area were carried out during an exploratory trip by Dr. J. Baade in 2000. Extended field work (four to seven weeks each) was carried out by Dr. J. Baade, R. Hesse and S. Kastner in 2002, by Dr. J. Baade, S. Kastner and S. Angelstein in 2004 and by Dr. J. Baade and R. Hesse in 2005 and 2006. Work concentrated on the following areas in the Palpa Valley (Figure 9): (a) near the archaeological site of La Muña in 2002 and 2004, (b) along and north-west of the Rio Palpa in 2002 and 2004 and (c) between Rio Palpa and Rio Vizcas as well as south-east of the Rio Vizcas in 2005 and 2006. Besides the localities in the Palpa Valley, a few profiles were investigated in the valley of the Rio Grande upstream from La Muña and downstream at the Rio Grande narrows near Cerro Corpante in 2005 and 2006.

A total of 50 sediment profiles (38 pits and 12 bank exposures) were investigated in detail using standard sedimentological methods (AG BODEN 2005; TUCKER 1982). Pits were usually excavated in fallow fields. As a general rule, excavation continued until basal gravel was reached. In a few cases, work had to be terminated earlier due to requests of landowners or groundwater. Profiles were cleaned and photographed. Stratigraphic sketches were drawn and written descriptions produced in the field on the basis of grain size, colour and sedimentary features (e.g. bedding). Sediment colour was determined visually by J. Baade and S. Kastner without the use of standard colour charts. In selected profiles, carbonate tests were performed. The positions of all profiles were measured using either centimetre-resolution differential GPS (Magellan Pro Mark X) or a Leica total station referenced to

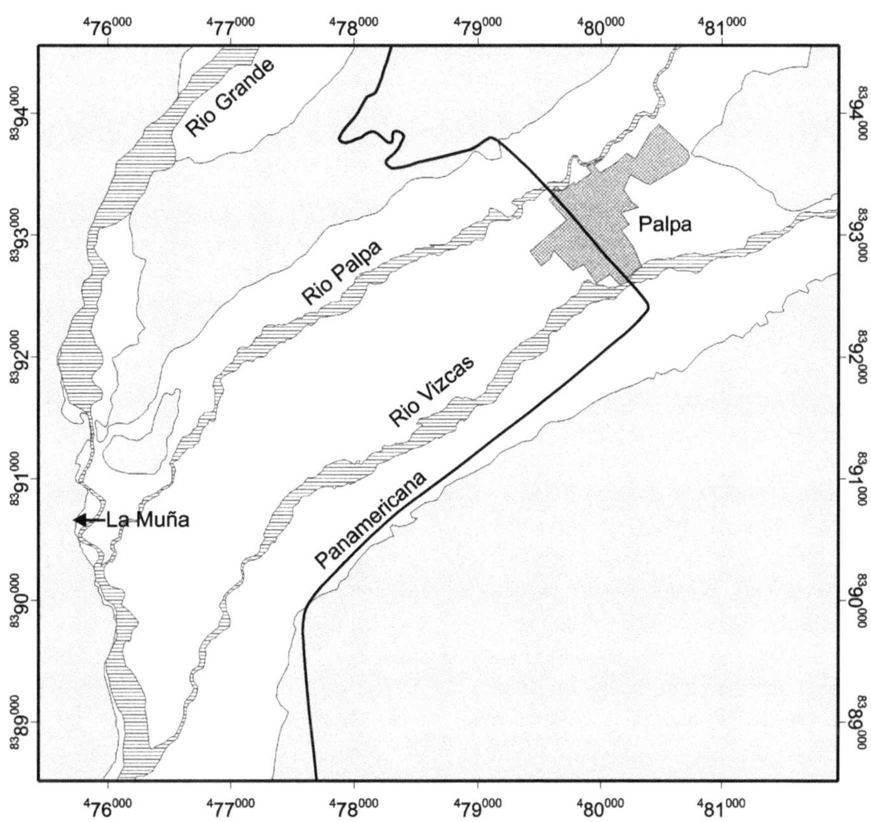

Fig. 9: Map of the Palpa Valley and the adjacent part of the Rio Grande valley.

points measured by centimetre-resolution differential GPS. From the profiles, a total of 487 samples were taken for grain size determination and chemical investigations and 407 samples for radiocarbon dating. Several hundred diagnostic archaeological finds (mainly ceramic shards) were recovered and, where possible, assigned to cultural phases by J. Isla Cuadrado and J. Palomino on the basis of the regional ceramic style chronology. Magnetic susceptibility was measured in profiles LM-1, LM-2, LM-5, LM-6, RPE, YN-1, SF-1, SF-2, SF-3 and SF-4. Results are shown in Appendix B. Since they proved inadequate for the intended purpose of aiding in the stratigraphical correlation of different profiles, they were not used in this thesis.

In March 2005, water table measurements were made in 33 wells in the Palpa and Grande valleys. From the end of February to early April 2006, sequential water table measurements were made in these and 16 additional wells to observe temporal changes of the groundwater level. Like the profile positions, well positions were measured using centimetre-resolution differential GPS.

3.2 Laboratory methods

Laboratory analyses were carried out in the geo-ecological laboratory at the Department of Geography, Friedrich Schiller University Jena. Grain size analysis was performed for all 487 soil and sediment samples by the combined sieve/pipette method after Köhn (DIN 19683, Deutsches Institut für Normung 1973). To reduce sample weight for transport, samples containing a large fraction >2 mm were dry-sieved in the field laboratory. Pre-treatment consisted of organic carbon destruction with H_2O_2. Samples were wet-sieved (6,300, 2,000, 630, 200, 63 µm); silt and clay fractions were determined by the pipette method. Sorting was calculated as the square root of the quotient of the 75% and 25% percentiles (cf. Füchtbauer 1988: 136). Clay-to-silt ratio (CSR) was calculated to investigate clay enrichment (Schaetzl & Anderson 2006: 579, after Van Wambeke 1962; cf. Noller 1993: 137). CSR values are only used for samples containing more than 10% silt and clay combined to avoid spurious values resulting from measurement error.

Carbonate content (346 samples) was determined using the volumetric method after Scheibler according to DIN 10693 (Deutsches Institut für Normung 1993) (measurement of CO_2 released upon reaction with 10% HCl). CO_2 volumes were converted to percent $CaCO_3$ under the assumption that calcium carbonate constituted the only carbonate present in the samples and using a correction for temperature and atmospheric pressure. Loss on ignition (LoI) was determined according to DIN 19684 (Deutsches Institut für Normung 1977) for all 487 samples by repeated heating to 430°C until stable sample weights were reached. It is assumed that LoI values reflect the organic matter content of the samples and that factors such as chemically combined water are of minor influence.

Measurement of chemical parameters using an atomic absorption spectrometer (AAS) Shimadzu AA-6800 after *aqua regis* extraction from the sample fraction <2 mm included the elements Cu, Cr, Fe, Mn, Ca, K, Mg and Na (224 samples) as well as Ni, Zn and Al (98 samples). However, due to calibration problems the results for the 98 samples measured in 2003 had to be discarded, leaving Cu, Cr, Fe, Mn, Ca, K, Mg and Na measurements for 126 samples. Furthermore, since significant correlations were found between grain size parameters and element content (suggesting that the choice of the total fine-grained fraction <2 mm was inadequate for chemical characterisation and intercomparison of the samples), the use of the AAS measurement results is very limited.

Further analyses were conducted for a limited number of samples: Bulk density (11 samples) and gravimetric water content (6 samples) were determined for the top 2.0 m of profile SR. H_2O extractable Cl^- content was measured in 21 samples (using a Dionex DX-120 ion chromatograph) as a proxy for salinity with the purpose of (a) determining the salinity of river and irrigation water and (b) observing down-profile changes of salinity in profile SF-2. Grain size measurements using a laser particle sizer Shimadzu SALD-2001 were conducted for samples of suspended sediment from river and irrigation water because in these cases the small sample sizes were not sufficient for sieving and pipette analysis.

3.3 Chronometric methods

Chronometric control of sedimentary layers was achieved mainly through radiocarbon dating. Out of the 407 collected samples, 74 were selected on the basis of (a) spatial coverage (distribution over the research area), (b) temporal coverage (stratigraphic position in a given profile: lowest suitable sample and additional samples in the sedimentary column) and (c) sample quality (preference of charcoal pieces over charcoal flitter). Samples were submitted to the Poznan Radiocarbon Laboratory. There, pre-treatment before AMS analysis was acid-alkali-acid treatment to remove carbonates as well as humic and fulvic acids (Poznan Radiocarbon Laboratory 2007). Results were reported by the Poznan Radiocarbon Laboratory as ^{14}C years BP. Four samples did not yield results due to very small sample size. The 70 radiocarbon dates were subsequently calibrated to calendar years BP using OxCal Version 3.10 (Bronk Ramsey 1995, 2001) and southern hemisphere correction (shcal04.14c, McCormac et al. 2004). No reservoir effects have been taken into account.

The possibility of hiatus or remobilised (older) carbon is acknowledged; however, unless additional information (e.g. archaeological findings or stratigraphically inconsistent dates) suggest otherwise, the radiocarbon ages are considered to reflect the age of deposition of the layer in which they were found.

Additional information regarding the age of sediment deposition comes from the style chronology for pre-Columbian ceramics in southern Peru (cf. Silverman & Proulx 2002: 14–39). Classification of ceramic shards into culture phases was done by J. Isla Cuadrado and J. Palomino. Most ceramic samples pertain to the Paracas and Nasca cultures. Calendar age range assignment to the phases follows Unkel (2006: 111) (cf. Table 2). Where both radiocarbon and ceramic ages or several ceramic ages were acquired from the same context, comparison of ages revealed additional information: while a good correspondence between ceramic and charcoal ages or between several ceramic ages is interpreted as an indicator for primary deposition, discrepancies between radiocarbon and ceramic ages or between different ceramic ages are interpreted as strong evidence for remobilisation (i.e. secondary deposition) or disturbance. In this case, the ages were not included in age-depth models. They can, however, be used in the temporal interpretation of the sediments insofar as the youngest of several conflicting ages represents a minimum age for the deposition.

Stratigraphic information was used to correlate corresponding layers in different profiles. However, the possibility of inter-profile correlation was limited by the ambiguous nature of many stratigraphic features. Sedimentation rates have been calculated mainly from the available radiocarbon dates and in a few cases from stratigraphically consistent and well-defined diagnostic archaeological finds.

3.4 Aerial photographs, satellite images and DTMs

Where applicable, geospatial data were used both as supplementary data and for independent analysis. Available data included aerial photographs, satellite images and data products as well as digital terrain models (DTM).

Aerial photographs of the Palpa Valley taken in the years 1944, 1970 and 1997 were made available by M. Reindel. They were mosaicked, georeferenced and resampled to a resolution of one metre per pixel using the 1:50,000 topographic map of the area. Unfortunately, only the 1944 images cover the entire Palpa Valley. The 1970 images cover the north-western and much of the central part of the Palpa Valley and the 1997 images only cover the north-western margin of the valley.

High-resolution satellite images (Quickbird, taken on Mach 25, 2005) were downloaded tile by tile from Google Earth (http://earth.google.com) on June 20, 2006 and mosaicked and georeferenced at a resolution of one metre per pixel using ground control points which were measured by cm-resolution GPS or transferred from the georeferenced aerial photographs.

Landsat images at a resolution of 28 m (taken by Landsat TM ca. 1990) and 14.25 m (taken by Landsat 7 ETM ca. 2000) (NASA Applied Research & Technology Project Office, https://zulu.ssc.nasa.gov/mrsid/mrsid.pl, download: 26.01.2005) as well as a Landsat 7 ETM scene of April 26, 2000 (Global Land Cover Facility, Institute for Advanced Computer Studies, University of Maryland, http://glcf.umiacs.umd.edu, download: 05.03.2007) were used for catchment-scale analysis and visualisation. The SRTM digital terrain model (DTM) (SRTM Version 2, ftp://e0srp01u.ecs.nasa.gov/srtm/version2, download 13.07.2006) was used for visualisation and terrain analyses on the Rio Grande catchment scale. A high resolution (2 m per pixel) DTM of the Palpa Valley and the surrounding bajadas derived from aerial photography (LAMBERS 2004) was made available by A. Grün and M. Sauerbier (Institute of Geodesy and Photogrammetry, ETH Zurich). For the purposes of this thesis, the DTM of the Palpa Valley was transformed from HAE (height above ellipsoid) to ALT (altitude above mean sea level) to conform to the GPS measurements. GOES weather satellite data of South America (NOAA Comprehensive Large Array-Data Stewardship System, http://www.class.noaa.gov, download: 10.–19.03 2007) are available for extended periods between 1998 and 2007 and were used for observation of synoptic conditions related to precipitation and coastal fog extent. MODIS Vegetation Continuous Fields data of 01.11.2000 to 01.11.2001 (data source: http://glcfapp.umiacs.umd.edu, download: 09.01.2006) were used for quantification of the altitude dependency of vegetation zones. The Global Seafloor Topography data set (SMITH AND SANDWELL 1997, topex.ucsd.edu//pub/global_topo_2min, download: 29.11.2003) was used to determine and visualise offshore topography (near-shore shelf slope, Nazca Ridge).

The geospatial data were mainly used for (1) change detection from 1944 to 2006 with focus on fluvial dynamics and anthropogenic changes, (2) detection of recent as well as pre-1944 features, (3) selection of sites for sedimentological work and (4) relating the results of the stratigraphical and sedimentological work to the surrounding landscape, in particular height relationships among profiles and between profiles and the rivers.

3.5 Discharge data

Annual discharge data for the Rio Grande de Nazca (1932–1969) have been published by ONERN (1971a: 192). Unpublished daily discharge data for the Rio Grande de Nazca and its main tributaries were provided by Junta de Usarios de Palpa for the years 1975 to 2007. However, since consistency of measurement (i.e. location of measurement station above or below tributary confluences for the old data set) between the two data sets was questionable, only the data from 1975 to 2007 are used in this thesis.

4. Late historic landscape dynamics

4.1 *Fluvial dynamics and anthropogenic change*

4.1.1 The near-natural state of the rivers

Aerial photographs taken in the years 1944, 1970 and 1997 and satellite images taken in 1990, 2000 and 2005 document changes related to both fluvial dynamics of and anthropogenic influences on the rivers Rio Palpa and Rio Vizcas since the middle of the 20th century. In particular, the erosion of a large tract of agricultural land in the central part of the Palpa Valley as well as the straightening and narrowing of river segments and the construction of artificial embankments are obvious. Natural fluvial dynamics of these allogenic rivers are dominated by extreme interseasonal and interannual discharge variability. For most of the year, Rio Palpa and Rio Vizcas are dry. For the years 1975 to 2007, the average annual number of days with discharge was 138 days for the Rio Grande, 77 days for the Rio Palpa and 79 days for the Rio Vizcas (data made available by Junta de Usarios Palpa). Interannual discharge variability in this period was characterised by ratios between highest and lowest annual discharge of 60 for the Rio Grande (5.4 to 323 Mio. m^3/a), 75 for the Rio Palpa (0.96 to 72.1 Mio. m^3/a) and 60 for the Rio Vizcas (1.24 to 74.4 Mio. m^3/a). Peak discharges were 280 m^3/s (Rio Grande), 160 m^3/s (Rio Palpa) and 260 m^3/s (Rio Vizcas).

The early aerial photographs document a near-natural state of the rivers in the year 1944 (see Figure 10). Over most of their course through the Palpa Valley, Rio Palpa and Rio Vizcas were typical braided gravel-bed rivers with first- and second-order flow channels and longitudinal bars. Within approximately one kilometre of its confluence with the Rio Grande and in the vicinity of the Panamericana bridge, the Rio Palpa showed a strong tendency towards meandering behaviour. The rivers had not been straightened or narrowed and there were almost no artificial embankments; however, some anthropogenic structures like irrigation canal intakes were present. This near-natural state of the rivers (which by 2005 was only observable on half of this stretch of the Rio Palpa and only as remnants outside of the artificial embankments for the Rio Vizcas) is characterised by high width-depth-ratios of >50 (width 50–180 m, average flow depth <1 m), division of flow into several channels and strong lateral activity. The valley slope is 1.2 to 1.6%, and river bed material is gravel to boulders with sandy to silty low-energy deposits on banks and in depressions. The above data describe the rivers as braided rivers of the categories D3 and D4 in the classification by Rosgen (1994). Such rivers are attributed with very high sensitivity to disturbances, poor recovery potential, low critical dimensionless shear stress, very high streambank erosion potential and moderate vegetation controlling influence (Rosgen 1994: 191–194). Valley slope and discharge data place Rio Palpa and Rio Vizcas at the diffuse threshold between meandering and split channels according to Alabyan & Chalov (1998) and slightly on the meandering side of the meandering-braiding threshold according to Leopold & Wolman (1957).

4.1.2 Pre-1944 fluvial dynamics

The 1944 aerial photographs as well as the 1997 high-resolution DTM were used to deduce features related to pre-1944 fluvial dynamics in the Palpa Valley. In the aerial photographs, such features include in particular arcuate to straight lineaments originating from the rivers and quebradas. From the DTM, erosional banks and terrace edges were extracted. Fluvial activity, e.g. channel avulsions or flood events, result in changes in surface and sub-surface properties of sediments and soils (e.g. colour, grain size, moisture capacity) as well as in the density and vitality of subsequent vegetation and have the potential of being recognisable on aerial photographs. The lineaments visible in the 1944 aerial photographs are therefore interpreted as being caused by preceding fluvial events. However, agricultural and other human activities may destroy or conceal such traces. Therefore, the recognised lineaments probably record only a fraction of the fluvial events. Furthermore, a chronological classification into earlier and later events is largely impeded by the paucity of crosscutting features. The visibility of features such as the 1944 river margins and the near-obliteration of earlier lineaments in the 1970 aerial photographs support the notion that the lineaments in the 1944 images record mainly events with an age depth of several years to decades.

Figure 10 shows the results of this mapping. Along-valley terraces can be recognised along the north-western margin of the Palpa Valley. Both Rio Palpa and Rio Vizcas are accompanied by a fringe of fluvial lineaments. The Rio Vizcas appears to be characterised by a stronger lateral activity than the Rio Palpa. This is attributed to the higher peak discharge of the Rio Vizcas (see 4.1.1) and the fact that the Rio Vizcas runs at a higher elevation (13.5 m at the crossings of the Panamericana) (Figure 11). While lineaments originating from the Rio Palpa are generally restricted to a narrow fringe along the river, lineaments from the Rio Vizcas sweep across a large part of the central valley. In the central part of the valley, a prominent swarm of lineaments extends from the Rio Vizcas to the north-west. In the south-western segment of the valley, a well-defined erosional bank related to a former river course is present for a length of 670 m at a distance of ca. 600 m from the present course of the Rio Grande. Lineaments originating from the mouths of quebradas are interpreted as traces of runoff caused by intense local precipitation.

4.1.3 Qualitative and quantitative changes: Rio Palpa

In 1944 (Figure 10), the only clearly identifiable anthropogenic change along the Rio Palpa was the narrowing of

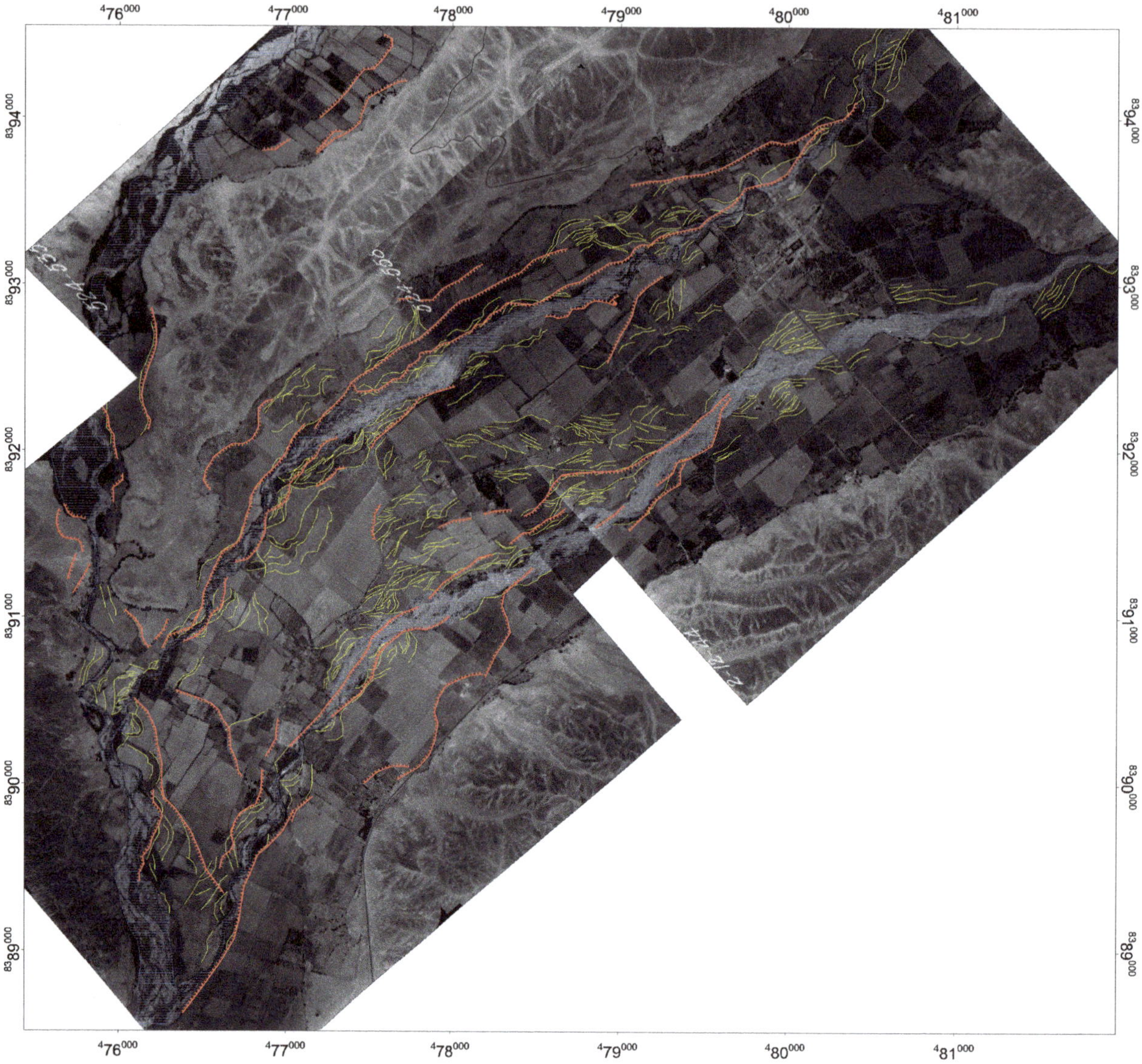

Fig. 10: Aerial photograph of the Palpa Valley from the year 1944 showing the near-natural state of Rio Palpa and Rio Vizcas. Riverbeds are shown in blue, fluvial lineaments in yellow and erosional banks and terrace edges (from the 1997 DTM) in red.

the riverbed at the bridge of the Panamericana. Even in this case, it can be assumed that engineering considerations have led to the construction of the road crossing at a site with minimal river width and dynamics. In a stretch upstream and up to ca. 0.2 km downstream of the Panamericana bridge, the Rio Palpa appears to have been on the threshold between braiding and meandering as indicated by a relatively narrow (up to ca. 40 m) bed with a tendency towards branching flow channels within the bed. From 0.2 to 3.8 km downstream of the bridge, the river was characterised by a wide bed (>100 m) with clearly distinguishable braided primary and secondary flow channels as well as longitudinal bars. The stretch from 3.8 km downstream of the bridge to the confluence with the Rio Grande again had transitional character between braiding and meandering. By 1970, a 0.6 km stretch downstream of the bridge had been straightened and narrowed and ca. 1.5 km of narrow artificial embankments (30–40 m wide) had been constructed between 1.0 and 2.5 km downstream of the bridge. However, most of these embankments had already been severely damaged. By 1997, the prominent meander ca. 0.2–0.4 km upstream of the bridge had been cut. By 2005, the entire length of a reach from >1.1 km upstream to 1.1 km downstream of the bridge had been straightened and enclosed in embankments, narrowing the river to a width of 20–30 m. However, at the time of the 2006 field work more than a hundred metres of coarse gravel embankments had already been swept away by the river despite only moderate discharge during the intervening period (peak discharge 27.0 m^3/s on 04.02.2006) (Figure 12).

Table 3 shows the results of quantitative analysis for changes from 1944 to 1970 in three segments of the Rio Palpa. While the total area occupied by the Rio Palpa in 1944 (39.2 ha) and 1970 (39.3 ha) remained almost con-

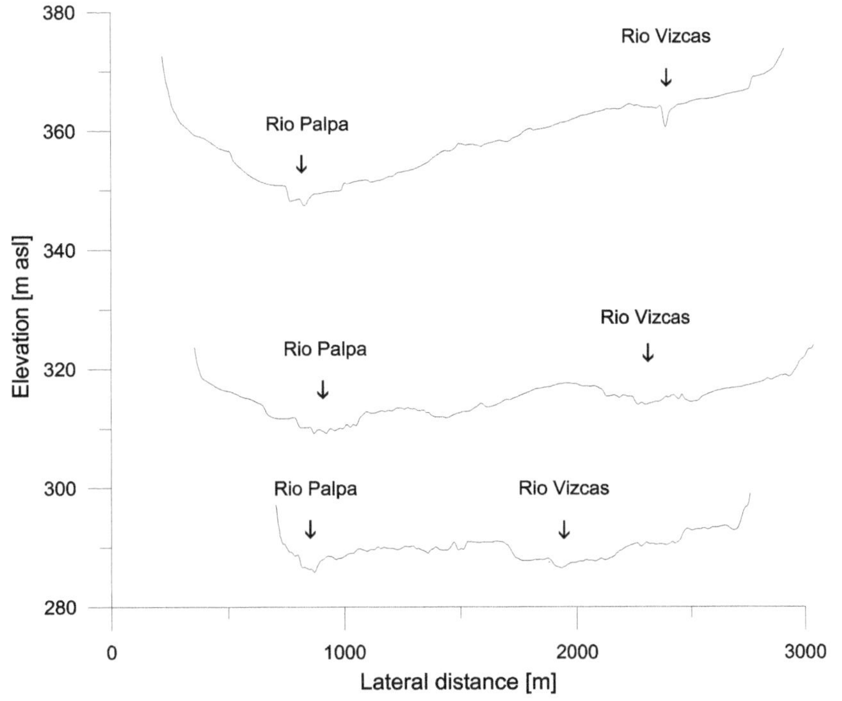

Fig. 11: Transverse elevation profiles of the Palpa Valley (approx. 0.1, 2.3 and 3.8 km down-valley from the Panamericana; data from DTM provided by ETH Zurich).

Fig. 12: Left: Quickbird image of the Rio Palpa downstream of the Panamericana (satellite image taken 25.03.2005) with photo position and field of view indicated by yellow lines; right: eroded artificial gravel embankments (photograph taken 11.03.2006).

Tab. 3: Quantitative changes of the Rio Palpa riverbed due to fluvial dynamics and anthropogenic impact.

	Riverbed 1944	Riverbed 1970	Riverbed 1944 and 1970	Abandoned by the river since 1944	Eroded by the river since 1944	Total affected area
Segment A [ha]	10.8	8.6	6.4	4.4	2.2	13.0
Segment B [ha]	20.7	21.2	15.8	4.9	5.4	26.1
Segment C [ha]	7.7	9.5	6.3	1.4	3.3	10.9
Total [ha]	39.2	39.3	28.5	10.6	10.8	50.0
Total [%]	78.4	78.6	57.0	21.2	21.6	100.0

stant, Figure 13 illustrates the strong lateral activity of the braided riverbed and the land use changes effected by it (cf. Figure 10). The greatest change from 1944 to 1970 occurred in a segment ca. 1.8 km downstream of the Panamericana where the right bank was dislocated by 140 m. Between 1944 and 1970, 10.8 ha were eroded and 10.6 ha abandoned by the river. It has to be assumed that areas abandoned by the riverbed and returned to agriculture require several years to a few decades until efficient agricultural production is possible. The fluvially affected areas (eroded and abandoned by the riverbed) total 21.4 ha and represent ca. 5% of the agricultural land along the Rio Palpa for the period 1944–1970.

4.1.4 *Qualitative and quantitative changes: Rio Vizcas*

The 1999 flood event

In the central part of the Palpa Valley, there is an area with abundant arcuate lineaments north-west of the Rio Vizcas (see 4.1.2). Along this reach of the river, modern erosion has widened the active riverbed from ca. 180 m in 1944 to up to 310 m in 2005. Furthermore, the left bank of the river has shifted by ca. 70–80 m towards the north-west. A strip of land 1.6 km long and up to ca. 160 m wide has been eroded on the right side of the river (Figure 14) where in 1944 there had been fields. The area is not covered by the available aerial photographs of 1970 and 1997. However, the high-resolution DTM based on aerial photography of 1997 reflects a situation with a 150 to 200 m wide riverbed with its left margin shifted by ca. 30 m to the north-west relative to the situation of 1944. A Landsat scene taken on 26.04.2000 shows that strong erosion has taken place between 1997 and 2000. A high-resolution satellite image of March 2005 (Quickbird) allows detailed and quantitative comparison with the earlier conditions. Given the temporal constraints discussed above, the erosion is attributed to the flood event of 02.03.1999 with a peak discharge of 260.0 m^3/s (the highest discharge of the Rio Vizcas on the 1975–2007 record). Between 1997 and 2000, probably on 02.03.1999, 17.8 ha of agricultural land were eroded by the Rio Vizcas. However, while this was the largest erosional event between 1944 and 2007, only a very small portion of the agricultural land in the Palpa Valley was affected to an extent that would inhibit agricultural use for several years.

Beside the erosion of the north-western river bank, the flood event also entailed the deposition of sediment on field surfaces in the central Palpa Valley. According to a local informant (G. Marcaquispe Vasquez), the flood deposited sediment as far away as 280 m from the Rio Vizcas at the location of profiles R-2 and R-4 (see Appendix A). The flood sediment consists of well-bedded and slightly cemented sand and coarse silt and is also present as a surface layer in profiles SF-2 and SF-3. While upstream straightening and narrowing of the Rio Vizcas after 1944 may have contributed to the amplitude of the erosional event by increasing flow velocity (see below), the fluvial lineaments visible in the 1944 aerial photographs indicate that this area has been particularly susceptible to flood events before.

River straightening and incision

Between 1944 and 2005, the entire reach of the Rio Vizcas within the Palpa Valley has been straightened and enclosed in artificial embankments. Two segments were analysed (Figure 15). River straightening has led to a reduction of the main flow channel length from 4.48 to 4.19 km for the lower reach (segment B) between the Llipata ford and the Panamericana bridge and a related reduction in sinuosity (calculated as main channel length divided by length of reach, Rosgen 1994: 183; Rust 1978: 188) from 1.11 to 1.04 for segment B (Table 4). For the upper reach (segment A) north-east of the Panamericana bridge (Figures 15 and 16), the reduction of main flow channel length has been from 1.84 to 1.71 km and the reduction in sinuosity has been from 1.10 to 1.02. Assuming that riverbed elevation paralleled the valley slope along the Rio Vizcas, main flow channel gradient would have increased from 1.07% to 1.15% in segment B and from 1.66% to 1.78% in segment A. However, while the Rio Vizcas today is deeply incised (ca. 5.5 m) and narrow (10–11 m) where it enters the Palpa Valley, its planform at this location was on the threshold between braiding and meandering (20–60 m wide) in 1944, suggesting that incision has since taken place. This incision has more than compensated the influence of the decrease in sinuosity on flow channel gradient. Besides the initial increase in channel slope effected by straightening, channel narrowing leads to an increase in shear stress due to increased flow depth and flow velocity, in particular where smooth artificial embankments have been built. Incision of the river in this reach has led to the abandonment of several irrigation canal intakes and the upstream construction of new intakes connected to the original canals by additional, river-parallel canal segments. Upstream canal intake migration from 1944 to 2005 was 200 m for Canal El Taro, 310 m for Canal Alameda, 95 m for Canal Valdivia and 70 m for Canal Chipiona; it is highest for the most deeply incised part of segment A (Figure 16).

The phenomenon of canal intake migration due to river incision has been described by Moseley (Moseley 1983: 774; Clement & Moseley 1991: 427) who has further argued that natural river incision driven by climatic change, extreme discharge events or tectonic movement could have led to the collapse of irrigation systems (Moseley 1983: 773–792). However, if, as stated by Clement & Moseley (1991: 426), the allogenic rivers crossing the Peruvian coastal were generally characterised by "degrading, erosional regimes", the continuously increasing depth of the riverbeds relative to the land surface should by now be prohibitive for irrigation. Contrary to this, even the rapid

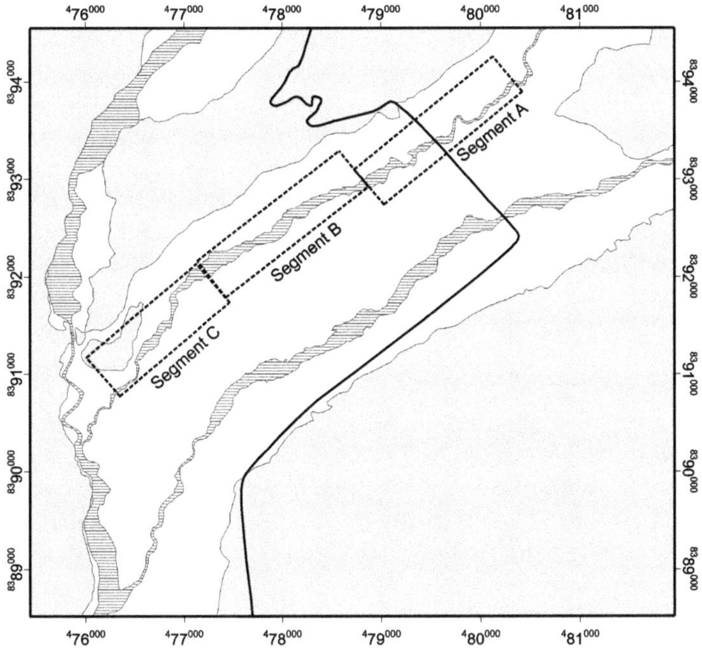

Fig. 13: Lateral changes of the Rio Palpa (deduced from aerial photographs from 1944, 1970 and 1997). Left: location of river segments; bottom: changes in river width.

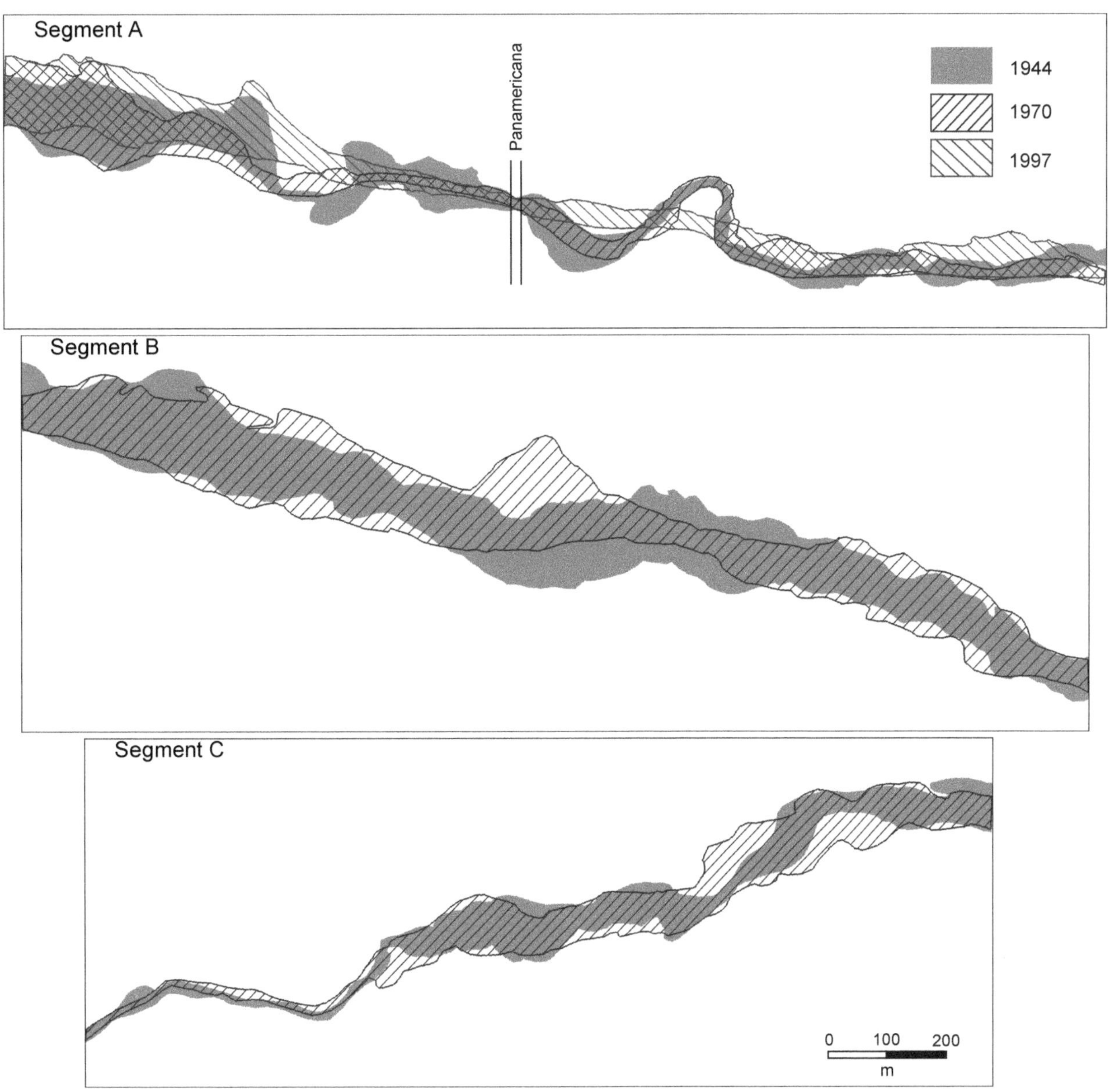

modern incision of this reach of the Rio Vizcas due to river straightening and narrowing has not led to the abandonment of fields.

While the problem of irrigation water access has been mitigated by modifications of the canals and flood risk along this reach has been reduced by the increase of the vertical distance between riverbed and land surface, downstream sections may now suffer higher risk due to increased flow velocity, erosional capacity or sediment transport (see above). Today, only limited meandering is possible within the artificial embankments. However, the high discharge variability and the high shear stress during discharge events let the anthropogenic measures appear to be of very limited efficiency. Despite only moderate discharge in the intervening year, several sections of gravel embankments visible on the satellite image of March 25, 2005 had been eroded by March 2006 both along the Rio Vizcas and the Rio Palpa (cf. Figure 12). This apparent failure of current hydraulic engineering measures calls for an adaptation of the present settlement and agricultural use of the valley bottom.

4.2 Irrigation systems

4.2.1 Present-day irrigation systems

Of the 15.9 km^2 valley floor area of the Palpa Valley analysed in this project, 13.7 km^2 (86%) are used for irrigation agriculture. Of the remaining area, riverbeds occupy 1.5 km^2 (9%) and settlements 0.7 km^2 (4%). Given the hyperarid conditions of the coastal desert and the fact that a high groundwater table would lead to rapid salinisation (as observed locally in low-lying areas near the Rio Palpa and in the Rio Grande valley), agriculture depends exclusively on irrigation. The irrigation system in the Palpa Valley consists of 25 main canals (11 originating from the Rio Palpa, 14 from the Rio Vizcas, see Table 5) which tap river water during the discharge season and distribute it to the fields through secondary and lower-order channels. They are generally simple, open channels without any specific measures to prevent evaporation or infiltration. However, while Clement & Moseley (1991: 427, quoting data from USDA 1955) assume that water losses of ≥50% due to evaporation and seepage are an important constraint on irrigation, it has to be noted that seepage from the channels likely benefits plant growth on adjacent fields, thus reducing the adverse effects. Accumulation of sediment deposited in the irrigation channels and excavation of this sediment in the course of channel maintenance regularly leads to the creation of topographic heights so that channels often run in slightly elevated positions relative to the surrounding land surface.

Irrigation is practised up to a vertical distance of ca. 15 m above the respective riverbed, with increasing height difference requiring increasing canal length and/or lower canal slope. The entire irrigation system of the Palpa Valley has remained almost unchanged since 1944 with the exception of the migration of some canal intakes (see 4.1.4) and the construction of three modern canals. The Canal La Peña Alta was constructed after 1944 and irrigates the lower slopes of the Pampa San Ignacio. In recent years, two additional large canals have been constructed to transfer water from the Rio Grande into the Palpa Valley. These two canals are larger than the traditional ones and are lined with concrete. However, as of 2006 they had not been in use. In late historic times, artificial wells and pumps have supplemented irrigation water outside of the discharge season and in times of drought (see 4.3).

The allocation of irrigation water to individual fields is regulated by the Junta de Usarios Palpa, the self-administration body for irrigation in the Palpa Valley. Agriculture certainly has constituted the predominant economic activity in the Palpa Valley throughout pre-Columbian and historic times and is still a major factor in the local economy. However, despite the fact that easily irrigable land is a valuable resource, the town of Palpa occupies 0.7 km^2 (ca. 4%) of the valley bottom. This clearly contradicts the assumption that pre-Columbian settlements and cemeteries may have been placed exclusively outside of the valley bottoms due to economic necessity (cf. Schreiber & Lancho Rojas 1995: 248 whose surface survey did not yield any sites).

4.2.2 Quantification of irrigation-related sedimentation

The use of river water with high concentrations of suspended sediment for irrigation poses the question of irrigation-related sedimentation. To answer this question, two case studies were undertaken in March 2006. Both sites under investigation are situated in the central Palpa Valley and receive water from the Rio Vizcas (the first site through Canal Jasmin and the second site through Canal San Fermin). The 2005–2006 discharge season was approximately average compared to the mean annual discharge for the period 1975–2007.

The site of profile SR (see Figure 22 on page 37; see Appendix A) had been under irrigation until a few days before excavation of the pit. Six sediment samples taken from profile SR were used to determine water content to a depth of 2.0 m. Since profiles on fallow, not recently irrigated fields were generally dry to this depth, water content in profile SR is assumed to be exclusively related to irrigation. The amount of percolation to depths below 2.0 m is unknown; therefore, the calculated 495 mm water column attributed to irrigation represents a minimum amount. Assuming a suspended sediment concentration between 3.7 and 6.6 g/l (measured approximately contemporaneously in the Rio Palpa and the Rio Grande, respectively) and a dry sediment density of 1.29 g/cm^3 (average value for the top 2.0 m of profile SR) results in a minimum sedimentation rate of 1.4–2.5 mm per irrigation season.

Fig. 14: Lateral changes of the Rio Vizcas in the central Palpa Valley (blue – riverbed according to the 1944 aerial photographs; red hatched – riverbed according to the 1997 DTM; yellow – area eroded during the 1999 flood event).

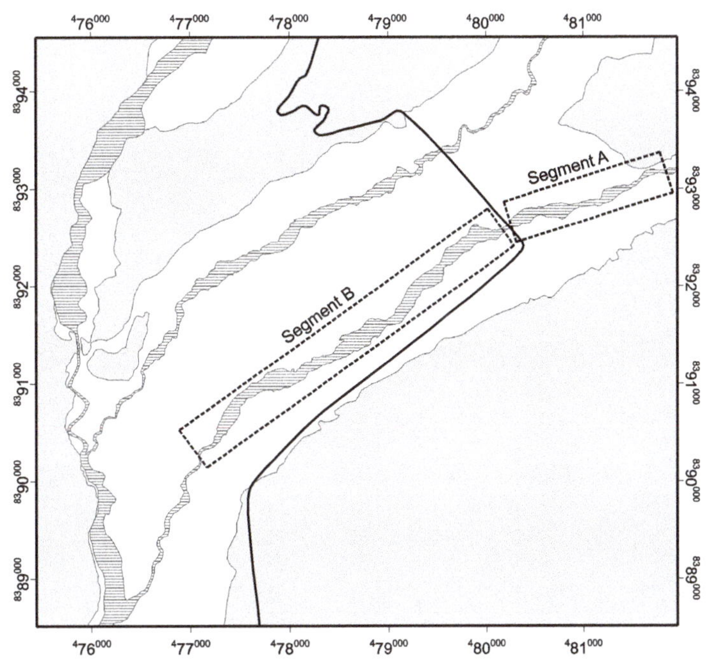

Fig. 15: Location of the lower and upper segment of the Rio Vizcas for which hydraulic parameters have been quantified.

Tab. 4: Quantitative changes in hydraulic parameters of the Rio Vizcas due to anthropogenic straightening.

	Segment B		Segment A	
	1944	2005	1944	2005
Length of reach [km]	4.03		1.67	
Length of main flow channel [km]	4.48	4.19	1.84	1.71
Sinuosity of main flow channel	1.11	1.04	1.10	1.02
Valley elevation difference [m]	48.0		30.5	
Riverbed elevation difference [m]		49.0		26.0
Gradient of main flow channel assuming valley slope [%]	1.07	1.15	1.66	1.78
Actual gradient of main flow channel [%]		1.17		1.52

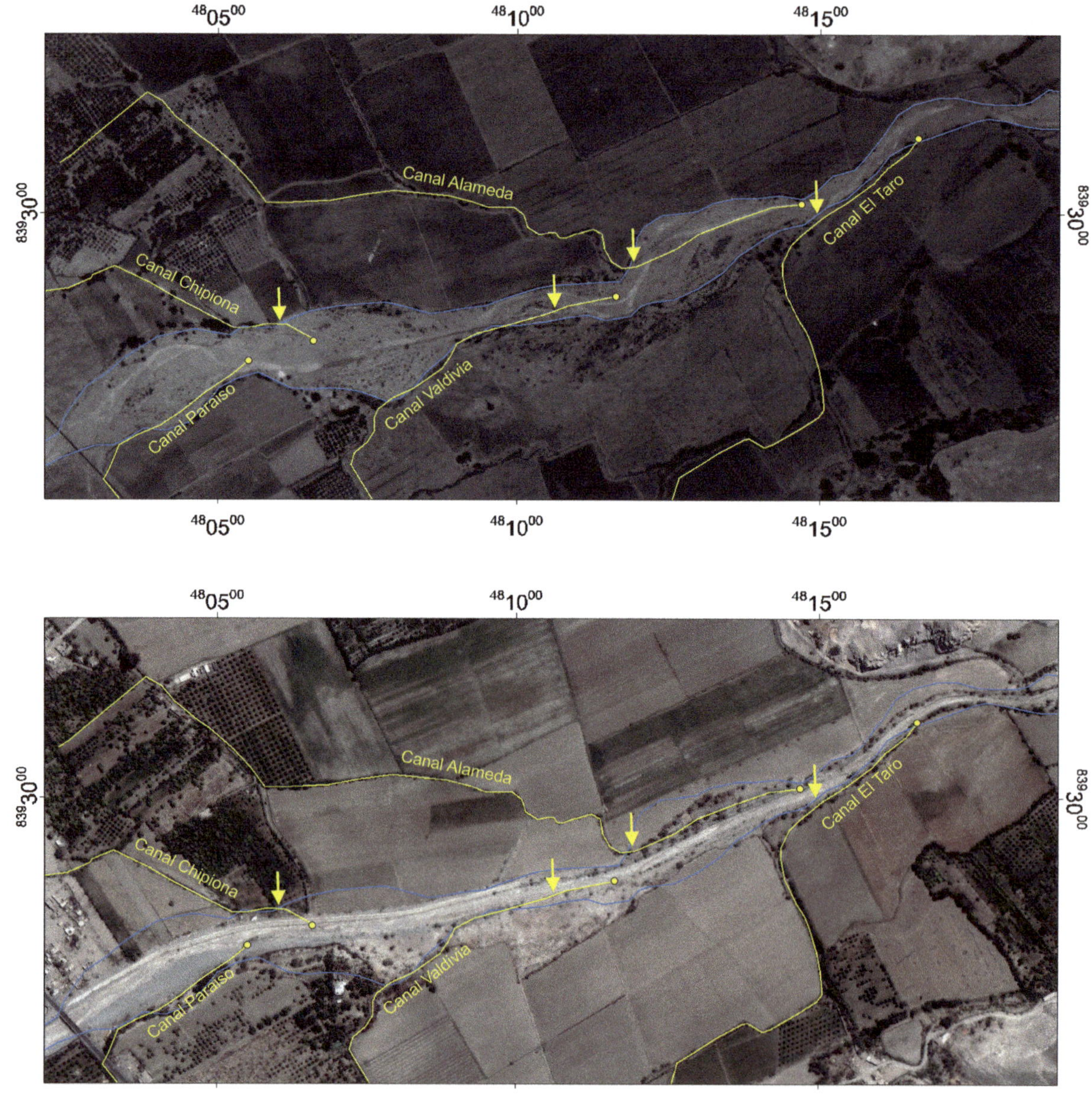

Fig. 16: Anthropogenic changes in river channel planform from 1944 (top, aerial photograph) to 2005 (bottom, Quickbird image) in segment A of the Rio Vizcas (cf. Figure 15). 1944 river margins are outlined in blue, irrigation canals are shown in yellow (information on irrigation canals provided by Junta de Usarios Palpa). Arrows indicate positions of canal intakes in 1944; circles indicate positions of canal intakes in 2006 (GPS measurements).

Tab. 5: Irrigation canals in the Palpa Valley (data provided by Junta de Usarios Palpa).

Location	Name	Maximum flow rate [l/s]	Length [km]	Irrigated area [ha]	ha/km
Rio Palpa (right side)	Canal La Maquina	1,203	3.35	57.10	17.04
Rio Palpa (left side)	Canal San Antonio	404	1.20	9.84	8.20
Rio Palpa (right side)	Canal Utua Alto	521	1.12	10.86	9.70
Rio Palpa (right side)	Canal Utua Bajo	982	1.70	2.35	1.38
Rio Palpa (right side)	Canal Pacae	1,325	2.10	28.30	13.48
Rio Palpa (left side)	Canal Jauranga	1,000	2.90	88.90	30.66
Rio Palpa (right side)	Canal Bolillo	1,000	3.47	56.75	16.35
Rio Palpa (right side)	Canal Donayre	242	2.65	29.12	10.99
Rio Palpa (left side)	Canal Mauleon	1,000	1.45	57.70	39.79
Rio Palpa (left side)	Canal Santo Domingo	1,500	1.35	60.86	45.08
Rio Palpa (left side)	Canal Rivera	86	1.16	5.50	4.74
Rio Vizcas (left side)	Canal Peña Alta	299	13.10	117.89	9.00
Rio Vizcas (left side)	Canal Peña Baja	521	4.30	44.89	10.44
Rio Vizcas (right side)	Canal Carapo	659	2.90	48.45	16.71
Rio Vizcas (left side)	Canal El Taro	912	7.47	108.58	14.54
Rio Vizcas (right side)	Canal Alameda	673	1.30	36.46	28.05
Rio Vizcas (left side)	Canal Valdivia	865	1.36	17.40	12.79
Rio Vizcas (right side)	Canal Chipiona	789	3.83	97.31	25.41
Rio Vizcas (left side)	Canal Paraiso	440	2.60	35.61	13.70
Rio Vizcas (right side)	Canal San Fermin	506	1.35	64.82	48.01
Rio Vizcas (left side)	Canal San Antonio Vizcas	882	4.03	55.90	13.87
Rio Vizcas (right side)	Canal Jasmin	142	2.32	49.96	21.53
Rio Vizcas (left side)	Canal La Victoria	767	5.22	138.02	26.44
Rio Vizcas (right side)	Canal Pacayal	1,199	1.42	53.08	37.38
Rio Vizcas (left side)	Canal Llipata	895	5.63	98.70	17.53
	Rio Palpa (right side)	5,273	14.39	184.48	12.82
	Rio Palpa (left side)	3,990	8.06	222.80	27.64
	Rio Vizcas (right side)	3,969	13.12	350.08	26.68
	Rio Vizcas (left side)	5,582	43.71	616.99	14.12
	Valley centre	7,959	21.18	572.88	27.05
	Valley sides	10,855	58.10	801.47	13.80
	Total	18,814	79.28	1,374.35	17.34

As a second case study, irrigation water input to a field close to the right bank of the Rio Vizcas was determined by measuring flow rate and suspended sediment concentration in the feeder channel as well as the area of the field (4.55 ha). Flow rate was 50 l/s and suspended sediment concentration was 3.0 g/l. Assuming an irrigation period of seven to fourteen days, an irrigation water input of 660–1,330 mm and a sedimentation rate of 1.5–3.0 mm per irrigation season was calculated.

Based on the methodology described and applied by Nordt et al. (2004), necessary irrigation water input for the cultivation of maize was calculated for profile SR. Crop evapotranspiration was calculated using the Blaney-Criddle Equation (Dunne & Leopold 1978: 139). The resulting water requirement of 380–530 mm per growing season compares with the 540 mm of irrigation water required for maize during a growing season under the conditions in northern coastal Peru found by Nordt et al. (2004: 39) and supports the quantitative estimates of irrigation water input derived for the Palpa Valley in the above case studies.

4.3 Groundwater use and dynamics

While the main source of irrigation water is seasonal river discharge, groundwater withdrawal has been playing an important role to satisfy irrigation water demand outside of the discharge season and during periods of prolonged drought. Groundwater withdrawal in the Palpa Valley has been ca. 4 Mio. m^3/a in the late 20th century (Klohn Crippen 1998: 13) compared with ca. 48 Mio. m^3/a mean annual river discharge of Rio Palpa and Rio Vizcas combined (period 1975–2007, data made available by Junta de Usarios Palpa). River flow from the Cordillera Occidental during the short (a few weeks up to four months) discharge season supports irrigation and in "good" years recharges the aquifer to within two metres below the surface as stated by local informants. This is supported by the frequent presence in sediment profiles of oxidation stains related to water table fluctuations (see Appendix A).

Groundwater table was measured in March 2005 in 33 wells in the Palpa Valley (Figure 17). Drought conditions with minimal river discharge had prevailed from 2001 through 2005 (cf. Figure 5). After four years of drought, the groundwater table was on average nine metres below the surface in March 2005. Groundwater table lowering therefore amounts to ca. 1.75 m/a during times of prolonged drought without significant groundwater recharge. Assuming a spatial extend of 15.9 km^2 (i.e. the area of the Palpa Valley) for the affected aquifer and a specific yield range of 0.08–0.27 (Healy & Cook 2002: 96), this corresponds to an aquifer depletion of 2.2–7.5 Mio. m^3/a. Assuming an average specific yield between the values for the most common grain sizes (silt, 0.08; fine sand, 0.21; Healy & Cook 2002: 96), aquifer depletion is 4.2 Mio. m^3/a. Within the range of uncertainty regarding the amount of groundwater withdrawal and the aquifer volume affected by it, the observed water table lowering can thus be explained by withdrawal. Natural causes can be considered of secondary importance. These results also imply that during droughts lateral groundwater inflow into the valley is low.

Recharge of the Palpa Valley aquifer by infiltration of surface water takes place both in the riverbeds which consist of coarse gravel with high infiltration rates and on irrigated fields. Hydraulic engineering (straightening and narrowing of the rivers) since the middle of the 20th century (see 4.1) has led to a reduction of riverbed area in the Palpa Valley by more than 50% which likely increases the severity of water shortages by minimising aquifer recharge.

The implication of these findings for archaeology – where droughts have been put forward to explain the decline of the Nasca culture (e.g. Silverman 2002: 24; Reindel et al. 2001: 95) – is that droughts of equal physical characteristics in terms of river discharge reduction had less severe consequences on the groundwater table in pre-Columbian times, because there was no large-scale withdrawal. Pukios (filtration galleries with outflow into irrigation channels or storage basins, cf. Schreiber & Lancho Rojas 1995: 234-235; Silverman 2002: 23) are not known from the Palpa Valley (Reindel et al. 1999: 354); however, it is common to construct lateral filtration galleries at the bottom of well shafts. Given the lack of surface water during most of the year, wells were probably dug to access groundwater for domestic and perhaps also irrigation use during pre-Columbian times. The high concentration of archaeological sites in the vicinity of the Laguna La Maquina (a natural well transformed into an artificial pond north-west of the Rio Palpa) indicates its use in pre-Columbian times.

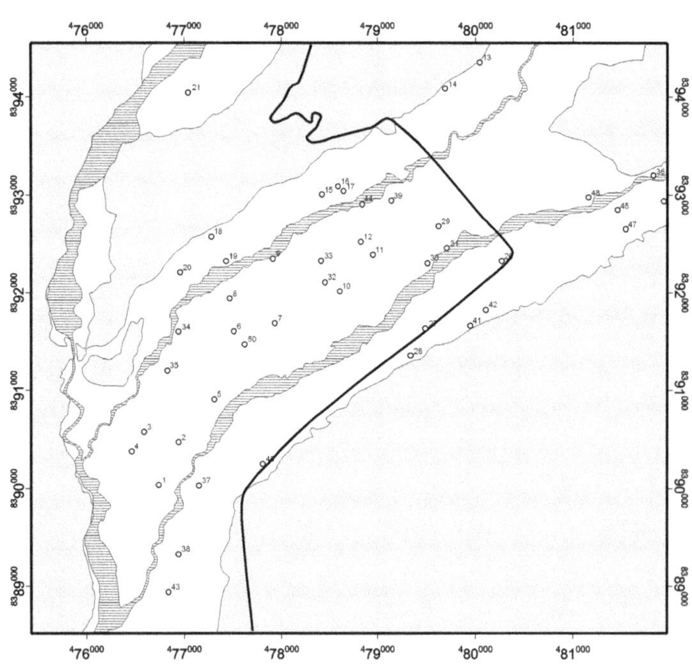

Fig. 17: Location of wells in the Palpa Valley.

5. Results of sedimentological studies

5.1 *Facies definition*

From the 50 sediment profiles investigated in detail, 487 sediment samples were taken and analysed in the laboratory (see 3.2). Together with the laboratory results, the sedimentological properties recorded in the field allow the definition of nine facies classes which are present in the Palpa Valley and in the valley of the Rio Grande. This facies definition is based on Miall (1996: 105–127) and FAO (1998); however, the finely resolved classification of fluvial facies (Miall 1996: 105–127) is adapted and simplified by combining several related and similar facies. Facies definition is based on grain size, sorting, CSR, LoI, sedimentary and pedogenic structures as well as the properties of layer boundaries. Of the 487 sediment samples analysed in the laboratory, 20 samples are from A_p horizons of recent agricultural fields, 112 samples are from layers of unequivocal fluvial origin as indicated by sedimentary structures such as bedding and 15 samples are from hyperconcentrated flow deposits. The laboratory results of these samples were used to characterise the defined facies. $CaCO_3$ content was not included in the facies definition because in the investigated profiles it generally had to be attributed to post-sedimentary enrichment. Nine facies classes have been defined (Table 6).

Facies F1 is limited to present-day surface or near-surface situations. It consists of poorly sorted silt to sand with variable clay content and crumbly structure. Bedding is absent. LoI values are mostly ≥2.0% and CSR is often relatively high. This facies is characterised by horizontal or undulating lower boundaries which are often diffuse and/or disturbed. Turbation with the underlying layer is common. Facies F1 frequently contains charcoal and occasionally artefacts and/or singular larger clasts whose presence within the fine-grained matrix can not be explained by fluvial processes. Clods of cohesive and often laminated silt are common and likely originate from sedimentation in irrigation furrows and subsequent turbation by tillage. Occasionally, plough traces are present. Facies F1 is interpreted as *irragric anthrosol* (FAO 1998; cf. 4.2). Figure 18a shows the grain size distribution for facies F1.

Facies F2 consists of poorly sorted silt to sand with variable clay content and mostly high CSR and high LoI values. It has crumbly or polyedric to columnar structure or is structureless. Layer boundaries are horizontal or undulating and often diffuse and/or disturbed. The lower boundary is often turbated with the underlying layer. Occasionally, plough traces can be observed. Charcoal, artefacts (in particular ceramic shards), larger clasts and cohesive clods of silt are frequently present. Ceramic shards are usually well-preserved with their edges not rounded and can therefore not have been transported over long distances by fluvial processes. Further traces of human activity include in particular marine mollusc shells. Facies F2 is very similar to facies F1 except for its more compacted nature (lack of crumbly structure); it is interpreted as *fossil irragric anthrosol* (FAO 1998; cf. 4.2). Figure 18b shows the grain size distribution for facies F2.

Facies F3 has a median grain size from clay to silt. It is characterised by low CSR and intermediate LoI values. Bedding or lamination is present; layer boundaries are generally sharp and horizontal. Facies F3 is interpreted as *still-water sediment* deposited as abandoned channel fill or in pond-like situations. Ponding of water may also occur on irrigated fields.

Facies F4 (median grain size silt to coarse sand) is characterised by good to poor sorting, variable CSR an generally low LoI. It is bedded or laminated. Occasionally, cross-

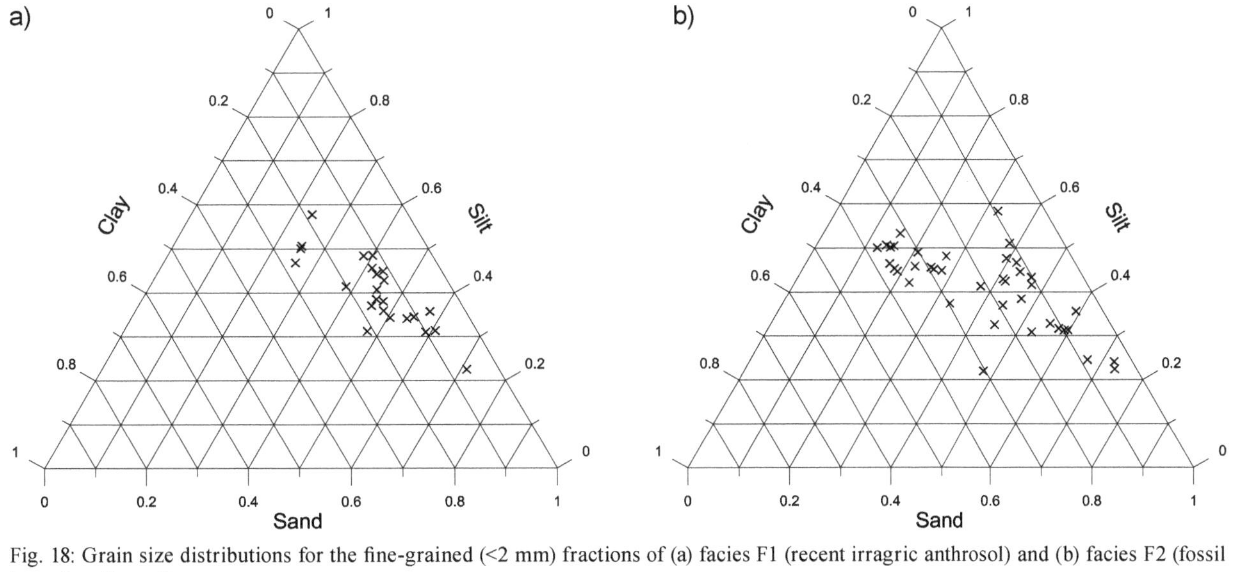

Fig. 18: Grain size distributions for the fine-grained (<2 mm) fractions of (a) facies F1 (recent irragric anthrosol) and (b) facies F2 (fossil irragric anthrosol).

Tab. 6: Facies classes of Palpa Valley sediments.

Facies	Median grain size	Sorting	CSR	LoI [%]	Sedimentary and pedogenic structures	Layer boundaries	Additional properties	Facies according to MIALL (1996)	Interpretation
F1	silt to sand	poor 2.6 – 8.7, mostly ≥3.0	0.2 – 0.6, mostly ≥0.25	1.1-5.6, mostly ≥2.0	crumbly; occasional plough traces	lower boundary horizontal or undulating, often diffuse and/or disturbed and often turbated with underlying layer	frequent presence of charcoal and cohesive clods of silt, occasional presence of artefacts and/or larger clasts	–	irragric anthrosol (A_P / irrigation horizon)
F2	silt to sand	poor 2.6 – 7.2, mostly ≥3.0	0.1 – 0.8, mostly ≥0.3	1.2 – 3.9, mostly ≥2.0	crumbly, polyedric to columnar structure or structureless; occasional plough traces	horizontal or undulating, often diffuse and/or disturbed; lower boundary often turbated with underlying layer	frequent presence of charcoal, artefacts, larger clasts or cohesive clods of silt	–	fossil irragric anthrosol
F3	clay to silt	good to poor 1.7 – 5.7	0.0 – 0.3	1.4 – 3.4	bedded or laminated	horizontal, generally sharp	–	Fl, Fsm, Fm	still-water sediment (abandoned channel or pond fill)
F4	silt to sand	good to poor 1.4 – 4.7	0.0 – 0.7	0.4 – 2.0	bedded or laminated; occasional crossbedding, ripple marks and graded beds	horizontal or undulating, generally sharp, lower boundary sometimes erosional	–	Sp, St, Sr, Sh, Sl, Se, Ss, Sm, Fl, Fsm, Fm	low-energy fine grained fluvial sediment (minor/abandoned channel fill, waning flow or overbank deposit)
F5	gravel	poor	–	–	clast- or matrix-supported, occasionally imbricated	horizontal or undulating, generally sharp, lower boundary often erosional	rounded clasts, layer thickness on the order of clast diameter	Gcm	high-energy, coarse grained fluvial sediment (overbank gravel splay)
F6	gravel to boulders	good	–	–	clast-supported, often imbricated	horizontal or undulating, generally sharp	rounded clasts, unweathered or slightly weathered	Gh, Gt, Gp, Gcm	high-energy, coarse grained fluvial sediment (channel deposit)
F7	gravel to boulders	good	–	–	clast-supported	horizontal or undulating, generally sharp	rounded clasts, strongly weathered (in particular granite), cemented	Gh, Gt, Gp, Gcm	older high-energy coarse grained fluvial sediment
F8	silt to gravel	poor 2.6 – 9.1	0.0 – 0.5	0.7 – 2.0	matrix-supported, structureless	horizontal or undulating, mostly sharp, lower boundary often erosional	angular to rounded clasts, generally strongly compacted	Gmm, Gmg (Dmm, Dmg)	hyperconcentrated flow deposit with fine-grained matrix (turbidity flow)
F9	gravel	poor	–	–	clast-supported, often imbricated	horizontal or undulating, mostly sharp, lower boundary often erosional	angular clasts	Gci, Gcm (Dci, Dcm)	matrix-poor debris flow deposits (grain flow)

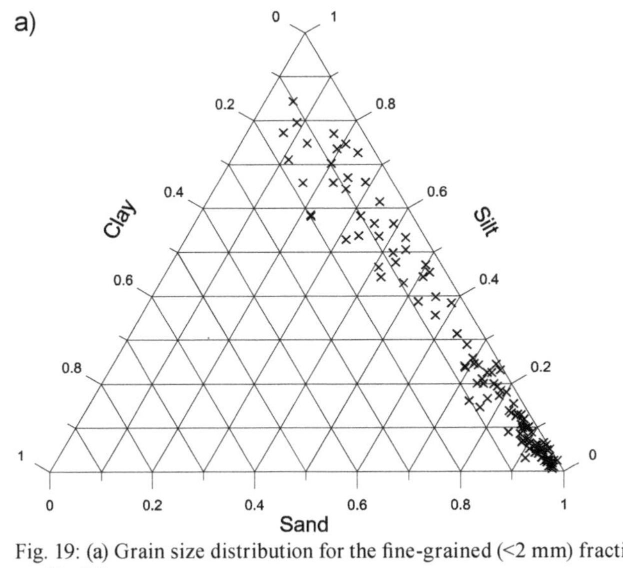

Fig. 19: (a) Grain size distribution for the fine-grained (<2 mm) fraction of facies F4 (fluvial sediment); (b) type example for facies F4 from profile AR.

bedding, ripple marks and graded beds are present. Layer boundaries are generally sharp and horizontal to undulating. Occasionally, the lower boundary is clearly erosional. Facies F4 is interpreted as low-energy fine-grained fluvial sediment deposited either as *overbank flood sediment, waning-flow or channel fill sediment.* Figure 19 shows the grain size distribution of facies F4 and a type example from profile AR (see 5.2.5).

Facies F5 consists of thin horizontal to undulating sheets of subrounded to well-rounded, clast- or matrix-supported gravel usually over- and underlain by fine-grained material. Thickness of these gravel sheets is usually on the order of clast diameter. Imbrication is occasionally observed. Layer boundaries are generally sharp and horizontal or undulating; the lower boundary is often clearly erosional. Sorting is poor; given the small layer thickness it is often impossible to determine whether the fine-grained matrix is of syn- or postsedimentary character. Facies F5 is interpreted as *high-energy overbank flood sediment* (overbank gravel splay) deposited in a channel-proximal situation.

Facies F6 is well-sorted and subrounded to well-rounded gravel to boulders. This facies is clast-supported; clasts are commonly imbricated. Layer boundaries are horizontal or undulating and generally sharp except for the vertical intrusion of fine-grained material from overlying layers. Facies F6 is interpreted as *fluvial sediment* deposited in a *high-energy* (main flow channel) environment.

Facies F7 consists of clast-supported, well-sorted and subrounded to well-rounded gravel to boulders. Layer boundaries are horizontal and undulating and generally sharp. A varying fraction of clasts (in particular granites) is strongly weathered to such a degree as to decompose to individual crystals when squeezed between fingers. Clasts and matrix are often cemented. Facies F7 is interpreted as *older high-energy fluvial channel sediment.* Similarity to exposed bajada sediments at the Cresta de Sacramento in terms of weathering and cementation suggests a pre-Holocene age for this facies; however, given the lack of numerical ages for these sediments this has to be considered a tentative interpretation.

Facies F8 consists of structureless, matrix-supported, poorly sorted silt to gravel. The material is generally massive and strongly compacted. Clasts are angular to rounded; there is a tendency that higher clast concentration occurs towards the top of layers. CSR is variable and LoI is generally low. Layer boundaries are horizontal to undulating and mostly sharp. The lower boundary is often erosional. Facies F8 is interpreted as *hyperconcentrated flow deposit* with abundant fine-grained matrix material. Figure 20 shows the grain size distribution for facies F8 and a type example from profile RP-06.

Facies F9 is clast-supported and often imbricated angular gravel. It occurs exclusively within or at the mouths of steeply descending quebradas (e.g. Quebrada La Muña, Quebrada Corpante). Layer boundaries are horizontal or undulating and mostly sharp. The lower boundary is commonly erosional. Facies F9 is interpreted as matrix-poor *hyperconcentrated flow deposit* (*debris flow / grain flow*). Figure 21 shows a type example from profile LM-1 (see 5.2.2).

5.2 *Sediment profiles*

The following pages contain short descriptive and interpretative summaries of the 50 sediment profiles investigated in detail. Extensive descriptions of these 50 profiles including their geographic coordinates, stratigraphic sketches as well as graphs of laboratory results appear in Appendix A. Magnetic susceptibility graphs can be found in Appendix B. The facies classification of chapter 5.1 is used throughout; however, for legibility reasons descriptive and inter-

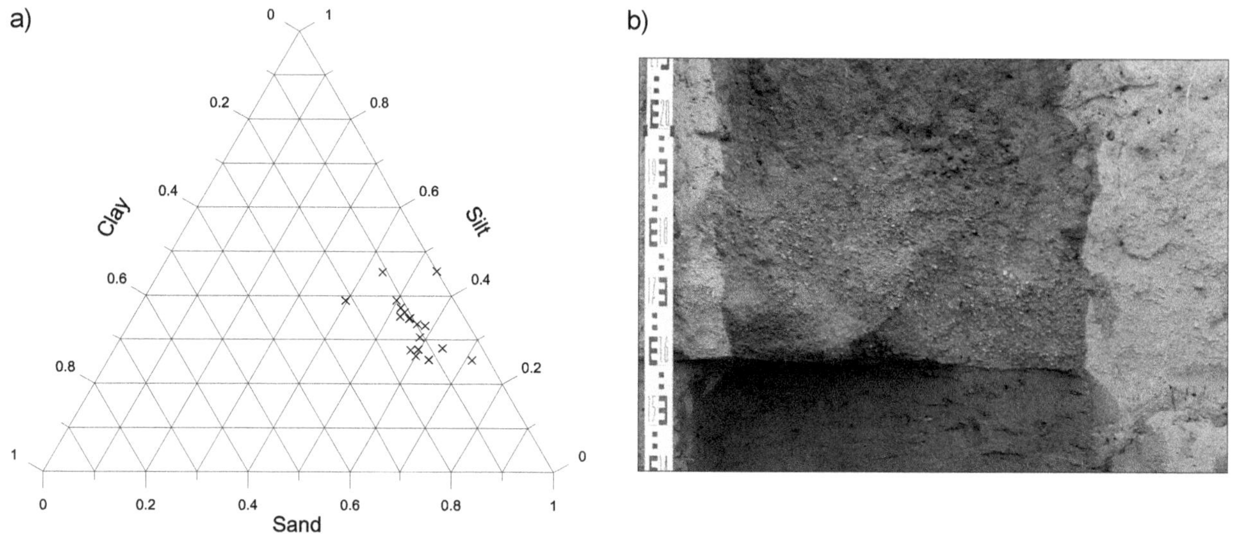

Fig. 20: (a) Grain size distribution for the fine-grained (<2 mm) fraction of facies F8 (hyperconcentrated flow deposit); (b) type example for facies F8 (top of photograph) from profile RP-06 (photograph by J. Baade).

Fig. 21: Type example for facies F9 (debris flow deposit) from profile LM-1 (photograph by J. Baade).

pretative terms rather than the facies codes of Table 6 are preferred. A further 12 profiles were only cursorily recorded and will be referred to in the discussion where applicable. Most of the profiles were located in the Palpa valley and a few were located in the Rio Grande valley just upstream of the confluence of Rio Grande and Rio Palpa (Figure 22). One profile was located at the narrows of the Rio Grande Valley near Cerro Corpante. The profiles are grouped spatially as follows: Rio Grande narrows (1 profile), La Muña (8 profiles), Rio Grande (7 profiles), north-west of the Rio Palpa (9 profiles), between Rio Palpa and Rio Vizcas (19 profiles) and south-east of the Rio Vizcas (6 profiles). Discussion and interpretation of the profiles at La Muña build partially on KASTNER (2005).

5.2.1 Rio Grande narrows

Profile RGE

Profile RGE records a sequence of erosion and autochthonous as well as allochthonous deposition ca. 15 km south-west of the Palpa Valley. A series of stratigraphically consistent radiocarbon dates constrains river valley incision to ≤6.0 m since ca. 30 ^{14}C ka BP and indicates a phase of fine-grained fluvial overbank sedimentation during the LGM followed by a sequence of (1) local debris flow accumulation, (2) quebrada incision to a base level ca. 6 m above the present riverbed and (3) river incision to its current depth.

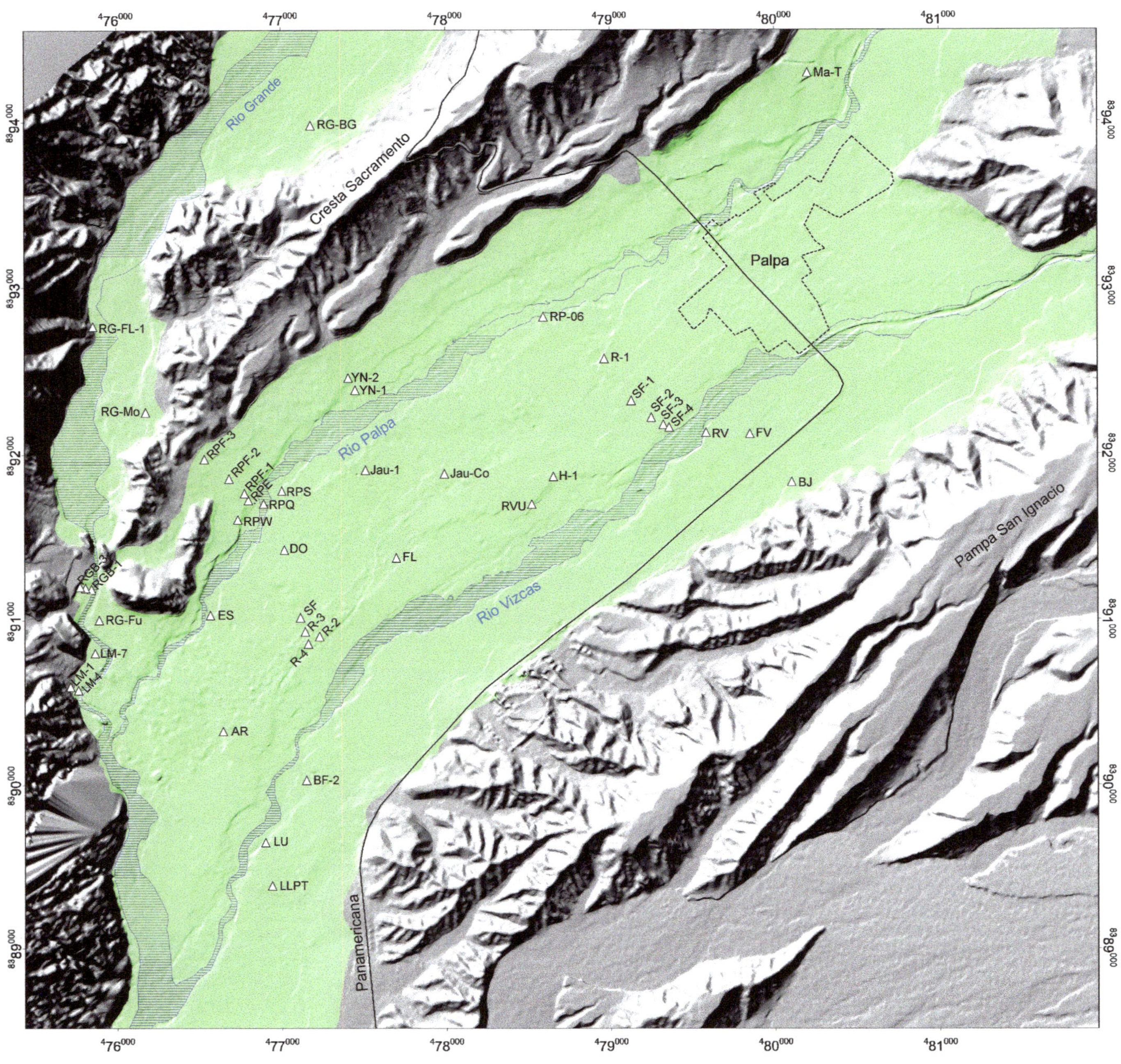

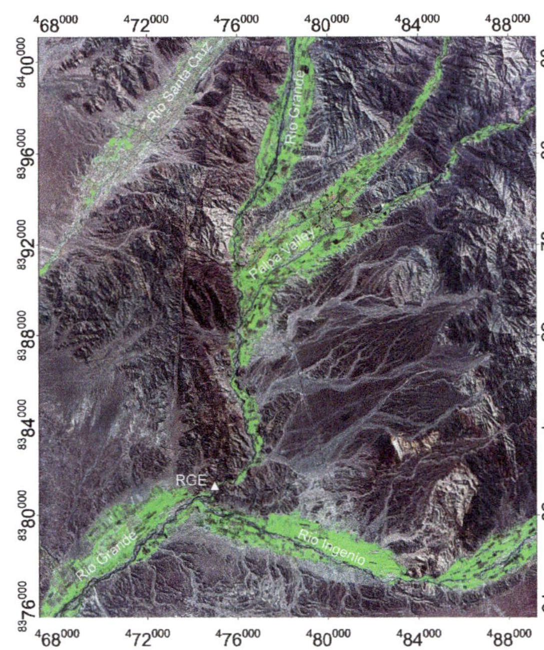

Fig. 22: Distribution of investigated sediment profiles. Left: profiles in the Palpa Valley (green: valley floor area; hatched blue: near-natural river courses from 1944 aerial photographs); below: overview and location of profile RGE. The maps are based on the high-resolution DTM of the Palpa Valley provided by ETH Zurich (left) and Landsat data (below).

Also available as an A3 download from: http://www.barpublishing.com/additional-downloads.html

5.2.2 *Profiles La Muña*

Profile LM-1

Profile LM-1 (see Figure 23) records a sequence of allochthonous and autochthonous sedimentation at the mouth of Quebrada La Muña. Fluvial gravel attributable to a former riverbed situation of the Rio Grande is present at a depth of 3.80 m. Fine-grained fluvial (allochthonous) sediment related to flood events is present between depths of 3.80 to 2.70 m. Evidence for a period of plant growth (indicated by root molds) and accumulation of disperse charcoal or organic matter on top of fine-grained fluvial sediment after 910–730 cal BP is overlain by the last secure allochthonous fluvial deposit and the first autochthonous debris flow deposit of coarse-grained, angular stones. Three further couples of fine sand overlain by thin carbon-rich bands with root molds extending downwards are present between depths of 2.45 and 1.68 m. While the fine-grained sediments can not be securely attributed to either allochthonous overbank flow of the Rio Grande or autochthonous fine-grained sedimentation from Quebrada La Muña, the diminution of layer thickness (30–25–20–12 cm) with increasing vertical accretion conforms with the pattern which would be expected from flood events (i.e. a decrease in flood sediment with increasing height above the riverbed). A relatively fine-grained hyperconcentrated debris flow deposit (layer 13) is present between the uppermost couples at a depth of 2.10–1.80 m. Above a depth of 1.68, all layers are attributed to coarse- and fine-grained debris flow or fluvial sedimentation from Quebrada La Muña, with six coarse-grained events recorded since 490–320 cal BP (depth 2.00 m). Human activity in the vicinity of the site is indicated by the presence of charcoal.

Profile LM-2

Profile LM-2 (see Figure 23) records both allochthonous and autochthonous sedimentation at the mouth of Quebrada La Muña. The bottom of the profile (depth 3.50–3.05 m) is characterised by low- to medium-energy fluvial sedimentation (sandy silt to gravel) overlying coarse gravel indicative of a high-energy fluvial environment (riverbed). Between depths of 3.05 and 1.43 m, several layers with high clay content are interpreted as results of still water deposition. Root molds present in most layers between depths of 1.75 and 0.65 m together with the presence of organic-rich bands indicate plant growth at the site. While autochthonous sedimentation has to be assumed for the top part of the profile due to the surface morphology of the site, the lack of clearly identifiable autochthonous material (angular clasts of local provenance; only present at the top of the profile) inhibits a definition of the transition from allochthonous to autochthonous dynamics. A series of four stratigraphically consistent radiocarbon dates for profile LM-2 indicates that accumulation of the sediments began before 680–560 cal BP (depth 2.00 m) and has since taken place at an average sedimentation rate of 2.72–3.25 mm/a.

Profile LM-3

Profile LM-3 (see Figure 23) records rapid sedimentation mainly by fluvial processes on the field between the mouth of Quebrada La Muña and the Rio Grande in historic times. High-energy fluvial channel deposits below a depth of 1.10–1.00 m are overlain by two thin fine-grained layers the upper of which is characterised by abundant charcoal (likely indicating nearby human activity). This layer was

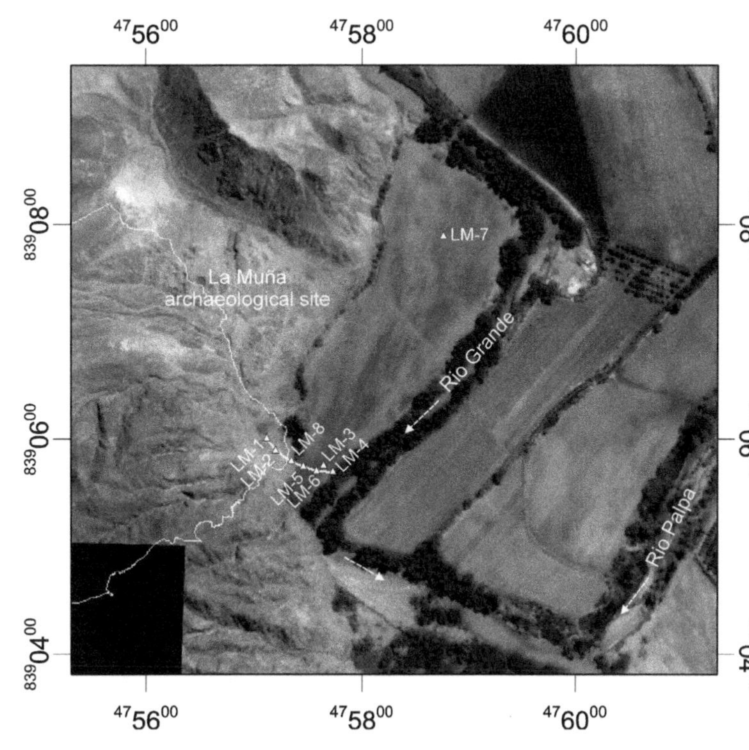

Fig. 23: Satellite image (Quickbird) showing the location of profiles in the vicinity of the archaeological site La Muña in relation to Rio Grande and Rio Palpa (arrows indicate flow direction; thin white line indicates the limits of the La Muña catchment).

radiocarbon-dated at ≤260 cal BP. The following layer (depth 0.95/1.00–0.70 m) is characterised by increased content of organic matter (LoI=2.2%) and likely represents agricultural disturbance of fluvial sediments and addition of irrigation sediment. The top 0.70 m of the profile represent young, agriculturally disturbed fluvial overbank sediment as indicated by low organic matter content (LoI: 0.5–1.4%), remnant bedding and broken-up layer boundaries and cohesive clods of broken-up material within layers. Average sedimentation rate for the top 0.98 m of the profile (since ≤260 cal BP) is ≥3.11 mm/a.

Profile LM-4

Profile LM-4 (see Figure 23) records rapid near-channel fine-grained aggradation of 2.00 m since ≤260 cal BP. The exclusively low-energy fluvial deposits include mainly bedded to laminated fine to coarse sand as well as structureless silty sand. Indicators of human impact and agricultural disturbance such as broken-up layer boundaries or mixed-in cohesive clods of silt, stones or abundant charcoal are absent. The average sedimentation rate for the top 2.00 m of the profile (since ≤260 cal BP) is ≥6.35 mm/a.

Profile LM-5

Profile LM-5 (see Figure 23) records a sequence of mainly fluvial low-energy deposition (abandoned channel fill and/or overbank flow) following the termination of a river channel situation. Erosional events and subsequent infilling are recognisable at a depth between 1.78 to 1.43 m. Increased LoI values (2.1–3.1%) in the lower portion of the stratigraphy indicate the accumulation of organic matter, probably in a context of riverine vegetation. Flood deposits are recognisable upwards to a depth of 0.69 m, followed by strongly turbated sediment with poorer sorting and increasing LoI (up to 2.4%) and indicators of human, in particular agricultural, activities such as the presence of cohesive clods of silt and charcoal. Figure 24 shows undisturbed fluvial sediment overlain by agriculturally turbated material as indicated by abundant cohesive clods of silt. However, given the proximity of the Rio Grande and the high sedimentation rate of ≥2.03 mm/a since ≤300 cal BP (depth 0.72 m), fluvial overbank sedimentation may have contributed to the build-up of the stratigraphy. Average sedimentation rate for the top 1.78 m of the profile is 4.73–8.66 mm/a.

Profile LM-6

Profile LM-6 (see Figure 23) records rapid aggradation from low-energy fluvial deposition on top of riverbed gravel which is present at a depth below ca. 1.10 m. Fine-grained sedimentation began after ≤260 cal BP as indicated by a radiocarbon-dated hearth on top of the basal gravel. Plant growth on very-low-energy (still water, likely abandoned channel fill) sediment (loamy silt, depth 1.00–0.74 m) is indicated by increased LoI (4.0%). The presence of charcoal and clods of silt indicates incipient agricultural use of the site which was followed by two further flood deposits between 0.74–0.36 m depth, the lower of which also shows traces of tillage. The top 0.36 m of the profile represent irrigation sediment likely augmented by flood deposits and strongly turbated by tillage. Average sedimentation rate for the top 1.12 m of the profile (since ≤260 cal BP) is ≥3.56 mm/a. A further radiocarbon date from the base of the trench (depth 1.05 m; 8.1 m from LM-5 / 11.9 m from LM-6) connecting profiles LM-5 and LM-6 is ≤300 cal BP (Poz-8016), resulting in an average sedimentation rate of ≥2.96 mm/a.

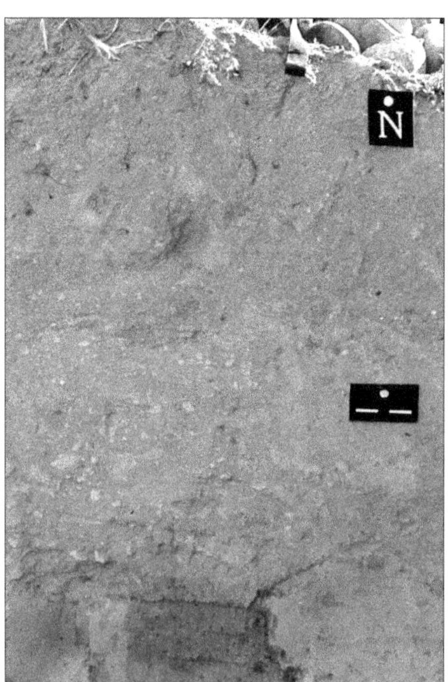

Fig. 24: Photograph of the cut between profiles LM-5 and LM-6 (distance between markers is 0.5 m; photograph by J. Baade)

Profile LM-7

Profile LM-7 (see Figure 23) records at least four fluvial overbank deposition events interspersed and capped by sediments indicative of agricultural activity. Basal gravel is present at a depth of 1.50 m. Like the top 0.40 m of the profile, two layers at depth of 0.40–0.83 m and 0.98–1.18 m overlie bedded to laminated fluvial sediments and are characterised by strong turbation (with cohesive clods of silt distributed throughout the layers). LoI values are low (0.7–1.5%) throughout the profile and indicate lacking accumulation of organic matter and hence a rapid build-up of the stratigraphy.

Profile LM-8

Profile LM-8 (see Figure 23) records sedimentation at a site which is today the edge of an irrigated field. Coarse fluvial gravel is present at a depth of 2.65 m and is overlain by a sequence of low-energy fluvial sediments up to a depth of 0.69 m. The top 0.69 m of the profile are characterised by traces of human impact (charcoal) and sedimentation related to the adjacent irrigation channel (proximal irrigation sedimentation and locally erosion) as well as the deposition of excavated material which had accumulated in the irrigation channel. Carbonate precipitation (2.7–3.4%) has affected the profile at depths between 1.77 and 0.60 m.

5.2.3 *Profiles in the Rio Grande valley*

Profile RG-FL-1

Profile RG-FL-1 records the occurrence of debris flows interspersed with human activity during the Middle Horizon indicated by the repeated presence of hearths in the stratigraphy. The precipitation of high amounts of carbonate since 1,290–1,140 cal BP indicates the possibility of a late-Holocene genesis of caliche layers in the study area.

Profile RG-Fu

Profile RG-Fu records a sequence of fine-grained fluvial sedimentation overlying coarse fluvial gravel. Rapid deposition is indicated by thick beds of sand with foreset structures and ripple marks. Although the profile is situated in an irrigated field, the entire stratigraphy between depths of 2.20 to ca. 0.70 m is of undisturbed fluvial origin. The upper 0.70 m of profile RG-Fu shows traces of agricultural activity; however, the sediment likely represents agriculturally reworked fluvial deposits with only a minor contribution of irrigation sediment.

Profile RGB-1

Profile RGB-1 records a river-proximal sequence of fluvial overbank sedimentation characterised by well-sorted and well-bedded to laminated loam to coarse sand. This sequence is capped by ca. 0.70 m of structureless fine-grained material interpreted to represent either irragric anthrosol or irrigation sediment turbated with flood deposits by agricultural activity. The upper 1.45 m of the stratigraphy were deposited since ≤270 cal BP. The high average sedimentation rate of ≥4.46 mm/a calculated from this age (≥2.15 mm/a for the agriculturally disturbed top portion if fluvial aggradation between depths of 1.45 and 0.70 m is assumed to have been instantaneous) strongly suggests that there has been a contribution of fluvial overbank sedimentation to the agriculturally influenced layers.

Profile RGB-2

Profile RGB-2 shows a sequence of sediments dominated by fluvial sedimentation in high-energy (basal gravel indicating a former riverbed situation) and low-energy (well-sorted and bedded fine-grained overbank sediments) environments. This sequence is interspersed with indicators of human impact (abundant charcoal). While some of the fluvial sediments both in the lower and upper part of the stratigraphy may have been influenced by agricultural disturbance and irrigation, the relative influence of irrigation sedimentation and fluvial sedimentation can not be discerned.

Profile RGB-3

Profile RGB-3 represents both fluvial and anthropogenic influences. At the base of the profile, well-rounded fluvial gravel is present. Most of the profile is characterised by well-sorted, fine-grained sediment with low CSR, low LoI and the presence of bedding. These layers (depth 1.95–0.67/0.58 m) are attributable to fluvial overbank deposition. Human impact is indicated by the presence of charcoal down to a depth of 1.55 m (radiocarbon-dated at 910–680 cal BP). Strongly turbated layers with poor sorting, increased CSR (≥0.27) and increased LoI (≥2.0%) as well as commonly occurring charcoal are interpreted as irragric anthrosol. Inception of irrigation agriculture is indicated by the turbation of a fluvial layer and the commencing sedimentation of irrigation sediments at a depth of 0.58–0.67 m after 480–310 cal BP (depth 0.78 m). Six flood events are recorded between 910–680 cal BP and 480–310 cal BP, i.e. floods reached this site at a rate of one to three events per century. Average sedimentation rates in profile RGB-3 are 1.28–3.85 mm/a for the lower (exclusively fluvial) and 1.46–2.14 for the upper (fluvial and irrigation) part of the profile.

Profile RG-Mo

Profile RG-Mo (Figure 25) represents a transition from fluvial dynamics controlled by the Rio Grande through debris flow dynamics from a nearby quebrada to anthropogenically dominated (irrigation) sedimentation. High-energy riverbed sediments (coarse, well-rounded gravel) are present at a depth of 3.75 m (i.e. 4.60 m above the

Fig. 25: Location of profile RG-Mo in the Rio Grande valley (flow direction from right to left, the archaeological site Los Molinos in the background and the present-day settlement Los Molinos to the right).

present riverbed), followed by medium- to low-energy fluvial sediments between 3.75 and 3.25 m. Between 3.25 and 1.25 m, at least seven major debris flow deposits are recorded at the site, most of them represented by ca. 20 cm of imbricated angular, platy stones overlain by a thin bed of sand containing similar clasts. The platy stones are of local origin. At a depth between 1.75 and 1.35 m, fluvial and debris flow facies interfinger, followed by strong human impact present as a filled-in pit (possibly a looted grave) and abundant Nasca period ceramic shards as well as charcoal. A further major debris flow set (depth 1.35–0.90 m) was deposited before the accumulation of ca. 0.90 m of irragric anthrosol with a basal date of 745–655 cal BP. Average sedimentation rate for the irragric anthrosol is 1.12–1.27 mm/a.

Profile RG-BG

Profile RG-BG represents both natural and anthropogenically-induced sedimentation. Coarse fluvial gravel corresponding to a riverbed situation was not reached in this profile, and the basal sand likely represents modern infilling of an undercut bank. Sediments at depths between ca. 3.20 and 1.41 m are interpreted as five separate low-energy fluvial overbank deposits and one (depth 1.64–1.72 m) overbank gravel splay deposit. The upper portion of the stratigraphy is characterised by 1.03 m of irragric anthrosol and 0.38 m of field-edge deposition. An age of 1,880–1,710 cal BP near the base of the irragric anthrosol indicates incipient irrigation agriculture at the site by the Early Nasca period (1,870–1,690 cal BP, cf. Unkel 2006: 111). The average sedimentation rate since 1,880–1,710 cal BP is 0.45–0.49 mm/a (referring to the recent field surface elevation) or 0.65–0.71 mm/a (including the field-edge deposit).

5.2.4 *Profiles northwest of the Rio Palpa*

Profile RPE

Profile RPE reflects a complex history of erosion and deposition of a terrace near the Rio Palpa. Well-weathered and cemented coarse gravel interpreted to be of pre-Holocene age is exposed locally at the base of the profile, suggesting that erosion and accumulation have remained within the limits of a few metres during the Holocene. High carbonate concentrations and well-developed caliche are interpreted as being due to preferential carbonate precipitation from ground or irrigation water. Human impact at the site is indicated by charcoal (both distributed throughout the stratigraphy and concentrated in charcoal-rich bands). A high age of the upper part of the stratigraphy is suggested by the Nasca 7 ceramic shards at positions 4.2 m (depth 0.55 m), 4.5 m (depth 0.25 m), 17.3 m (depth 0.40 m) and 17.6 m (depth 0.90 m). However, this high age is contradicted by two radiocarbon assays of charcoal from position 4.90 m (depth 1.50 m) and position 5.10 m (depth 0.45 m) which both gave ages of ≤260 cal BP (Poz-5270 and Poz-5271). The presence of even earlier Nasca ceramic (position 8.9 m, depth 0.82 m: Nasca 3 or 5; position 15.9 m, depth 0.60 m: Middle Nasca; position 22.6 m, depth 0.75 m: Nasca 4) in layers stratigraphically younger than historic period artefacts (position 13.8 m, depth 1.30 m: glass shards in gravel layer) further shows that the upper ca. 1.50 m of profile RPE consist of massively disturbed material. It is probable that this disturbance was caused by either fluvial reworking and short-distance hyperconcentrated flow transport (e.g. slumping of upstream river banks) or mechanised agricultural activity. In any case, despite their secondary deposition during late historic times, the abundance of ceramic shards (108 pieces were recovered from profile RPE) from a range of periods including Early, Middle and Late Nasca as well as the Middle Horizon (Loro) and possibly Initial Nasca strongly suggests long-term and perhaps uninterrupted settlement activity in the vicinity of the site. Two Nasca 3 shards found at position 20.9 m (depth 3.56 m) may reflect primary deposition as judged from their placement very low in the stratigraphy. While traces of fluvial activity (bedded overbank deposits) are present throughout profile RPE and indicate that flooding of the lower terrace must have occurred since 260 cal BP, the strongly disturbed character and chronologically inconsistent findings do not allow a detailed interpretation of the sedimentary history at this site.

Profile RPF-1

Profile RPF-1 reflects significant anthropogenic disturbance throughout the stratigraphy. The surface of older (perhaps pre-Holocene) fluvial gravel has been affected by erosion and locally been covered by younger fluvial gravel. Between depths of 1.5 and 0.5 m, the stratigraphy reflects repeated horizontal deposition of poorly sorted material as well as the excavation and infilling of steep-sided pits, with most of the deposited material containing charcoal and/or ceramic shards pertaining to the Middle to Late Nasca period (1,620–1,300 cal BP, cf. UNKEL 2006: 111). The chronologically inconsistent position of Nasca 5 and Nasca 7 shards in the stratigraphy indicates that much of the disturbance took place during or after Nasca 7. The abundance of ceramic shards (in total, 161 pieces and fineware and ca. 6 kg of plainware were recovered from the pit) strongly suggests the presence of a nearby settlement.

Profile RPF-2

Profile RPF-2 reflects mainly anthropogenic influences on the main terrace north-west of the Rio Palpa. Riverbed sediments (coarse, well-rounded gravel) are present at a depth of 1.60–1.70 m below the surface. This fluvial gravel is overlain by structureless silty to sandy sediments generally containing disperse gravel-sized clasts which are interpreted as irrigation sediments. Anthropogenic influence is evident from abundant ceramic shards which were recovered from the excavated material (from depth ≥1.45 m: 9 pieces Middle Nasca, 1 piece Late Paracas; from depths ≤1.45 m [mainly from 1.07–0.63 m]: 40 pieces Middle Nasca, 2 pieces Nasca 7). Below 1.45 m, three mollusc shell fragments were recovered. Charcoal is present mainly at 0.85–1.07 m and at ca. 1.50 m. A Nasca period pit grave extends from a charcoal- and ceramic-bearing paleosurface at 0.90 m to a depth of 1.48 m in the left (southern) part of the south-western side. The stratigraphy on the opposite (north-eastern) side of the pit was not affected by the grave. From charcoal recovered from a depth of 1.50 m, an age of 2,070–1,820 cal BP (Poz-13540) was determined. This age pertains to the Initial Nasca or Early Nasca period (2,150–1,690 cal BP, cf. UNKEL 2006: 111). Sedimentation related to irrigation agriculture therefore had commenced by Initial Nasca or Early Nasca or perhaps as early as Late Paracas times. Human use of the site (possibly settlement-related) of a paleosurface at ca. 0.90 m depth and deposition of the grave probably occurred during the Middle Nasca period (1,620–1,520 cal BP, cf. UNKEL 2006: 111). Average sedimentation rates are 1.09-3.00 mm/a between 2,070–1,820 cal BP and the Middle Nasca period and 0.54–0.57 mm/a since the Middle Nasca period.

Profile RPF-3

Profile RPF-3 (Figure 26) reflects geomorphic processes close to the Cresta Sacramento. While high-energy (channel) fluvial deposits characterised by well-rounded coarse gravel are present at the base of the pit at a depth of only 1.35 m, most of the profile is characterised by poorly-sorted hyperconcentrated flow deposits with angular clasts derived from the nearby Cresta Sacramento. At least six such events are recorded at the site. Low- to medium-energy fluvial sediments (bedded, well-sorted sand to fine gravel) occur in the lower third of the profile. Charcoal and other indicators of human activity at the site are missing except for a single fragment of Nasca 5 ceramics recovered from the excavated material. A possible incipient soil development indicated by a slight increase of LoI values (1.4–1.6%) is present below the hyperconcentrated flow deposits at a depth of 1.35–0.95 m. Irragric anthrosol is present only in the top portion of the profile from a depth of 0.65–0.52 m to the surface. Assuming average sedimentation rates of 0.54–0.57 mm/a as derived at the nearby profile RPF-2 indicates that irrigation agriculture at the site may have been initiated upon the uppermost hyperconcentrated flow deposit between 1,150 and 860 cal BP (i.e. during the Middle Horizon, 1,300–950 cal BP, or Late Intermediate Period, 950–550 cal BP, cf. UNKEL 2006: 111).

Fig. 26: Stratigraphy of profile RPF-3: coarse-grained fluvial sediments overlain by partially eroded fine-grained fluvial sediments and coarse-grained debris flow deposits in the lower portion; irragric anthrosol in the upper half of the profile (photograph: J. Baade).

Profile RPQ

Profile RPQ consists of more than two metres of fluvial sediments and serves as a type example for the material deposited in the riverbeds of Rio Palpa and Rio Vizcas. The observed grain size parameters and sedimentary structures (bedding, imbrication, cut-and-fill features) are characteristic for braided, gravel-bed rivers. Coarse, imbricated gravel sheets with horizontal or inclined bedding indicate sedimentation in high-energy flow environments such as channels and bar structures. Finer-grained (silt to fine gravel) and commonly well-bedded and/or upwards-fining sediments indicate waning-flow or bar accumulation (a) in primary channels close to the end of the discharge season, (b) in secondary and higher-order channels or (c) on inter-channel flow surfaces. The presence of loam with high LoI values at stations RPQ-1 and RPQ-2 which is interpreted as eroded soil hints at lateral erosion affecting fine-grained riverbanks upstream. Repeated erosion and accumulation of sediment during discharge periods and possibly the movement of sediment pulses through the reach within late historic times (second half of the 20th century) are shown by the presence of a plastic bottle and a piece of plastic bag at a depth of 1.43–1.53 m below the surface of the riverbed.

Profile RPW

Profile RPW records mainly anthropogenic impact, i.e. the accumulation of irragric anthrosol, turbation of fluvial sediment and/or a combination of irragric anthrosol and fluvial sediment and the incorporation of charcoal and ceramic shards into the sediment. The only undisturbed fluvial layer is characterised by very good sorting, low LoI and the presence of bedding. The presence of charcoal dated to 1,690–1,410 cal BP almost 0.2 m below the upper boundary of the coarse and partially well-weathered (and therefore presumably very old) basal gravel is somewhat enigmatic. It is tentatively assumed that either the charcoal may have been turbated into the gravel from the overlying, very-poorly sorted and ceramic-bearing stratum (which appears likely because material from the basal gravel also appears to have been turbated into the overlying layer) or that old, weathered gravel may have been remobilised and may have incorporated the charcoal at that time. The Middle Nasca date of 1,690–1,410 cal BP is a maximum age for the deposition of the overlying sedimentary column. Assuming that this age reflects the begin of fine-grained sedimentation, the average sedimentation rate is 0.93–1.11 mm/a.

Profile YN-1

Profile YN-1 records river-proximal sedimentation on the right side of the Rio Palpa since the Late Intermediate Period. After the retreat of the course of the river (indicated by riverbed gravel) from the site, ca. 0.60 m of well-sorted fluvial sand were deposited. The maximum age for this deposition is 750–665 cal BP. After this time, mainly irrigation sediments were accumulated, interspersed by two minor flood sediments at depths of 1.60–1.50 m and 1.43–1.40 m. Given the proximity of the Rio Palpa, further fine-grained flood deposits may have affected the site; however, if this did occur the associated sediments have been completely incorporated into the irragric anthrosol by tillage. The average sedimentation rate has been 2.55–2.85 mm/a since 750–665 cal BP.

Profile YN-2

Profile YN-2 records high- and low-energy fluvial sedimentation (gravel and bedded silt to sand) as well as anthropogenic influences (strongly turbated coarse- and fine-grained material, a charcoal-rich layer and a filled-in depression into which large stones appear to have been intentionally set). The complex lower portion of the profile (depth between 2.20 and 0.60 m) is overlain by 0.60 m of irragric anthrosol.

Profile Ma-T

Profile Ma-T shows that the large upper irrigated terrace in this part of the Palpa Valley is constructed exclusively of fine-grained material. There are no dating constraints except for the assumption that the strongly-weathered basal gravel (below 3.20 m) is of pre-Holocene age while the unweathered stone layer at a depth between 2.80 and 3.20 m is of Holocene age. Given the proximity of the Laguna La Maquina, a source of water of high importance in pre-historic times, the lack of artefacts in this excavation is surprising and may have to be interpreted in terms of an early deposition of much of the terrace before human impact. The lack of clearly identifiable irrigation sediment can be explained with the use of sediment-free water from the natural spring, allowing agricultural use without sediment accumulation. This can also explain the presence of five conspicuous layers of well-cemented caliche. The erosional activity of the Rio Palpa creating the edge of the upper terrace apparently stripped away the fine-grained material and was halted by the underlying gravel base. The thin (a few decimetres) irrigation sediment with agriculturally turbated underlying gravel on the adjacent lower terrace hints at a young age (on the order of a few hundred years) of this erosional event.

5.2.5 *Profiles between Rio Palpa and Rio Vizcas*

Profile AR

Profile AR records a sequence from basal, strongly-weathered (likely pre-Holocene) coarse gravel, five low-energy (silt to sand) and one high-energy (gravel) fluvial sediments and five likely anthropogenically dominated layers (irragric anthrosol and charcoal-rich layers). Carbonate content peaks at a depth of 1.25–0.88 m with 9.2%, suggesting preferential evaporative enrichment from irrigation-induced groundwater seepage from the field above the exposure. The situation of this large (670 m long) fluvially cut bank in combination with the horizontally layered and bedded sediments in the upper part of the profile and an age of 530–330 cal BP from a depth of 1.37 m in the recorded profile indicates that a strong erosional event of the Rio Grande de Nazca has taken place probably during the historic period. Assuming a sedimentation rate of 3.0–5.0 mm/a (cf. profiles ES and DO) to estimate the time required for the accumulation of the top 1.37 m of the profile constrains the age of this event to ≤260 cal BP.

Profile R-2

Profile R-2 is characterised by long-term human impact since at least the Paracas period, following the deposition of fine-grained fluvial overbank sediments and a massive (depth 3.76–4.87 m, 1.11 m thick) hyperconcentrated flow deposit. A minor hyperconcentrated flow deposit (depth 1.69–1.80, 0.11 m thick) separates a Paracas midden (likely indicating nearby settlement activity) and irragric anthrosol from a later, Early to Middle Nasca irragric anthrosol which is overlain by three further anthrosols separated by diffuse boundaries, recording an apparently almost continuous agricultural use of the site since Early to Middle Nasca times. The top of the profile is marked by the 1999 flood deposit. No other sediment comparable with this deposit is present within the 3.67 m of the profile. Sedimentation rates at this river-distal site are 0.50–0.94 mm/a for the period from 2,700–2,340 cal BP (depth 2.28 m) to 1,720–1,540 cal BP (depth 1.70 m) and 0.82–0.91 mm/a for the period from 1,720–1,540 cal BP (depth 1.70 m) to AD 1999 (depth 0.25 m).

Profile R-3

Profile R-3 records a sequence of sediments from both fluvially and anthropogenically dominated environments. All eleven layers at the bottom of the profile (depth 4.57–6.10 m) are interpreted as low-energy overbank deposits due to their grain size, sorting and bedding characteristics. At a depth of 4.57–3.60 m, a massive (0.97 m thick) hyperconcentrated flow deposit is present. Incipient irrigation and agricultural use of the site are indicated by the diffuse upper boundary of the hyperconcentrated flow deposit and the apparent turbation of coarse-grained material from this deposit into the overlying fine-grained irrigation sediment. The sharply defined top of this irragric anthrosol is marked by ceramic shards and abundant charcoal, possibly indicating nearby settlement activity at 2,300–1,950 cal BP; probably during the Initial Nasca period (2,150–1,890 cal BP, cf. UNKEL 2006: 111). A ca. 25 cm thick fluvial layer separates this usage horizon from a second likely irragric anthrosol at a depth between 2.24 and 1.80 m. Another fluvial layer with a turbated upper boundary containing Late Intermediate Period ceramic shards (950–555 cal BP, cf. UNKEL 2006: 111) is overlain by a third irragric anthrosol which in turn is capped by a complex sequence of irrigation sediment and material excavated from the nearby irrigation channel (Figure 27). The two top layers of profile R-3 are interpreted as agriculturally reworked flood deposit possibly deposited by the 1999 event. Average sedimentation rate for the upper portion of profile R-3 is 1.05–1.23 mm/a, however, the complex situation near the top of the profile has to be taken into account.

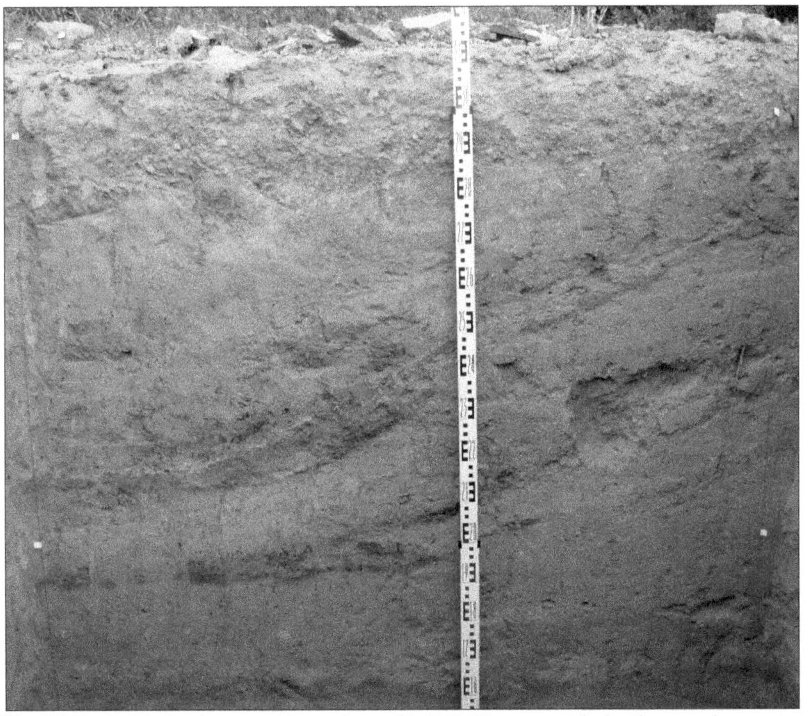

Fig. 27: Buried irrigation canal levee in profile R-3. The present irrigation canal runs at a distance of ca. 10 m to the right.

Profile R-4

Profile R-4 records a sequence of mainly anthropogenic influence interspersed by a few fluvial events. A lower irragric anthrosol (depth below 3.07 m) containing ceramic shards of unknown age was capped by two layers of fine-grained fluvial sediment following erosional truncation. Re-initiation of agriculture took place by irrigation sedimentation and turbation of the upper portion of the fluvial sediment. A sediment interpreted as irragric anthrosol with poor sorting and high LoI values is present between depths of 2.85 and 0.05 m interrupted by three fluvial layers which all show turbation of their top with the base of the overlying layers. The profile is capped by a thin fluvial deposit attributed to the 1999 flood event.

Profile ES

Profile ES records repeated fine-grained overbank deposition in a river-proximal situation during the historic period. At least six separate events can be discerned, five of which occurred since ≤270 cal BP, i.e. at an average rate of ≥2.2 events per century. Anthropogenic influence can be assumed for the top 0.54 m of the profile (irragric anthrosol and/or agriculturally turbated overbank deposit) and for the charcoal-rich layer at a depth of ca. 1.45 m. A massive layer of poorly sorted material at a depth between 0.76 and 1.27 m is interpreted as a hyperconcentrated flow deposit. Average sedimentation rate since ≤270 cal BP has been ≥4.86 mm/a and has to be attributed mainly to fluvial deposition, with irrigation sediment accounting for ≤34% of the stratigraphy above the datum. Assuming that the land use change from field to plantation between 1944 and 1970 (as seen in aerial photographs) has led to changes visible in the sedimentary record, the boundary between layers 1 and 2 may be attributed to this land use change. Under this assumption, the sedimentation rate by late historic plantation irrigation can be estimated to 3.03–3.71 mm/a.

Profile DO

Profile DO is characterised by a sequence of initial deposition of coarse fluvial gravel followed by river-proximal flood events and late anthropogenic influence. Twenty-two low- to medium-energy fluvial layers alternating between sand and silt have been identified, five of them with internal layering / bedding. Eight of the fluvial events occurred between 900–670 cal BP and 490–320 cal BP, i.e. at an average rate of 1.4 to 4.4 events per century. Twelve events occurred after 490–320 cal BP, i.e. at an average rate of 2.4 to 3.8 events per century. From the bottom of the profile upwards, there is a tendency towards thinner layers as would be expected for overbank deposits as the bank rises relative to the riverbed due to sedimentation. Sorting is generally good in the lower, fluvially dominated portion of the profile; however, LoI reaches ca. 2.0% for several of the fluvial layers (generally in the more silty sediments). Since charcoal is present only in very small amounts throughout the fluvial portion, these relatively high LoI values are best explained by accumulation of organic material through plant growth during periods between flood events. The top 0.95 m of profile DO differ significantly from the lower part in that they are characterised by poor sorting, relatively high CSR and LoI, diffuse layer boundaries and lack of bedding. These differences are attributed to both a different mode of sedimentation (irrigation instead of floods) and strong turbation by agriculture. These irragric anthrosol layers have been deposited since to 490–320 cal BP. Average sedimentation rates are 1.01–4.17 mm/a for the period from 900–670 cal BP to 490–320 cal BP and 3.00–4.37 mm/a for the period since 490–320 cal BP.

Profile RPS

Profile RPS shows alternating high- and low-energy fluvial sedimentation similar to profile DO as well as some human impact (as indicated by the presence of abundant charcoal) in a river-proximal environment.

Profile Jau-1

Profile Jau-1 reflects an alternation of fluvial and anthropogenic influences. Basal fluvial gravel is present at a depth of 3.40 m. Since 1,490–1,300 cal BP (depth of 3.20 m) and until historic times, the site was repeatedly impacted by fluvial events which deposited well-sorted silt, sand and fine gravel. Some layers are of definite fluvial origin as indicated by bedding characteristics and were deposited in low- to medium energy flow environments (ripple marks and cross-bedding in silty sand to fine gravel). It is likely that there are hiatus in the stratigraphy since several of the layer boundaries probably have erosional character. Some fluvial layers are characterised by diffuse or pocketed upper boundaries overlain by well-turbated, structureless fine-grained sediment containing disperse charcoal. These features are interpreted as indicating incipient agricultural use and irrigation on fluvial sediment. Other anthropogenically dominated layers as well as the surface A_p irrigation horizon are characterised by increased LoI values and the presence of charcoal. Sedimentation rates are 0.91–1.20 mm/a for the period 1,490–1,300 cal BP (depth 3.20 m) to 550–505 cal BP (depth 2.30 m), 0.57–26.0 mm/a for the period 550–505 cal BP (depth 2.30 m) to 500–320 cal BP (depth 1.00 m) and 1.80–2.67 mm/a for the period since 500–320 cal BP (depth 1.00 m). Possible hiatus and the strongly episodic nature of deposition in profile Jau-1 have to be taken into account when interpreting these sedimentation rates.

Profile FL

Profile FL records the results of a high energy event of erosion and deposition followed by a sequence of fine-grained flood deposits and anthropogenically induced sedimentation (irragric anthrosol). Fluvial overbank deposits are clearly distinguishable from irrigation sediments on the basis of sorting, LoI and bedding characteristics. The ten fine-grained fluvial layers possibly represent only three separate events with respective internal stratigraphic sequences as suggested by the recurrent capping of these three sedimentary packages by well-defined silty layers. Average sedimentation rates are 1.0–1.9 mm/a for the lower, fluvially dominated part of the profile (depth 1.04–1.80 m, 1,060–920 cal BP) and 1.8–2.8 mm/a for the upper part of the profile (above 1.04 m, since 510–320 cal BP).

Profile H-1

Profile H-1 is characterised by basal gravel of probably pre-Holocene age as judged from their strong weathering, overlain by poorly sorted, younger (Holocene) gravel which in turn is overlain by 1.25–1.40 m of irrigation sediment and/or turbated fine-grained overbank flood sediment. A very large boulder (diameter >1.5 m) indicates the occurrence of an extreme flood event of unknown age. A former irrigation channel filled with coarse sand and capped by silty sand was identified at a depth of 1.05 m.

Profile Jau-Co

Profile Jau-Co is interpreted as a succession of riverbed and irrigation sediments with interspersed fluvial overbank deposits (layers 6 and 3). The minimum age of riverbed gravel deposition is given by a radiocarbon date of 1,700–1,420 cal BP from the base of the overlying fine-grained sediment. A major flood event later deposited 15 to 30 cm of coarse sand to silt. The top 0.75 m of the profile contain Nasca shards as well as 20th century refuse and are therefore considered to represent material related to nearby 20th century settlement activity (landfill / discarded artefacts brought to the site). Excluding the top 0.75 m of the profile and assuming a plausible age of AD 1900–2000 (50 to -50 cal BP), fine-grained sedimentation of fluvial and irrigation sediment occurred at an average rate of 0.57–0.58 mm/a before the deposition of modern refuse.

Profile RP-06

Profile RP-06 (Figure 28) is a sequence from natural to human-dominated sedimentation. There is an increase in carbonate content and LoI from a depth of 2.00 m to the surface. The increase in carbonate content is explainable by evaporative processes. While pore water was available underground from a nearby irrigation channel, there was no irrigation (and hence no downward percolation flushing out soluble salts) on the site since at least 1944. Carbonate and LoI also peak in near-basal layer which is interpreted as a natural paleosol developed in a riverine situation. The lowest LoI values (0.6–0.8%) of profile RP-06 were measured in the hyperconcentrated flow deposits and fluvial layers. At the base of the profile, strongly weathered gravel of probably pre-Holocene age is exposed. A shallow soil developed during the mid-Holocene in a hyperconcentrated flow deposit. Fluvial overbank deposits and hyperconcentrated flow deposits characterise the period from 6,190–5,920 cal BP to 3,080–2,840 cal BP, followed by deposition of irrigation sediment (irragric anthrosol) during the late Initial Period (3,080–2,840 cal BP) and human use of the site during late Paracas times (Ocucaje 9, 2,300–2,150 cal BP, cf. Unkel 2006: 111). Possible and actual anthropogenic alteration of the stratigraphy (e.g. by fire, deposition of material and erection of buildings) inhibit detailed interpretation of the upper 0.85 m of the profile. Average sedimentation rates are 0.43–0.51 mm/a

Fig. 28: Profile RP-06. Left: profile photograph; top right: overview of bank exposure with persons indicating the positions of wall foundations (**A** – mid-Holocene palesol; **B** – mid-Holocene hyperconcentrated flow deposit; **C** – layer containing abundant Late Paracas ceramic shards). Photographs by J. Baade.

for the mid- to Late Holocene sequence of fluvial and hyperconcentrated flow deposition and paleosol development (6,190–5,920 to 3,080–2,840 cal BP), 0.32–0.56 mm/a for the incipient irrigation agriculture (3,080–2,840 to 2,300–2,150 cal BP) and 0.47–0.50 mm/a for the anthropogenically dominated upper part of the profile (after 2,300–2,150 cal BP).

Profile SF-1

Profile SF-1 records a sequence of natural sedimentation from riverbed gravel followed by fluvial channel or proximal overbank deposits (sand and sand with fine gravel), distal or waning overbank deposits (sandy silt) and hyperconcentrated flow deposits which is capped by 3.55 m of irrigation sediment. An age of 6,280–5,990 cal BP for incipient fine-grained sedimentation places the underlying hyperconcentrated flow deposit into the early to mid-Holocene. The low average sedimentation rate between 6,280–5,990 cal BP and 440–140 cal BP (0.24–0.27 mm/a) and the very high average sedimentation rate since 440–140 cal BP (4.0–10.3 mm/a) are anomalous in comparison with other profiles. Possible explanations are reworking of older material, turbation of younger material, a pre- or early historic sedimentary hiatus or mechanised anthropogenic deposition in late historic times.

Profile R-1

Profile R-1 records a sequence of fluvial and irrigation sedimentation. Two radiocarbon dates (1,530–1,350 cal BP and 1,870–1,630 cal BP) place this layer into the Nasca period (1,870–1,300 cal BP, cf. UNKEL 2006: 111). Incipient irrigation agriculture in Early Nasca times at this site is interpreted from the turbation of clods of irrigation sediment into underlying fluvially deposited material, however, given the poorly defined character of some of the underlying layers (similar to other irragric anthrosols but lacking artefacts), an earlier period of irrigation agriculture is possible. The recent ploughing horizon likely contains reworked material from the 1999 flood event. Sedimentation rates are 0.79–4.1 mm/a for the Nasca irrigation sediment, 0.58–0.81 mm/a for Late Nasca to historic time and 2.6–3.8 mm/a for historic times.

Profile SF-2

Profile SF-2 represents a succession from fluvial gravel (riverbed) dynamics followed by the deposition of a fine-grained fluvial layer, a hyperconcentrated flow deposit and long-term accumulation of irrigation sediments. While all layers except the three fluvially deposited layers are characterised by high CSR and LoI values and poor sorting, the layer pertaining to Early/Middle Paracas through early Nasca (using linear age interpolation, cf. UNKEL 2006: 111) has the highest CSR (0.64) and almost the highest LoI (3.0%). With two exceptions, flood events did either not occur or are masked by agricultural activity. The top layer probably is the reworked deposit of the 1999 flood event. Some sand lenses could be the result of flood events; however, their very limited size lets an interpretation as results of irrigation channel flow appear more likely. Besides abundant plainware and fineware ceramic shards pertaining to the Middle Horizon and Late Intermediate Period, definite anthropogenic structures are present in this profile: a filled-in pit and a hearth with associated ceramic shards dating to the early Late Intermediate Period (Figure 29). With the exception of one uncertain piece of ceramic from the filled-in pit, ceramic and radiocarbon ages are stratigraphically consistent. Sedimentation rates are 0.81–1.00 mm/a for the lower part of the profile (3,220–2,950 cal BP to 1,080–930 cal BP) and 1.01–1.16 mm/a for the upper part of the profile (1,080–930 cal BP to AD 1999).

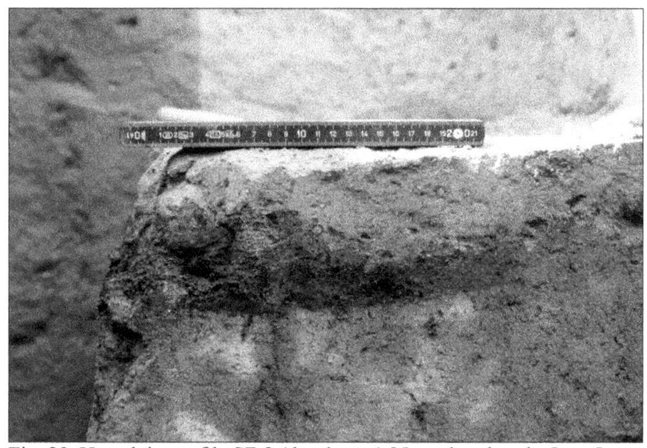

Fig. 29: Hearth in profile SF-2 (depth ca. 1.35 m, dated to the Late Intermediate Period by radiocarbon dating and cermic chronology). Photograph by J. Baade.

Profile SF-3

Profile SF-3 records a history of human impact at least since the Late Intermediate Period (950–550 cal BP, see UNKEL 2006: 111) and possibly since the Middle Horizon (1,300–950 cal BP, cf. UNKEL 2006: 111) as evidenced by the presence of ceramic shards, foundation-like structures (Figure 30), marine mollusc shells, charcoal and strongly turbated sediment. Among other properties, poor sorting and high LoI values together with anthropogenic traces suggest agricultural use of the site. Fluvial impact is evidenced by river-bed gravel at the base and overbank flood sediment at the top of the profile.

Profile SF-4

Profile SF-4 records a sequence of fluvial activity and human impact. Very coarse river bed sediments were deposited at the site probably before 1,520–1,300 cal BP. The abundance of ceramic shards suggests that during the Middle Horizon (Loro, 1,300–950 cal BP, cf. UNKEL 2006: 111) and perhaps already during Late Nasca period (1,520–1,300 cal BP, cf. UNKEL 2006: 111), there was a set-

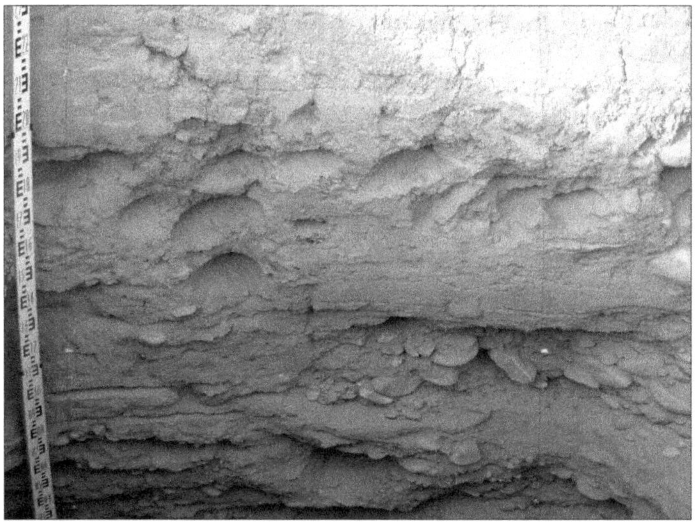

Fig. 30: Profile SF-3. Moulds of stones set in a foundation-like manner. Note the cone-shaped incision of mainly fine-grained material into bedded sand and gravel below the moulds of stacked boulders. Photograph by J. Baade.

tlement near this site. Given the mixture of coarse gravel and well-preserved ceramic shards, the site likely was situated at the edge of the river at that time. A later more distal position of the river allowed the deposition of poorly sorted, fine-grained sediment likely related to irrigation agriculture as indicated by strong turbation, increased LoI values and the presence of charcoal. The following layers are interpreted as results of repeated fluvial overbank deposits (fine-grain as well as gravel) which were used for agriculture in the intervening periods. The top of the profile is marked by the 1999 flood sediment (well-sorted, laminated sand with low LoI).

Profile SR

Profile SR can be divided into two parts: While the top part (above a depth of 1.20 m) is interpreted to be accumulated by irrigation, several unambiguous fluvial layers identified on the basis of their grain size, sorting, CSR, LoI and bedding are present in the lower part. Irrigation sediment is recognisable due to its grain size and sorting at a depth of 2.25–1.70 m. Possible traces of tillage at a depth of 1.50–1.20 m hint at agricultural use of fluvial deposits after flood events. Deposition of the upper 1.70 m occurred since 470–310 cal BP, i.e. during historic times. The high average sedimentation rate of 3.2–5.5 mm/a is interpreted to be due to both rapid aggradation by a small number of overbank flood events and intensive historic to modern irrigation with high water and sediment input.

Profile RVU

Profile RVU (Figure 31) records a complex (albeit incomplete) pattern of erosion and deposition in the proximity of the Rio Vizcas since the mid-Holocene (5,500–4,950 cal BP). During the mid-Holocene, structureless and homogenous fine-grained sediment (silty sand) with a minimum thickness of 0.95 m was deposited, the origin of which is unknown. None of the possible interpretations (turbated fluvial overbank / low energy flow deposit, irrigation sediment or hyperconcentrated flow deposit) can be securely rejected or supported on the base of available

Fig. 31: Profile RVU. Left: two coarse-grained channel deposits overlying mid-Holocene fine-grained sediment and capped by irragric anthrosol; right: partially excavated erosional surface and coarse-grained fluvial channel fill. Photographs by J. Baade.

data, although the presence of charcoal weakly suggests a possible early human influence. The mid-Holocene was followed by repeated erosion and deposition as evidenced by erosional surfaces as well as several coarse gravel sheets and fine-grained sediments. The earliest definite human impact appears in the form of Loro ceramic shards and coeval (1,260–1,060 cal BP) charcoal, giving a minimum age for the last gravel sheet deposition at this site. Repeated fluvial as well as anthropogenic influence of probably mainly historic age is evidenced throughout a large part of the profile by well-bedded or laminated fluvial sediments on one hand and strongly turbated and charcoal-rich layers locally containing ceramic shards on the other hand. Bedded fluvial overbank deposits likely related to the 1999 erosional event are present along several parts of the exposure.

5.2.6 *Profiles south-east of the Rio Vizcas*

Profile LLPT

Profile LLPT is characterised by basal fluvial gravel overlain by a gravel-sand sequence interpreted as a gravel/sand bar complex which in turn is overlain by an upwards-coarsening layer which can be interpreted as an overbank splay sheet. A conspicuous change in sedimentation separates the lower from the upper portion of the profile. The deposition of 3.30 metres of silty sand is only interrupted by three thin lenses or layers of silt, silty clay and sand, respectively. The entire upper sequence is interpreted as irragric anthrosol indicating the inception of irrigation agriculture by early to middle Paracas times (2,360–2,720 cal BP, cf. Unkel 2006: 111). Quasi-continuous sedimentation lasted at least until early to middle Nasca times (1,610–1,830 cal BP, cf. Unkel 2006: 111) after which three undated interruptions (as shown by silt / silty clay deposition and layer boundaries) occurred until late historic times (260 cal BP to the present) without a strong change in average sedimentation rates (0.68–1.41 mm/a for the lower and 0.82–1.15 mm/a for the upper portion of the profile). The anomalously high sedimentation rate for the top 1.00 m (≥3.85 mm/a) in late historic times may be due to very high irrigation water input for modern high yield – high demand plants, an intentional sedimentation in the adobe pits or mechanised earth movement.

Profile BF-2

The bottom 1.60 m of profile BF-2 expose strongly weathered and cemented coarse-grained material reminiscent of the road cuts through the bajadas (e.g. the Cresta Sacramento). This material is probably of pre-Holocene age and likely corresponds to the early period of bajada deposition. It is overlain by a sequence of well-sorted material with low LoI values which is interpreted as resulting from fluvial channel and overbank deposition during the mid-Holocene. The top 3.57 m of the profile are characterised by increasingly poorer sorting and higher LoI values as well as the inclusion of dispersed small stones in a matrix of structureless to columnar silty sand / sandy silt. Therefore, these layers are interpreted as irragric anthrosol.

Profile BJ

At the base of profile BJ, 3.90 m of strongly weathered and cemented gravel of probably pre-Holocene age have been recorded. It is overlain by a sedimentary sequence of fluvial overbank deposition (ten layers) and hyperconcentrated flow / debris flow deposition (7 layers) originating from a nearby quebrada in the flank of the Cresta San Ignacio. At a depth of 3.55 to 2.90 m, beginning human impact at the site is indicated by the appearance of charcoal. Given the still low LoI values and the proximity of a quebrada mouth, the turbation of gravel with silty sand may be due to a hyperconcentrated flow event. The layer at a depth of 2.80 to 1.40 m is interpreted as irragric anthrosol with a quasi-continuous (seasonal) sedimentation from shortly after 2,720–2,360 cal BP until at least 1,610–1,410 cal BP with a mean sedimentation rate between 0.46 and 0.80 mm/a. The upper section of profile BJ records a complex sequence of erosion, deposition and human impact. At an unknown time between 1,610–1,410 cal BP and 470–300 cal BP, deposition or irrigation sediment was terminated and followed by erosion of at least 0.70 m in the northern and north-eastern sector, perhaps related to the current position of the profile close (ca. 5 m) to a terrace edge. Neither river- nor quebrada related erosional processes are plausibel given the lack of erosion in profile FV and the orientation of the erosional flank perpendicular to the nearby hillslope. The partial erosion was followed by the deposition of either colluvial or irrigation sediment. A 90 cm deep pit (probably a grave) was dug into this sequence in historic times (after 470–300 cal BP). Apparently, this grave was later looted since human bones fragments as well as glass and ceramic shards were found in the fill of the pit as well as in the layer which represents the concurrent paleosurface. A low dam then separated a sector of low-energy fluvial or still-water deposition from a turbated sector into which shallow channels were incised. This is interpreted as being related to the position near the lower terrace/field edge with a shallow pond situation, a field edge and a collector channel for excessive irrigation water. The top 15 cm represent the A_p horizon of an irragric anthrosol.

Profile FV

Profile FV represents a transition from natural dynamics (hyperconcentrated flow deposits) to anthropogenic influence (irrigation agriculture). In the upper 2.70 m, ony one layer shows unequivocal fluvial influence. Structures interpreted as planting holes are evidence for agricultural activity at this site by 2,800–2,150 cal BP after a fluvial deposition event which left bedded sand to fine gravel. The planting holes as well as low CSR and LoI values and high sand content of the overlying material suggest that a significant part of the fluvial sediment was turbated into the overlying irrigation sediment. Between 2,800–2,150 cal BP and 1,710–1,540 cal BP, irrigation sediment was deposited at a rate of 0.68–1.93 mm/a. The sedimentation rate for the top 0.80 m is 0.45–0.52 mm/a.

Profile RV

Profile RV records a succession of fluvial and irrigation sediments as well as several phases of direct anthropogenic influence as seen repeatedly as heaped-up material of low field-edge banks. Interspersed laminated silt layers are interpreted as low-energy deposits related to irrigation. Oxidation stains are present below 1.78 m and indicate the influence of a strongly fluctuating ground water level. The entire sequence was deposited since 490–310 cal BP; the average sedimentation rate was 6.13–10.8 mm/a.

Profile LU

Profile LU represents both anthropogenic and fluvial influence. Besides Late Intermediate Period ceramic shards in the recorded profile, Nasca and Wari shards were found laterally in the bank exposure at depths between 2.0 and 1.0 m. While a reworked or colluvial nature of some of the sediments can not be excluded, layers above a depth of 1.85 m are interpreted as irrigation sediment. The bedding present in the tops of two layers is interpreted as low energy / still water sedimentation related to irrigation because of the fine (silt, silty sand) grain size atypical for proximal flood deposits. Assuming primary deposition of the Late Intermediate Period shards, average sedimentation rate since that time has been 0.69–1.28 mm/a.

5.3 *Transects in the Palpa Valley*

5.3.1 *Transect La Muña*

Transect La Muña (Figures 32 and 33, Table 7; see also Figure 23) is located at the mouth of Quebrada La Muña at the western edge of the Palpa Valley. It allows the reconstruction of autochthonous and allochthonous dynamics as well as of anthropogenic influences near the archaeological site of La Muña. From profile LM-1 on the alluvial/debris fan, a row of five pits (two of which were connected by a 20 m long cut) extends across a field surface to the right bank of the Rio Grande (profile LM-4). In all profiles, coarse basal gravel is overlain by fluvial sediments of varying thickness. Autochthonous sedimentation from debris flows is restricted to profiles LM-1 and LM-2; allochthonous fluvial sediments in profiles LM-8, LM-5, LM-6 and LM-4 were deposited by the Rio Grande. Fine-grained fluvial sediment in profiles LM-1, LM-2 and LM-8 is partially of autochthonous and partially of allochthonous origin. Profiles LM-8, LM-5 and LM-6 are capped by irragric anthrosol. Sediments in the river-bank exposure LM-4 are exclusively of fluvial origin.

The uppermost profile LM-1 records seven coarse grained debris flows which have occurred since 910–730 cal BP (Poz-8010). In profile LM-2, which is situated only eight metres downslope from LM-1, only fine-grained sediments are present in the stratigraphy with the exception of the surface layer. This likely has to be attributed to the rapid exhaustion of energy in the runout zone of debris flows due to flow expansion and slope reduction, i.e. the down-slope limit for most coarse-grained flows is near profile LM-1. Profile LM-1 records an increase in the frequency of debris

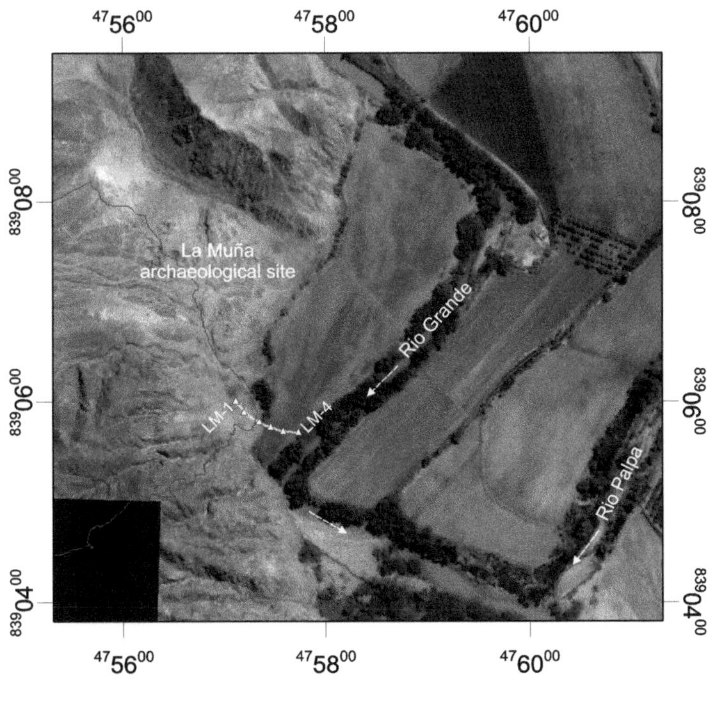

Fig. 32: Satellite image (Quickbird) showing the location of transect La Muña (line LM-1 – LM-4) in relation to Rio Grande, Rio Palpa and the archaeological site of La Muña. The thin grey line indicates the limits of the La Muña catchment.

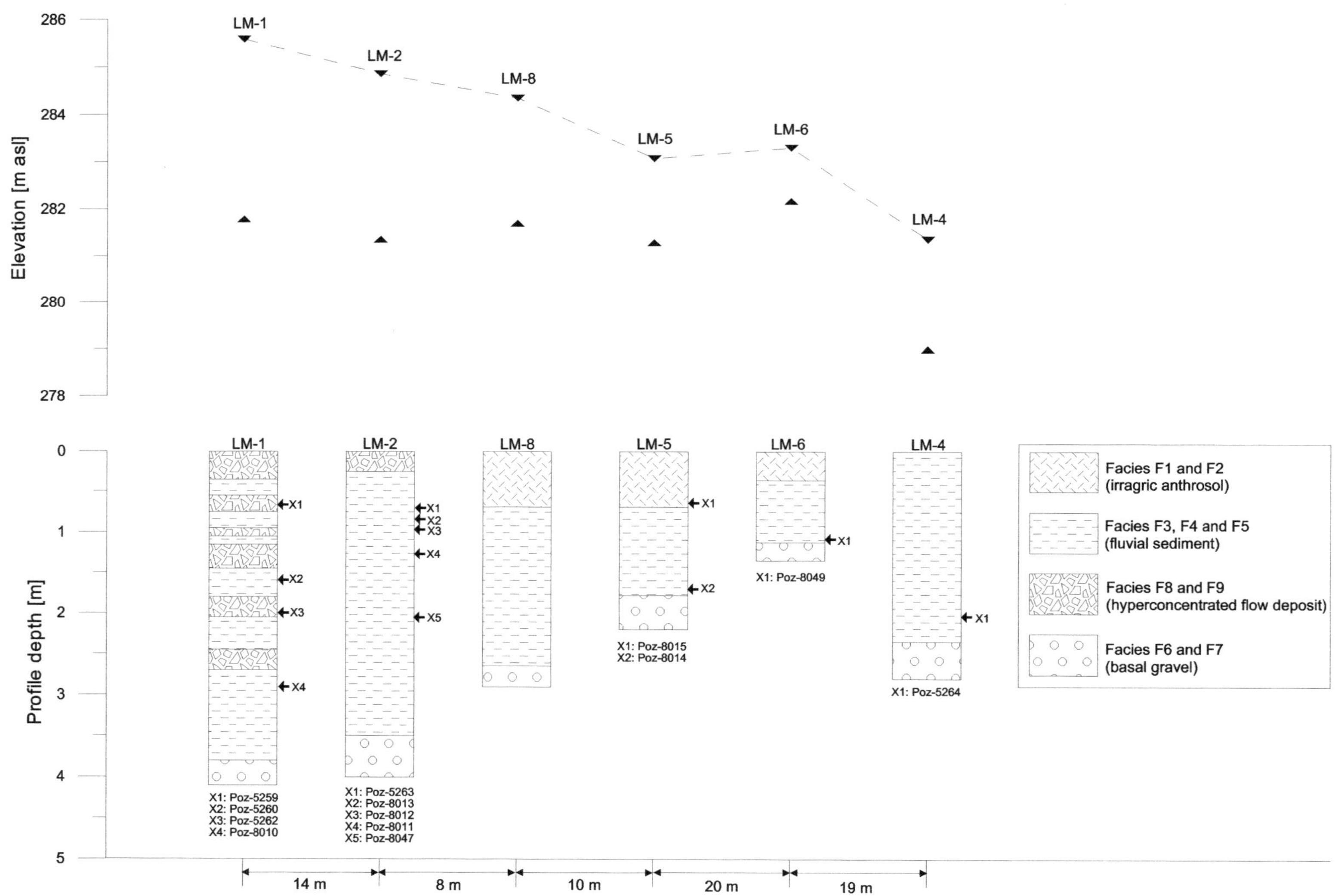

Fig. 33: Transect La Muña. Top: elevation of land surface (downward-pointing triangles) and basal gravel (upward-pointing triangles); bottom: simplified stratigraphic sketches with approximate depths of ^{14}C samples (ages shown in Table 7).

Tab. 7: Radiocarbon dating results for transect La Muña (* = sample excluded from analysis due to stratigraphic inconsistency).

Sample-ID	^{14}C BP	Location	Depth below surface [cm]	cal BP (2σ)
Poz-5259	28 ±530	LM-1	70	450 – 150
Poz-5260*	525 ±30	LM-1	160	545 – 495
Poz-5262	385 ±30	LM-1	200	490 – 320
Poz-5263	195 ±30	LM-2	68	≤300
Poz-5264	85 ±30	LM-4	200	≤260
Poz-8010	925 ±25	LM-1	293	910 – 730
Poz-8011	520 ±30	LM-2	122	545 – 495
Poz-8012	450 ±25	LM-2	97	520 – 330
Poz-8013	335 ±25	LM-2	83	450 – 300
Poz-8014	255 ±25	LM-5	175 – 180	320 – 150
Poz-8015	205 ±25	LM-5	72	≤300
Poz-8016	210 ±25	LM-5/6	105	≤300
Poz-8047	725 ±35	LM-2	200	680 – 560
Poz-8049	110 ±30	LM-6	112	≤260

flows from one depositional event between 910–730 cal BP (Poz-8010) and 490–320 cal BP (Poz-5262) to six events after 490–320 cal BP. Radiocarbon dates of 910–730 cal BP (Poz-8010) and 450–150 cal BP (Poz-5259) bracket four couples of fine-grained fluvial sediment overlain by organic-rich material which indicate repeated plant growth and burial by floods, likely in a river-proximal situation.

The pits on the field between the quebrada mouth and the Rio Grande (profiles LM-8, LM-5 and LM-6) are characterised by rapid aggradation in historic times (1.8 m since 320–150 cal BP, Poz-8014) after river channel migration from near the mouth of the quebrada to its present location. Exclusively fluvial sediments in the lower part of the stratigraphy of the area abandoned by the river are evidence that channel migration was not caused by accumulation of debris flow sediment originating from Quebrada La Muña but rather by the fluvial dynamics of the Rio Grande itself. Agricultural use of the area abandoned by the river due to channel migration commenced after the deposition of ca. 0.7–2.0 m of fine-grained fluvial sediments as evidenced by a change in sedimentary properties.

5.3.2 *Transect Rio Palpa*

Transect Rio Palpa (Figures 34 and 35, Table 8) consists of a row of three pits (RPF-3, RPF-2 and RPF-1) running perpendicular to the Palpa Valley from near the foot of the Cresta Sacramento to a bank exposure (RPE) on the right side of the Rio Palpa. It further includes four profiles recorded in a ditch across the riverbed (RPQ-1 to RPQ-4) as well as a pit (DO) on a field on the left side of the Rio Palpa. In profiles RFF-1, RPF-2 and RPF-3 to the northwest of the Rio Palpa, basal gravel is present at a depth of 1.4–1.7 m, i.e. ≥5.3 m higher than the present-day riverbed. In profile DO on the opposite side of the river, basal gravel was reached at a depth of 2.9 m, i.e. 1.1 m above the adjacent riverbed.

Autochthonous events are recorded only in profile RPF-3 near the foot of the Cresta Sacramento. At least six debris flows from the nearby hillslope reached the site. There are no radiocarbon dates for this profile, but the application of the average sedimentation rate for the accumulation of irragric anthrosol in nearby profile RPF-2 provides an estimate of 1,150 to 860 cal BP for the minimum age of the last debris flow. Irrigation agriculture commenced much later at the site of profile RPF-3 (1,150–860 cal BP, i.e. during the Middle Horizon or Late Intermediate Period) than at profile RPF-2 (ca. 2,070–1,820 cal BP, Poz-13540, i.e. Initial Nasca or Early Nasca). This is likely due to the late construction of an irrigation canal (equivalent to today's Canal Bolillo) reaching the position of profile RPF-3 on a terrace 11.5 m above the riverbed. The amount of ceramic shards (Middle to Late Nasca) recovered from RPF-1 and the presence of a Middle Nasca burial and associated paleosurface suggest the existence of a nearby settlement. This was corroborated by findings in a nearby trench during canal construction in 2005 and 2006 (profile RPW) which included abundant Middle to Late Nasca artefacts and a radiocarbon date of 1,690–1,410 cal BP (Poz-8026, depth 2.58 m).

The bank exposure RPE is characterised by strong disturbance and stratigraphic inconsistencies: Nasca ceramic shards occur above and below glass fragments and rusted iron and the two radiocarbon dates of ≤260 cal BP (Poz-5270 and Poz-5271). While some unambiguous fluvial sediment is present in the profile, most of the material appears to have been remobilised, possibly by mechanised

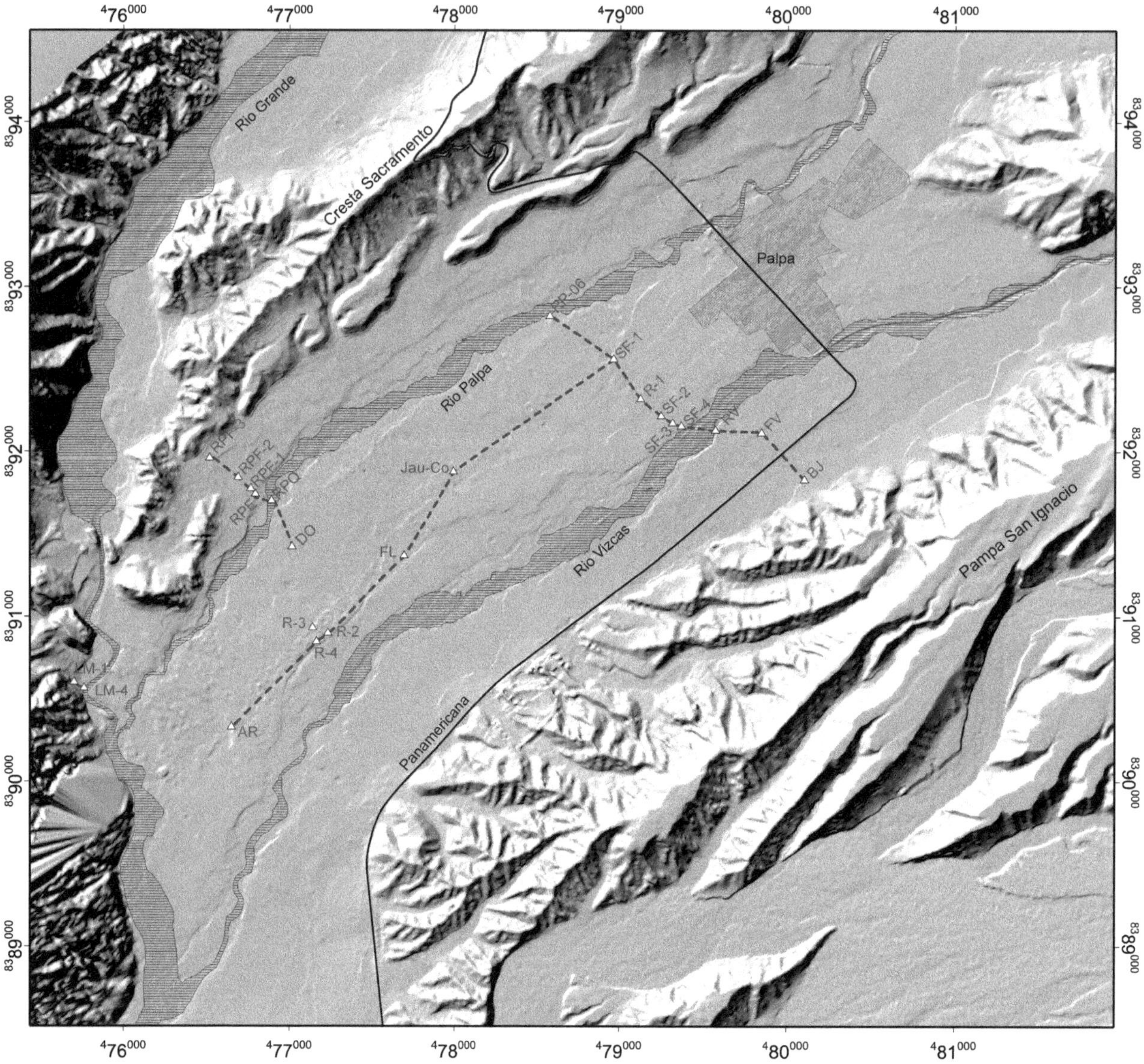

Fig. 34: Location of transects in the Palpa Valley (DTM provided by ETH Zurich).

agricultural activities or by fluvial reworking. The intensity of erosion and deposition within the nearby riverbed is exemplified by the presence of plastic artefacts at depths of ca. 1.5 m within coarse fluvial gravel.

Profile DO on the opposite side of the Rio Palpa and at a significantly lower elevation relative to the riverbed is characterised by very different sediments: The basal gravel is overlain by a sequence of 22 fluvial layers with varying grain size but generally alternating between sand and silt. There is a general upward trend to thinner layers due to increasing elevation relative to the river. Eight such layers have been deposited between 900–670 cal BP (Poz-15235) and 490–320 cal BP (Poz-13546) and a further twelve after 490–320 cal BP (Poz-13546). Contrary to profile RPF-2 and RPF-3, at this site irrigation agriculture as indicated by the accumulation of irragric anthrosol began after 490–320 cal BP (Poz-13546), i.e. during the historic period.

5.3.3 *Transect San Fermin*

Transect San Fermin (Figure 34 and 36, Table 9) is located approximately one kilometre south-west of the town of Palpa transverse to the Palpa Valley. The profiles RP-06, SF-4 and RV were located in banks of Rio Palpa and Rio Vizcas; however, due to the construction of artificial embankments after the 1999 flood event, SF-4 and RV are not adjacent to active river channels. The other profiles in this transect were pits. In all profiles except RV and FV basal gravel was reached at depths between 2.40 and 5.52 m. In the lower parts of all profiles, fluvial sediments and/or hyperconcentrated flow deposits are present. The upper 1.60 to 3.60 m of all profiles consist of irragric anthrosol which locally contain fluvial sediments of minor thickness interspersed within the stratigraphy or capping the profiles.

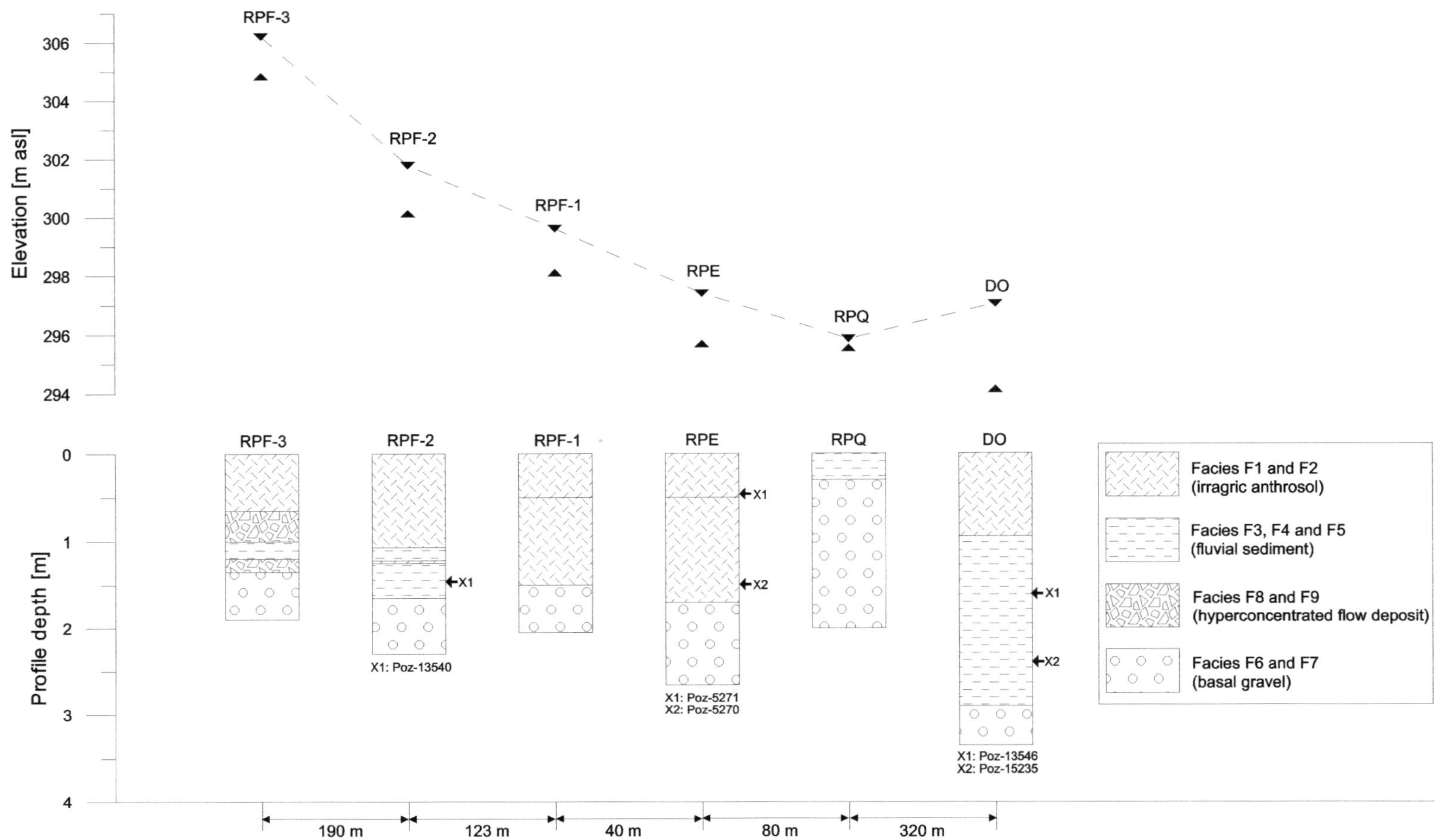

Fig. 35: Transect Rio Palpa. Top: elevation of land surface (downward-pointing triangles) and basal gravel (upward-pointing triangles); bottom: simplified stratigraphic sketches with approximate depths of ^{14}C samples (ages shown in Table 8).

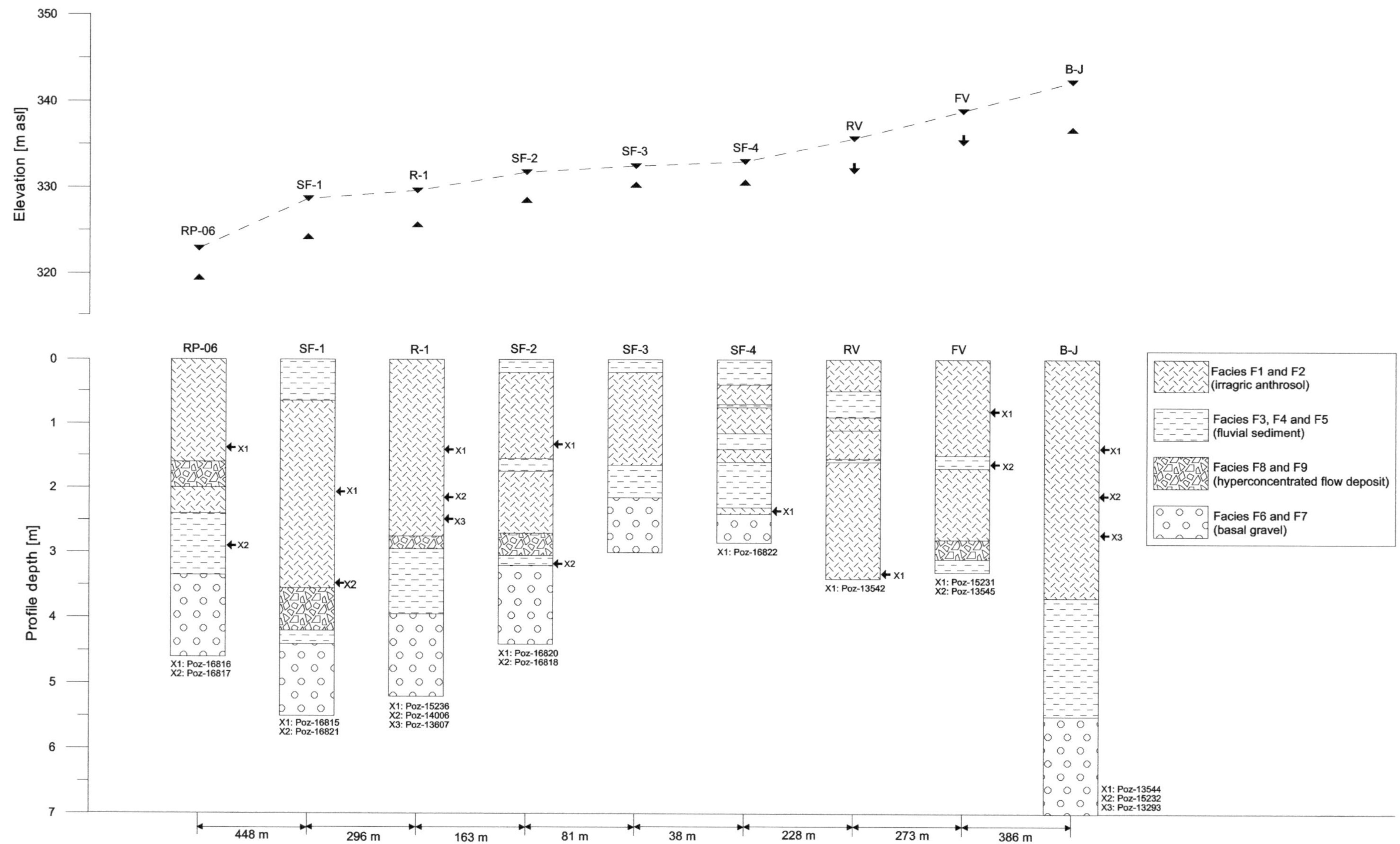

Fig. 36: Transect San Fermin. Top: elevation of land surface (downward-pointing triangles) and basal gravel (upward-pointing triangles); bottom: simplified stratigraphic sketches with approximate depths of ^{14}C samples (ages shown in Table 9).

Tab. 8: Radiocarbon dating results for transect Rio Palpa.

Sample-ID	^{14}C BP	Location	Depth below surface [cm]	cal BP (2σ)
Poz-5270	115 ±30	RPE (4.9 m)	150	≤260
Poz-5271	100 ±30	RPE (5.1 m)	45	≤260
Poz-8026	1,705 ±30	RPW	258	1,690 – 1,410
Poz-13540	2,050 ±40	RPF-2	150	2,070 – 1,820
Poz-13546	380 ±30	DO	164	490 – 320
Poz-15235	880 ±40	DO	239	900 – 670

Tab. 9: Radiocarbon dating results for transect San Fermin.

Sample-ID	^{14}C BP	Location	Depth below surface [cm]	cal BP (2σ)
Poz-13293	2,525 ±30	BJ	275	2,720 – 2,360
Poz-13542	375 ±35	RV	334	490 – 310
Poz-13544	355 ±30	BJ	142	470 – 300
Poz-13545	2,510 ±110	FV	165	2,800 – 2,150
Poz-13607	1,885 ±30	R-1	251	1,870 – 1,630
Poz-14006	1,595 ±30	R-1	210	1,530 – 1,350
Poz-15231	1,765 ±30	FV	80	1,710 – 1,540
Poz-15232	1,690 ±30	BJ	215	1,610 – 1,410
Poz-15236	385 ±30	R-1	140	490 – 320
Poz-16815	255 ±35	SF-1	200	440 – 140
Poz-16816	2,890 ±35	RP-06	140	3,080 – 2,840
Poz-16817	5,320 ±40	RP-06	285	6,190 – 5,920
Poz-16818	2,980 ±35	SF-2	320	3,220 – 2,950
Poz-16820	1,165 ±30	SF-2	134	1,080 – 930
Poz-16821	5,410 ±40	SF-1	350	6,280 – 5,990
Poz-16822	1,510 ±40	SF-4	232	1,420 – 1,290

Anthropogenic traces are present in layers interpreted as irrigation sediments in all profiles. The most prominent examples were found in profiles RP-06, SF-2, SF-3 and B-J. In profile RP-06, a layer containing ceramic shards pertaining to the Paracas culture (phase Ocucaje 9, ca. 2,350–2,150 cal. BP, ceramic phase assigned by J. Palomino, numerical age assigned after Unkel 2006: 111) as well as marine mollusc shells and worked stone fragments was present at a depth of 0.85–1.10 m. Furthermore, foundations of adobe walls associated with this layer were identified by J. Palomino and are tentatively interpreted as being of similar age. Therefore, this may have been the site of a Paracas culture settlement like the one recorded by Isla C., Reindel & de la Torre Z. (2003) near the southeast bank of the Rio Palpa and located 1.5 km downstream from profile RP-06. In profile SF-2, a hearth with associated ceramic shards pertaining to the Late Intermediate Period (ca. 950–550 cal. BP, ceramic phase assigned by J. Palomino, numerical age assigned after Unkel 2006: 111; radiocarbon dating result for charcoal from the hearth: 1,080–930 cal. BP, Poz-16820) was found at a depth of 1.34 m. In profile SF-3, foundation-like structured as well as ceramic shards of the Middle Horizon and the Late Intermediate Period (ca. 1,300–550 cal. BP, ceramic phase assigned by J. Palomino, numerical age assigned after Unkel 2006: 111) at a depth of 1.10 to 2.50 m document the use of the site by humans. Thick irrigation sediments are also present in profile B-J, 650 m southeast of the Rio Vizcas and ca. seven metres above the recent riverbed elevation. The lower part of these sediments (up to 2.80 m below surface) was dated to the time of the Paracas culture (2,360–2,720 cal. BP, Poz-13293). The presence of these sediments at this location, in particular at this elevation relative to the riverbed, points to the early existence of a high irrigation canal close to the foot of the slope of the Pampa de San Ignacio during the Paracas period. The intake of the modern Canal La Peña Baja which irrigates this site today is located almost 3.8 km up-valley.

Strong fluvial influence on the sedimentation is only discernible in the river bank exposures SF-4 and RV. The stratigraphy of these profiles is characterised by repeated alternation between fluvial and irrigation sediments. According to the embedded ceramic shards, profile SF-4 represents several phases of sedimentation since the Middle Horizon (ca. 1,300–950 cal BP, ceramic phase assigned by J. Palomino, numerical age assigned after Unkel 2006: 111; ^{14}C age: 1,420–1,290 cal BP, Poz-16822), albeit very likely with erosional discontinuities. Profile RV represents a comparatively young infilling of an area which had previously been eroded by the river. While noticeable differences are present in the stratigraphy along transect San Fermin, marked similarities between profiles exist. Age and thickness of fine-grained anthropogenic sedimentation is not related to the vertical position relative to the riverbed as exemplified by profiles FV and B-J.

5.3.4 Transect Jauranga

Transect Jauranga (Figure 34 and 37, Table 10) begins at profile SF-1 and runs down-valley along the valley axis, perpendicular to transect Sam Fermin. All profiles were pits except profile AR which was recorded at a natural escarpment. Irragric anthrosols in transect Jauranga have thicknesses between 1.43 and 3.76 m. Like in transect San Fermin, basal gravel was reached in most profiles (with the exception of R-2 and R-4). Where basal gravel was reached, its upper boundary is between 1.43 and 4.40 m below the surface. Profile R-2, which in the upper three metres exhibits a stratigraphy similar to that in profile R-4, can be parallelised with the nearby profile R-3 by means of a hyperconcentrated flow deposit at a depth of ca. 4 m. Below this deposit, only sandy, well-bedded and unequivocally fluvial sediments are present in the 6.10 m deep profile R-3, suggesting a similar situation in profiles R-2 and R-4.

Profiles Jau-Co and FL represent an area in the central Palpa valley which has been fluvially reworked by the Rio Vizcas as recently as during the Middle Horizon (1,300–950 cal BP after UNKEL 2006: 111) as shown by a radiocarbon date of 1,060–920 cal BP (Poz-13547) a few centimetres above the basal gravel in profile FL. Whether this fluvial reworking happened during a single, large-magnitude event, several episodic events or during a phase of permanent displacement of the river can not be determined on the basis of the available data. Profile R-2 is capped by a 0.25 m thick flood deposit which is attributable to the March 1999 flood event. At a depth of 1.90 to 2.00 m, a layer containing a large amount of anthropogenic material (ceramic shards, marine mollusc shells, worked stone fragments, charcoal) was identified in profile R-2. Both the ceramic style chronology (ceramic phase assigned by J. Palomino) and the radiocarbon age of charcoal (2,700–2,340 cal BP, Poz-13609) place this layer into the Paracas culture (2,750–2,150 cal. BP, numerical age after UNKEL 2006: 111). The amount and variety of anthropogenic material in this midden-like layer indicates the close vicinity of a Paracas culture settlement. In the nearby profiles R-3 (at a depth of 2.45 to 2.50 m) and R-4 (at a depth of 3.07 to 3.15 m), ceramic shards were found in a similar stratigraphic position, however, in lower quantity and without pieces suitable for the application of the ceramic style chronology. The radiocarbon age of a charcoal sample from this layer in profile R-3 is somewhat younger (2,160–1,950 cal BP, Poz-16825). Profile AR is characterised by two distinctive, charcoal-rich layers. The age of charcoal in the lower layer (1.37 m below surface) is 530–330 cal BP (Poz-16829). Analyses of the high resolution DTM of the Palpa valley provided by the ETH Zurich and of aerial and satellite images clearly show that this profile represents a former cut bank of the Rio Grande which today flows at a distance of 620 m. At the time of deposition of the horizontal layers in profile AR, this cut-bank situation can not have existed, therefore an event or phase of lateral erosion by the Rio Grande must have occurred after 330–530 cal. BP.

5.4 Generalised stratigraphy of the Palpa Valley

Based on the facies classification (5.1) and the investigation of 50 sedimentary profiles (5.2) and their combined interpretation in four transects (5.3), a generalised stratigraphy for the Palpa Valley has been developed (Figure 38, based on a draft by J. Baade). In general, the valley bottom of the Palpa Valley is underlain by coarse fluvial gravel of varying age as indicated by the varying degrees of weathering and cementation. Strongly weathered and more or less cemented basal gravel comparable to the material of the bajadas has been reached in several excavations. Because of the intense weathering, in particular of the granites, a pre-Holocene age is tentatively assumed. The

Tab. 10: Radiocarbon dating results for transect Jauranga.

Sample-ID	^{14}C BP	Location	Depth below surface [cm]	cal BP (2σ)
Poz-13547	1,125 ±35	FL	180	1,060 – 920
Poz-13608	1,780 ±30	R-2	170	1,720 – 1,540
Poz-13609	2,460 ±30	R-2	228	2,700 – 2,340
Poz-15238	415 ±30	FL	104	510 – 320
Poz-16814	1,730 ±35	Jau-Co	155	1,700 – 1,420
Poz-16815	255 ±35	SF-1	200	440 – 140
Poz-16821	5,410 ±40	SF-1	350	6,280 – 5,990
Poz-16825	2,150 ±35	R-3	247	2,300 – 1,950
Poz-16829	460 ±30	AR	137	530 – 330

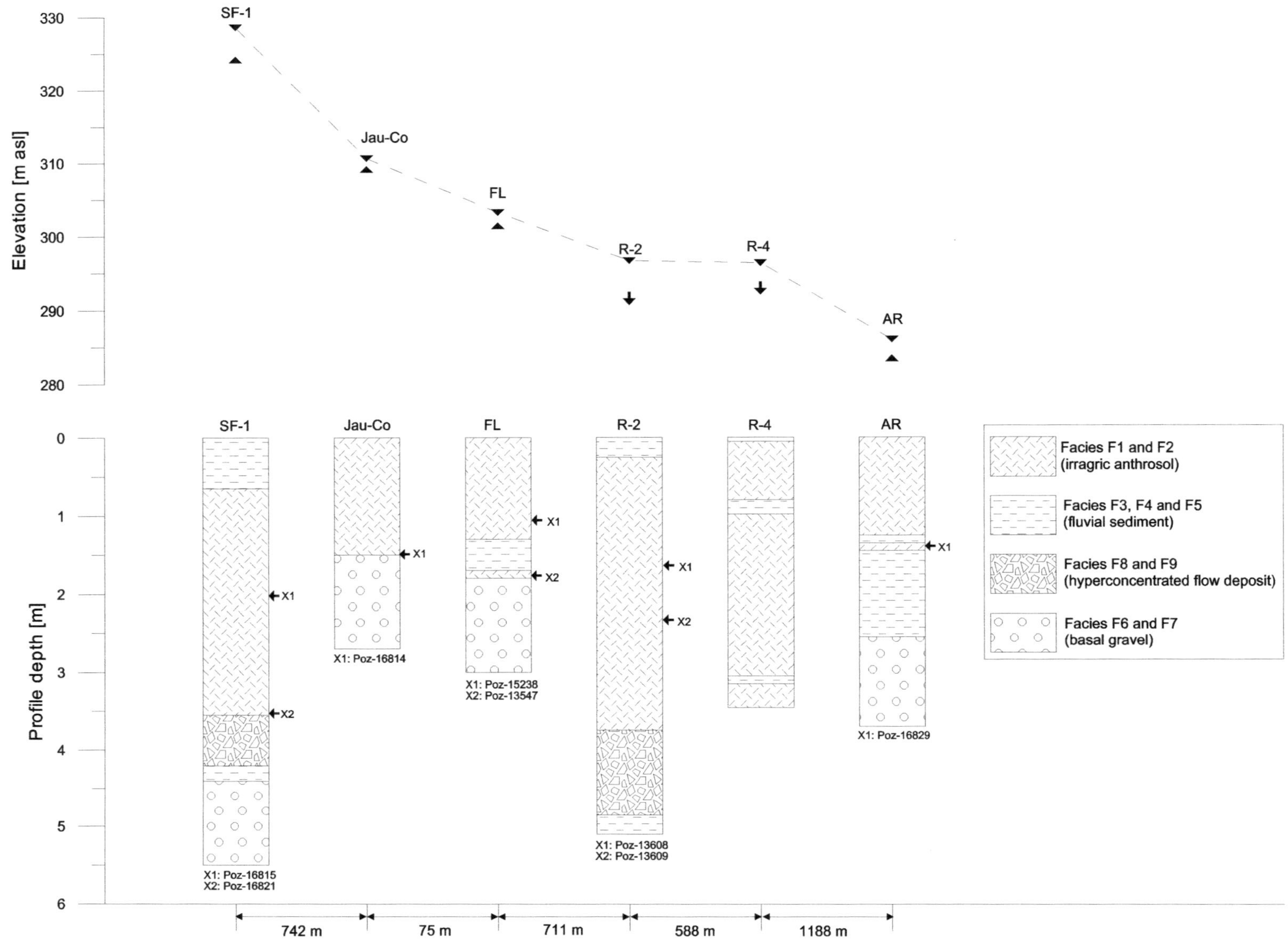

Fig. 37: Transect Jauranga. Top: elevation of land surface (downward-pointing triangles) and basal gravel (upward-pointing triangles); bottom: simplified stratigraphic sketches with approximate depths of ^{14}C samples (ages shown in Table 10).

fact that older and younger gravel was found at similar depths suggests that Holocene valley incision and/or re-deposition of fluvial gravel occurred within a few metres of the pre-Holocene limit of valley incision. The Palpa Valley can therefore not be considered to be rapidly incising or aggrading. Basal gravel is often overlain by fine-grained fluvial sediments and/or hyperconcentrated flow deposits. In most profiles, between one and more than three metres of irragric anthrosol have been deposited either directly upon the basal gravel or upon hyperconcentrated flow deposits or low-energy fluvial sediments. Occasionally, fluvial overbank sediments are interspersed within the stratigraphy. The A_p (ploughing) horizon is generally well defined mainly by its crumbly structure. However, no general pattern can be discerned for clay content, CSR and LoI when comparing the recent A_p horizon with the underlying sediment. Lessivation or similar soil development processes have not been detected within the irragric anthrosols. This is attributed to (1) the quasi-continuous (seasonal) addition of irrigation sediment and turbation by tillage and (2) the use of river water for irrigation which does not have the strong potential of rainwater for reducing soil pH.

Average sedimentation rates have been calculated from mean calibrated sample ages (Figure 39). They vary between 0.26 and 3.04 mm/a for the identified irragric anthrosols which is comparable to sedimentation rates established for recent irrigation agriculture in the Palpa Valley (see 4.2.2). Sedimentation rates of fluvial sediments are higher (1.04 and 7.34 mm/a). Differences in sedimentation rates of irragric anthrosols (e.g. some relatively high recent values) may be attributable to one or several of the following factors: (1) Irrigation water input for modern high-yield / high-demand crops likely has to be higher than for earlier, less demanding and less productive crops. (2)

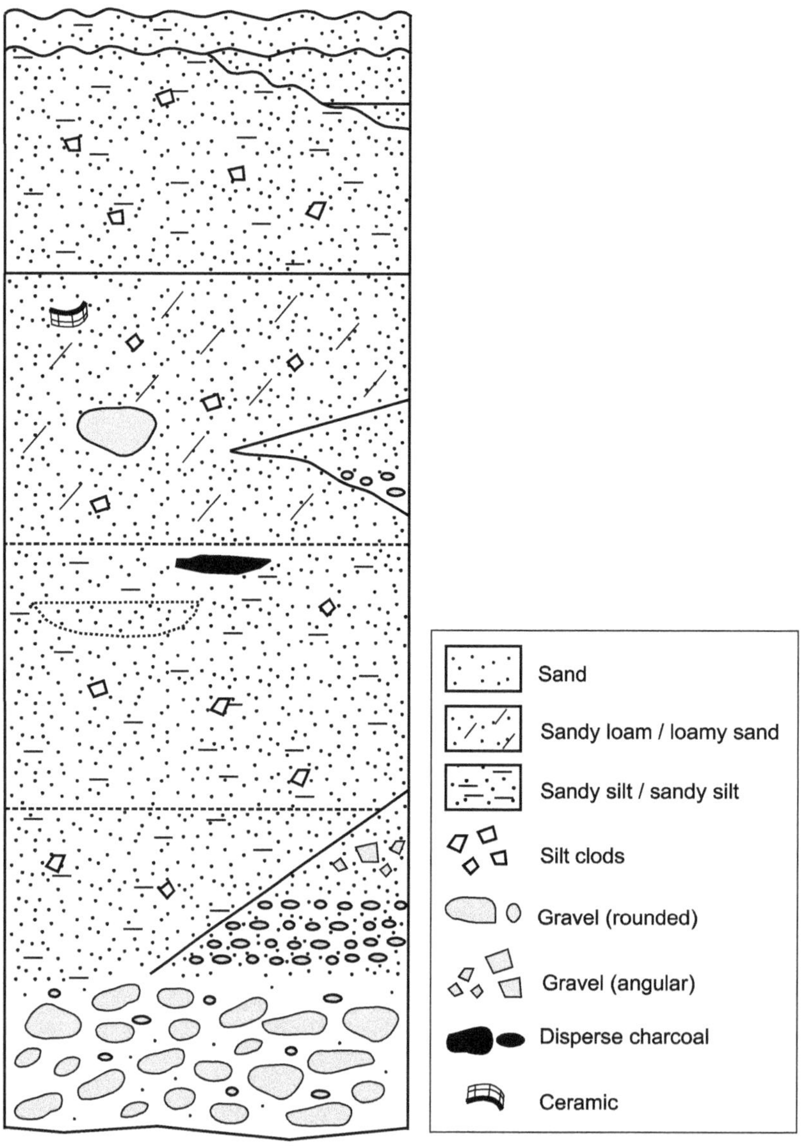

Fig. 38: Generalised stratigraphy of the Palpa Valley bottom (drawing: R. Mendler, based on a draft by J. Baade).

Quantified recent irrigation (cf. 4.2.2) took place under optimal conditions in terms of water availability. Since discharge variability is very high and years with very low discharge are common, the long-term average sedimentation rate likely is lower than the sedimentation rates determined in 2006. (3) The quantification of recent sedimentation rates did not incorporate losses (e.g. due to solution by percolating water or uptake in plants). (4) Reconstructed sedimentation rates may include hiatus (e.g. periods of fallow or drought). Interruptions or qualitative and quantitative changes in the deposition or irrigation sediment can be recognised in the stratigraphic record by the presence of layer boundaries and/or changes in sedimentary properties. However, the existing dating and analytical constraints (low temporal resolution, problematic detection of hiatus / fallow / land use change under the conditions of generally strongly turbated layer boundaries) inhibit detailed interpretation regarding the timing and character of these changes.

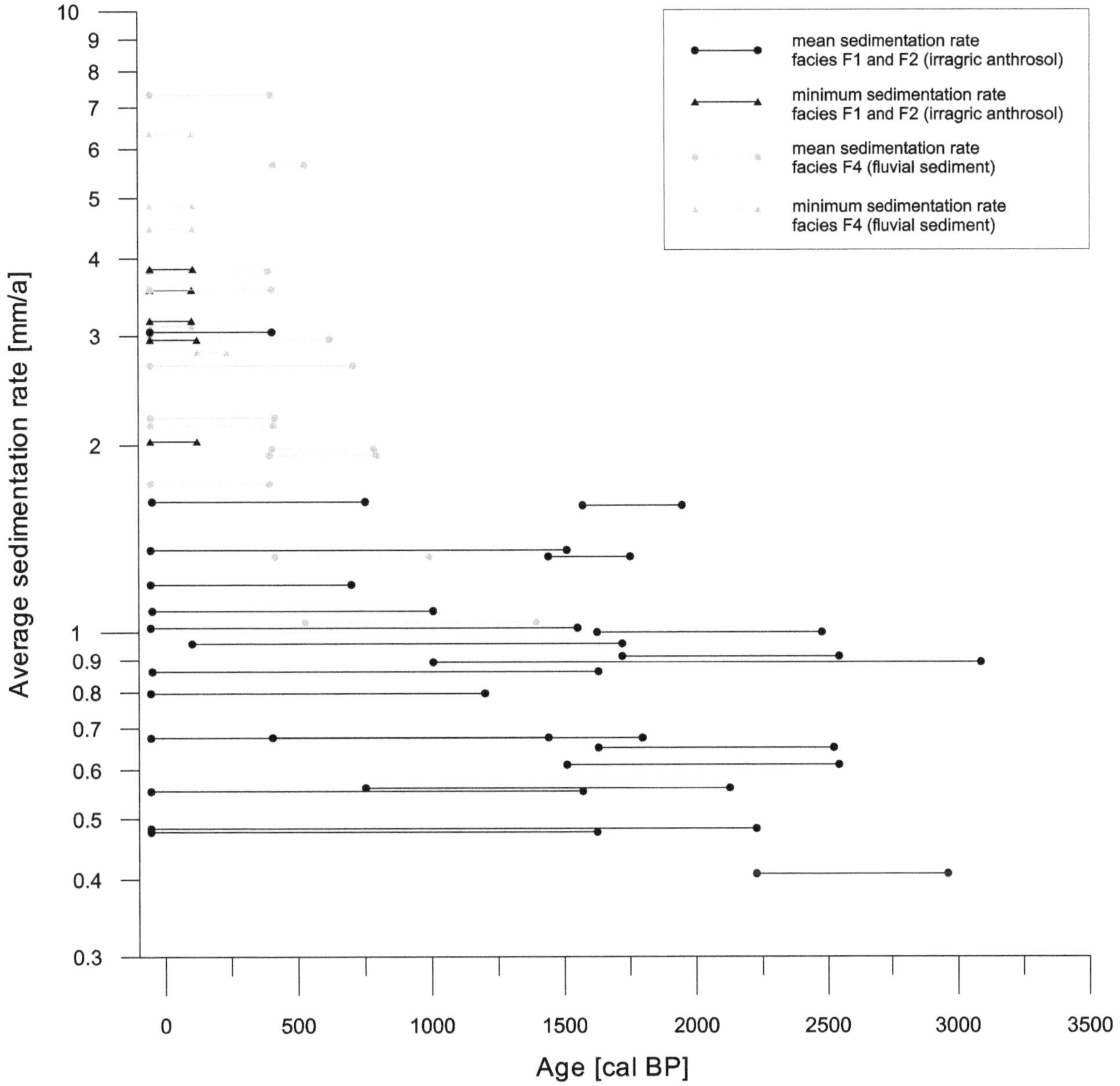

Fig. 39: Mean sedimentation rates (log scale) for predominantly fluvial sediments and predominantly irragric anthrosols (circles and triangles indicate mean cal BP ages; data: Table A-10 in Appendix B).

6. Late Pleistocene to mid-Holocene landscape evolution

6.1 Late Pleistocene to Holocene autochthonous and allochthonous dynamics

Marine terraces record Quaternary coastal uplift and sea level changes (Hsu 1992); lake sediments and shorelines, ice cores and rodent middens record long-term climatic fluctuations in the high Andes (e.g. Placzek et al. 2001; Baker et al. 2001; Betancourt et al. 2000; Bobst et al. 2001; Grosjean et al. 1995; Thompson et al. 1995; Thompson et al. 1998). Local conditions in the coastal desert however remain obscure for much of the Quaternary due to a paucity of suitable sedimentary environments for the formation of geo-archives as well as dating problems outside of the river valleys. Out of the considerable number of studies in which sedimentary sequences have been used to reconstruct flood and debris flow events in coastal southern Ecuador / northern Peru (e.g. Craig & Shimada 1986; Rodbell et al. 1999; Wells 1987; Wells 1990; Wells & Noller 1999) as well as southern Peru / northern Chile (deFrance & Álvarez 2004; Fontugne et al. 1999; Keefer et al. 2003; Magilligan & Goldstein 2001; Satterlee et al. 2000; Usselmann et al. 1999; Vargas & Ortlieb 1998; Steffen et al. 2007), only few (Vargas & Ortlieb 1998; Keefer et al. 2003; Steffen et al. 2007) extend beyond the Pleistocene-Holocene transition and confirm the occurrence of intensive local precipitation in the coastal desert during the Last Glacial.

6.1.1 Reconstruction of the erosional/depositional sequence

The results of the analyses of profile RGE at the Cerro Corpante narrows of the Rio Grande (Figure 40; see 5.2.1 and Appendix A) allow the reconstruction of autochthonous and allochthonous erosion and accumulation in the lower Rio Grande valley. Temporal constraints are provided by four radiocarbon dates and are applied directly as well as through recurrence interval estimates for local debris flow accumulation. Based on the investigations at profile RGE and taking into account glacial/interglacial cycles which globally play a dominant role in cyclic landscape changes, a model for landscape evolution in this part of the Peruvian coastal desert (Table 11, see Appendix A for a more detailed discussion) is developed.

Autochthonous dynamics in the form of debris flow accumulation which characterised the site at an unknown time before the LGM (unit 1) was interrupted by fluvial erosion. If debris flow activity is characteristic for glacial periods and erosion by the allothonous Rio Grande is characteristic for interglacial periods (as indicated below for the Last Glacial and Holocene), unit 1 can tentatively be attributed to the penultimate Glacial. Under this assumption, fluvial erosion and the accumulation of slope debris (unit 2) pertain to the last interglacial period. The early Last Glacial (before 23.1–24.1 ^{14}C ka BP) was characterised by debris flow aggradation only (unit 3). No fluvial sediments are present; therefore either the riverbed elevation or the maximum flood height must have been low to inhibit overbank deposition.

The LGM (unit 4) was characterised by repeated deposition from fluvial overbank events interspersed with debris flow events. LGM conditions in the Andes are considered to have been wet (e.g. Bobst et al. 2001; Baker et al. 2001) or dry bracketed by wet phases just before and after the LGM (e.g. Grosjean et al. 2001). While the timing of maximum glacial advances in different regions of the Andes was not synchronous and the role of the driving factors precipitation and temperature varies regionally (cf. 2.3.2), LGM conditions can generally be expected to have led to "flashier hydrographs" (Clapperton 1993: 194). Riverbed elevation must have been higher than today to allow the deposition of material by fluvial events in unit 4. Quebrada incision into the sediments took place after the end of unit 5 debris flow aggradation, i.e. during the late Glacial or

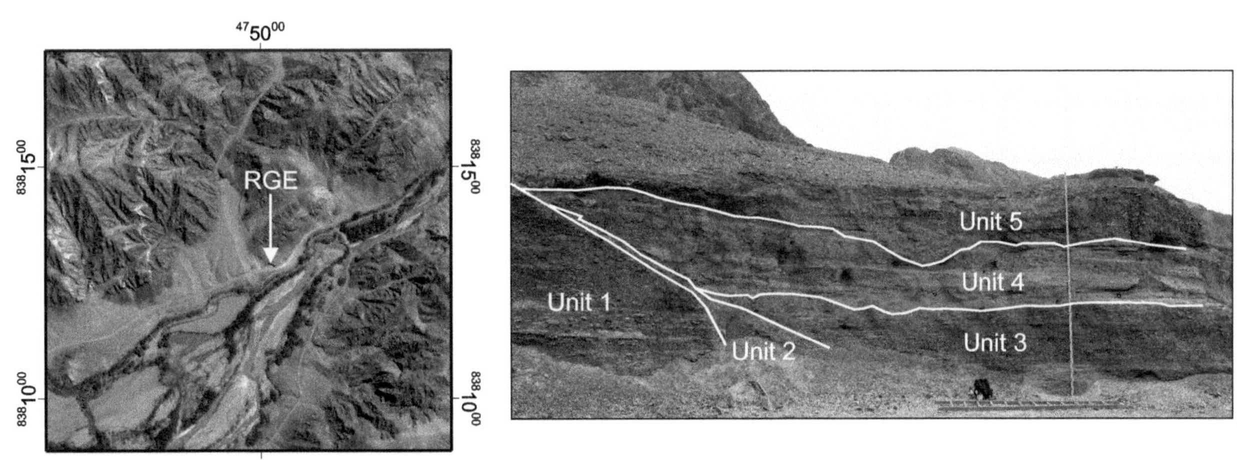

Fig. 40: Profile RGE. Left: satellite image (Quickbird) showing position of segment A of profile RGE; right: photograph of segment A with the sedimentary units discussed in the text (photograph by J. Baade).

Tab. 11: Model for Late Quaternary dynamics deduced from profile RGE.

Unit	Begin	End	Phase	River bed elevation	Activity
unit 1	duration of deposition: 4.2 – 16.6 ka		penultimate Glacial, last Interglacial or early Glacial	≤6 m higher than present	debris flows (≥14 events) fluvial sedimentation (1 event)
erosion	before 32.3 – 25.2 ka BP		last Interglacial or early Glacial		fluvial erosion
unit 3	≥32.3 – 25.2 ka BP	23.1 – 24.1 ka BP	early Glacial	≤6 m higher than present	debris flows (≥8 events)
unit 4	23.1 – 24.1 ka BP	20.6 – 17.0 ka BP	LGM	≤8 m higher than present	fluvial sedimentation (≥10 events) debris flows (≥4 events)
unit 5	20.6 – 17.0 ka BP	18.9 – 9.9 ka BP	LGM or late Glacial		debris flows (≥8 events)
	before 2,040 BP		late Glacial or early Holocene	ca. 6 m higher than present	quebrada incision
	after quebrada incision		Holocene		river incision

the Holocene. This incision left the surface of the debris fan as a relict surface unaffected by further activity in the quebrada and therefore must have commenced before 2,040–1,820 cal BP (Poz-13541, 2,040 BP) as evidenced by the lack of debris flow deposits over a radiocarbon-dated pre-Columbian hearth. The incised quebrada mouth is adjusted to a base level approximately six metres above the recent riverbed. Therefore, riverbed elevation must have been higher than today during quebrada incision. Fluvial incision of the Rio Grande by theses six metres to its recent riverbed elevation thus took place after the end of quebrada incision. Similar to the quebrada at the Rio Grande narrows, other quebradas mouths like the one near Ciudad Perdida de Huayuri in the Rio Santa Cruz valley are adjusted to levels several metres above the recent riverbeds. This indicates that the observed sequence of quebrada debris flow accumulation followed by quebrada incision to a higher-than-recent base level and followed in turn by fluvial incision of the allogenic rivers is a regional phenomenon. However, tighter chronological controls for this sequence need to be established.

6.1.2 *Deposition and dissection of bajadas and debris terraces*

Eitel et al. (2005) have suggested that the end of bajada accumulation and the inception of river valley incision into the bajadas at the foot of the Cordillera Occidental occurred ca. 45 ka BP. Their argumentation is based on a single OSL date of a silty vesicular layer under desert pavement near the town of Palpa, 13 km north-north-east of the location of profile RGE. However, since vesicular layers are considered to be the result of additive processes incorporating dust into the sub-surface (McFadden et al. 1998; Noller 1993: 133, 137), the 45 ka BP (OSL) date does not represent the end of bajada accumulation as assumed by Eitel et al. (2005) (cf. Hesse & Baade 2007). River incision by approximately 140 m after 45 ka BP (OSL) is at odds with the findings from profile RGE discussed above and with the oldest Palpa Valley bottom date in this project of 6,750–6,490 cal BP (Poz-13548, 5,880 ±50 BP, 4.98 m below the surface at profile BF-2). On Cerro Piedra Gorda near profile RGE, remnant bajada sediments occur at elevations of up to 350 m, 140 m above the recent riverbed. However, at the site of profile RGE, incision of the Rio Grande de Nazca was less than eight metres since the LGM (see 5.2.1 and Appendix A), with most of the incision attributable to the Holocene. Extrapolation based on debris flow recurrence interval estimates allows the inclusion of the time for the accumulation of units 1 and 3. Taking into account the time estimated for this accumulation and the intervening period of fluvial erosion – all of which must have occurred with riverbed elevation not more than six metres above its recent position –, incision of the Rio Grande de Nazca has been at most six metres during the past 30 ^{14}C ka, i.e. over the course of the entire LGM and Holocene. This strongly indicates that river incision into the bajadas began much earlier than 45 ka (OSL). Assuming an average river incision rate of ≤6 m per 30 ^{14}C ka results in an estimated ≥700 ^{14}C ka BP age for the begin of the incision.

Throughout the Nazca-Palpa sector of the Peruvian coastal desert, stacked terraces are set in large sediment bodies of poorly consolidated, coarse-grained, angular material of local origin (facies F9, see 5.1) along quebradas with small catchment areas. They are related to debris flow and fluvial activity caused by rare intensive local precipitation events in the otherwise hyperarid desert. The observed sequence of deposition and erosion of quebrada sediments appears to be the youngest episode of repeated cycles during the Quaternary. Sébrier & Macharé (1980: 12) describe a sequence of four autochthonous terraces in the coastal desert of central Peru and attribute it to climatic changes representing the entire Quaternary. Likewise, Paskoff (1978: 18f.) interprets the accumulation and dissection of fans in coastal northern Chile (21°42'S–23°45'S) as indicators for climatic changes. While fluvial terraces in the mid-latitudes are generally associated with greatly increased fluvial transport due to decreased vegetation cover, increased physical weathering and increased peak runoff during glacial periods, this can not necessarily be assumed for the quebrada terraces in the coastal desert of southern Peru. Factors influencing the accumulation–incision threshold for a given quebrada runoff event include base level relations, the total amount and peakedness of precipitation, the availability of erodible material and possible vegetation cover.

The presence of autochthonous terraces both within and outside of the area of Nazca Ridge influence (HAMPEL 2002; HAMPEL et al. 2002; cf. 2.2) indicates that the Nazca Ridge is not the dominant controlling factor. While base-level changes in this area of Quaternary coastal uplift (HSU 1992) must be important for long term changes (over the Quaternary period of uplift as well as over glacial–interglacial cycles of sea-level changes), the influence of base-level change is not straightforward: Local riverbed elevation is the base level for the quebrada at whose mouth profile RGE is situated. However, at present the quebrada is not adjusted to this base level, strongly indicating that other factors have to be considered. Furthermore, while the incision of the Rio Grande by up to six metres since the LGM is in agreement with a coastal uplift of approximately nine metres since the LGM (based on an uplift rate of 0.4 m/ka according to HSU 1992), riverbed elevation at this location does not respond directly to eustatic base level changes at the mouth of the Rio Grande: During the LGM period of global sea level minimum, the riverbed elevation was higher than today. The slope of the shelf at the mouth of the Rio Grande is shallow (Global Seafloor Topography, SMITH & SANDWELL 1997) and similar to the slope of the lower reach of the Rio Grande according to SRTM data; therefore, incision due to eustatic sea level changes is not to be expected (cf. BLUM & TÖRNQUIST 2000: 6).

Since an influence of base-level changes can be excluded, incision of the Rio Grande has to be attributed to changes in fluvial transport/erosion capacity possibly combined with a time lag due to progressing deposition and incision (cf. TUCKER & SLINGERLAND 1997). Therefore and under the conditions of a persistence of a generally arid climate throughout the Quaternary (CLARKE 2006) without sufficient vegetation cover to significantly influence weathering and erosion, the creation of the quebrada terraces has to be explained on the basis of a combination of the amount and peakedness of precipitation and the availability of erodible material. The existence of stacked terraces which clearly indicates a cyclic nature of deposition and erosion therefore has to be interpreted in terms of cyclic changes in the frequency and/or intensity of local precipitation events possibly in combination with changes in the intensity of physical weathering.

While the phases of accumulation and incision are thus related to climatic changes, direct inference of precipitation changes is impossible since both sedimentation of and incision into the coarse-grained sediments requires substantial runoff. This fact has been overlooked by KEEFER et al. 2003), resulting in the interpretation that a lack of sediments is attributable to a lack of El Niño events while it could plausibly be the result of incision caused by such events. For the Last Glacial, the minimum number of 20 authochthonous debris flow events between 32.3–25.2 ^{14}C ka BP and 18.9–9.9 ^{14}C ka BP recorded in profile RGE compares with a minimum number of 10 events represented by the Lower Series at Quebrada Tacahuay in the Ilo region (530 km to the south-east) for the time between 38.2 ±4.3 ka (IRSL) and 12.88–11.97 cal ka BP (radiocarbon) (KEEFER et al. 2003). Comparison between the two areas shows that the occurrence of these events during the Last Glacial followed a similar pattern as during the Holocene, i.e. they occurred less frequently with increasing distance from the equator. This pattern is in accordance with a southward decrease in El Niño influence (cf. 2.3.1; cf. NOLLER 1993: 130ff.).

6.2 *The mid-Holocene landscape*

6.2.1 *Mid-Holocene sediments in the Palpa Valley*

In the present study, four profiles in the Palpa Valley yielded mid-Holocene sample ages. In profile RVU (see 5.2.5 and Appendix A), charcoal from 0.1 m below the top of a massive (thickness >0.95 m) layer of structureless silty sand was dated to 5,500–4,950 cal BP (Poz-16826). The available data do not allow a definitive classification of this sediment; it may be either a turbated fluvial overbank deposit, irrigation sediment or a hyperconcentrated flow deposit. However, the presence of charcoal weakly suggests the possibility of an early human influence. The mid-Holocene deposition of this fine-grained sediment was followed by several episodes of erosion and deposition of coarse gravel sheets until 1,260–1,060 cal BP (Poz-16827).

In profile RP-06 (see 5.2.5 and Appendix A), charcoal from a depth of 2.85 m below the present surface was dated to 6,190–5,920 cal BP (Poz-16817). The layer from which the sample was retrieved is interpreted as a truncated A_H horizon developed in a riverine situation in a loamy hyperconcentrated flow deposit. It is overlain by low-energy fluvial sediment with minor soil development and two prominent hyperconcentrated flow deposits interspersed by low-energy fluvial sediments. A radiocarbon date of 3,080–2,840 cal BP (Poz-16816) represents the lower age boundary for the hyperconcentrated flow deposits in profile RP-06. In profile SF-1 (see 5.2.5 and Appendix A), a radiocarbon date of 6,280–5,990 cal BP (Poz-16821) (depth 3.50 m) provides both a maximum age for the inception of irrigation agriculture at the site and a minimum age for the accumulation of two hyperconcentrated flow deposits which overlie unambiguous fluvial sediments. Assuming that the prominent pairs of hyperconcentrated flow deposits in profiles RP-06 and SF-1 which are characterised by similar sedimentological properties indeed represent the same events, the radiocarbon dates bracketing these layers can be used to constrain the time of their deposition to between 6,210–6,010 and 6,170–5,940 cal BP by processing the ^{14}C dates as an ordered sequence (Figure 41).

In profile BF-2 (see 4.2.6 and Appendix A), charcoal from a depth of 4.98 m dates the top of a fluvial sediment (coarse sand to medium gravel) to 6,750–6,490 cal BP (Poz-13548). This is overlain by a 1.44 m thick sequence of fine-grained fluvial overbank deposits and a thin gravel splay followed by 3.55 m of probable irrigation sediment. Eight samples in profiles RPF-2, R-3, R-1, FV, LLPT, BJ, RP-06 and SF-2 were dated to the time interval between ca. 3,000 and 2,000 cal BP. Sediment below these samples was classified mainly as unambiguous fluvial sediment (facies F3 to F7, see 5.1) or hyperconcentrated flow deposit (facies F8, see 5.1). A massive hyperconcentrated flow deposit in profile R-3 (depth 3.60–4.57 m; also present in profile R-2) is overlain by a 1.15 m thick irragric anthrosol whose top (depth 2.47 m) was dated at 2,300–1,950 cal BP. Considering the necessary time for the accumulation of this anthrosol and a possible hiatus, the hyperconcentrated flow deposit could perhaps be ascribed to one of the two events recorded in profiles RP-06 and SF-1. Similarly, the hyperconcentrated flow deposit in profile FV 1.1–1.47 m below a radiocarbon sample dated to 2800–2150 cal BP (Poz-13545) could pertain to one of these events. In any case, hyperconcentrated flow deposits have to be considered a typical component of the mid-Holocene record. Judging from the spatial distribution of the deposits and the topography of the Palpa valley, the hyperconcentrated flows originated in the valley of the Rio Vizcas and deposited their sediments in the zone of flow expansion after entering the wide Palpa Valley.

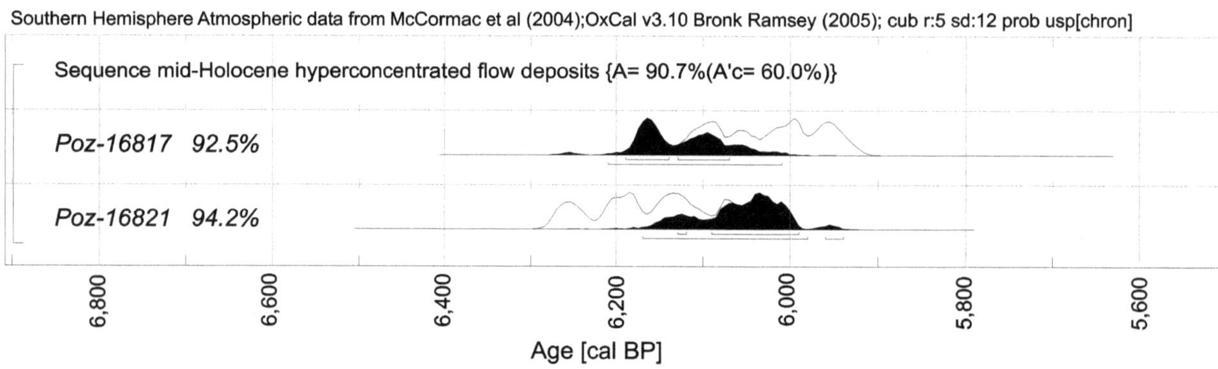

Fig. 41: Radiocarbon calibration of samples Poz-16817 and Poz-16821 as ordered sequence.

6.2.2 *Paleoenvironmental implications*

The mid-Holocene environment in the Palpa Valley was characterised by (1) the occurrence of large hyperconcentrated flow events, two of which occurred between 6,210–6,010 and 6,170–5,940 cal BP and likely affected a large part of the central Palpa Valley, (2) the lack of irragric anthrosols except in profiles SF-1, R-2 and R-3 (and possibly RVU and FV), (3) mainly natural fluvial dynamics (deposition of unambiguous high- and low-energy fluvial sediments) and (4) the lack of natural soil development except in river-proximal situations (RP-06 and possibly RVU). Hyperconcentrated flow events in the coastal desert of Peru have generally been ascribed to intense local El Niño precipitation. However, the apparent occurrence of such an event in conjunction with the March 1999 flood in the Rio Vizcas unrelated to El Niño is indicated by the presence of hyperconcentrated flow deposits together with fluvial sediments. This highlights the possibility that seasonal highland precipitation related to the SASM may trigger such events. Climatological interpretation of the hyperconcentrated flow deposits in the Palpa Valley is thus hampered by ambiguity. However, despite generally arid conditions in the Andean highlands during the mid-Holocene, the evidence for repeated major storms of regional significance in the Atacama Altiplano of northern Chile during that period (Grosjean et al. 1997: 239, 245) indicates high-amplitude, short-term excursions of precipitation amounts. Short-term intense precipitation periods punctuating a generally arid mid-Holocene highland climate are a plausible scenario for the release of large hyperconcentrated flows in the Cordillera Occidental.

For the early and mid-Holocene (10,950–3,750 cal BP), Eitel et al. (2005: 149) have suggested the formation of two fluvial terraces in the major valleys and quebradas of the research area under semi-arid conditions with precipitation of >200 mm/a at elevations between 250 and 700 m. However, they provide no information as to how they arrived at their quantitative precipitation estimates and present only a single mid-Holocene OSL date of 3.98 ±0.54 ka BP for silty alluvium from the lower (ca. 2.5 m above riverbed) "river terrace wall of the Rio Palpa" and one OSL date of 6.42 ±0.53 ka BP for a loess sample from an elevation of 1,154 m in the Rio Vizcas valley (EITEL et al. 2005: 148). The single mid-Holocene OSL date from the Palpa Valley is at odds with two radiocarbon dates of 450–150 cal BP (ET436, 305 ±49) and 440–140 cal BP (ET352, 280 ±45 BP) presented by UNKEL (2006: Anhang page 38-39) for the same site (distance 5.0 m) and with a date of 750–665 cal BP (Poz-8020, depth 2.05 m) from nearby profile YN-1 (this study, distance 42.8 m) (see 5.2.4 and Appendix A; cf. 7.1).

The findings discussed above contradict the assertions presented by EITEL et al. (2005). Rather than recording "open grassland" conditions (EITEL et al. 2005: 149) which should have been accompanied by soil development and increased LoI values, mid-Holocene sediments in the Palpa Valley are characteristic of fluvial and hyperconcentrated flow deposits practically devoid of vegetation with the exception of riverine situations and early irrigated fields. The pre-agricultural mid-Holocene environment of the Palpa Valley was characterised by shifting, braided, gravel-bed rivers with some fine-grained overbank deposits but no distinct hydraulic floodplain (cf. NANSON & CROKE 1992: 460). The regular occurrence of monsoonal precipitation in the coastal desert during the mid-Holocene suggested by EITEL et al. (2005) is contradicted by (1) a generally dry mid-Holocene in the Andean highlands (see 2.3.2) coincident with the *silencio arqueológico* on the Altiplano (ca. 8.0–4.8 ^{14}C ka BP, GROSJEAN et al. 1997: 239; GROSJEAN et al. 2005: 162; MESSERLI et al. 1993: 125, after Nuñez unpubl.), (2) a more northward position of the mean latitude of the Atlantic ITCZ during the mid-Holocene (HAUG et al. 2001; TEDESCO & THUNELL 2003: 1) and (3) the implausibility for significant amounts of Amazonian spill-over precipitation to reach the ground in the coastal desert (WEISCHET 1966: 1) as well as for the implied shift in the altitudinal zonation of precipitation by ≥2,100 m (see 2.3.2 and 2.4; cf. HOLMGREN et al. 2001: 245ff. for the altitudinal stability of the cactus belt in the Arequipa region throughout the Holocene; cf. MESSERLI et al. 1993: 126 for the persistence of arid conditions below 3,500 m; see also HESSE & BAADE 2007 for a further discussion).

Comparison with other valleys in the area helps to define, visualise and interpret the pre-agricultural environment of the Palpa Valley. Surface and near-surface sediments of Quebrada Yapona (Figure 42), a tributary to the Rio Ingenio with a catchment area of 352 km^2 and elevations from 460 to 3,660 m asl, exhibit many characteristic features of mid-Holocene sediments in the Palpa Valley: coarse gravel with well-rounded clasts and lenses or bars of bedded and massive sand to gravel as well as hyperconcentrated flow deposits. Several terrace-like levels of accumulation and incision are present along the valley. While Quebrada Yapona occasionally carries water from seasonal highland precipitation, this valley never had sufficient water for agricultural use in pre-Columbian or historic times due to the small extent of the watershed at elevations with significant precipitation. Therefore, natural fluvial dynamics during discharge periods and related sediments are not masked by anthropogenic impacts. The sedimentological similarities between the mid-Holocene Palpa Valley and the present Quebrada Yapona may be interpreted as representing typical pre-agricultural fluvial dynamics without human interference. Taking into account the significantly lower discharge in Quebrada Yapona compared to the Palpa Valley permits the paleoenvironmental interpretation as representing the typical fluvial dynamics of a stream with lower (and possibly more variable) discharge. Therefore, comparison of Quebrada Yapona and the Palpa Valley supports the notion of a generally dry mid-Holocene in the high Cordilleras associated with reduced allogenic river discharge.

An increase and westward expansion of monsoonal precipitation as suggested by EITEL et al. (2005) is not compatible with the evidence for a dry mid-Holocene in the Andes. On the contrary, it is compatible with an increase in fog conditions at the base of the atmospheric inversion (see 2.3.2) suggested to be underlying the mid-Holocene occurrence of land snails (cf. HESSE & BAADE 2007).

Fig. 42: Left: satellite image (Landsat, ca. 2000) showing the location of major rivers and Quebrada Yapona (**1** – Rio Santa Cruz; **2** – Rio Grande; **3** – Rio Palpa; **4** – Rio Vizcas; 5 – Rio Ingenio; **6** – Quebrada Yapona). Right: photograph of Quebrada Yapona.

Probable irragric anthrosols had already reached thicknesses of 1.35 m by the Late Paracas period (profile R-2) or 1.15 m by the Initial Nasca period (profile R-3) or overlie a basal date of 6,280–5,990 cal BP (Poz-16821) (profile SF-1). They indicate human presence in and agricultural use of the Palpa Valley during the Early to Middle Paracas period and possibly as early as the Initial and Archaic Periods. However, no artefacts have been recovered from the sediments predating the Paracas period.

The assumption explicit or implicit in a number of publications (e.g. West 1981: 60ff.; Moseley 1992: 16; Wells & Noller 1999: 775; Silverman 1996: 100) that early agriculture in the coastal valleys of Peru was based on floodwater farming (i.e. agricultural use of seasonally inundated areas) is not compatible with the fluvial characteristics of Rio Palpa and Rio Vizcas. Insofar as their near-natural state has not been transformed by hydraulic engineering Rio Palpa and Rio Vizcas are braided, gravel-bed rivers which laterally adjust to changes in discharge (see 4.1). This is corroborated by the properties of sediments deposited before the inception of irrigation agriculture. These sediments are attributable to braided river dynamics as well as hyperconcentrated flows, with low-energy fluvial sediments present in most profiles but generally accounting for only a minor part of the stratigraphy. Despite the fact that discharge variability is very high, overbank flood events are rare. Due to lateral adjustment of the riverbed they do not occur on a regular, annual basis (e.g. the two last overbank flows occurred in 1972 and 1999). However, a regular occurrence would be necessary for survival based on floodwater farming. Therefore, agriculture in the Palpa Valley must have been based on irrigation from the beginning.

7. Late Holocene landscape evolution in the Palpa Valley

7.1 Late Holocene fluvial dynamics

The results of chapters 4, 5 and 6 are combined in a geomorphological-chronological map of the Palpa Valley (Figure 43 and Table 12). Basal radiocarbon dates and diagnostic archaeological finds provide temporal constraints while sedimentological information from the profiles described in detail in Appendix A, height relationships and terrain edges extracted from the 1997 high-resolution DTM (provided by the ETH Zurich) and fluvial lineaments mapped on the 1944 aerial photographs allow spatial extrapolation. The resulting map indicates approximate minimum ages for the deposition of fine-grained sediment upon the basal gravel present throughout the valley. However, the reliability of the depicted information may decrease with distance from sampled locations (see Figure 22). Basal ages indicate that large tracts of the valley bottom have been outside of the influence of fluvial events for more than 1,500 years. In most profiles in the areas classified as >1,500 cal BP, basal ages are >2,000 cal BP. In these areas, overbank sediments are particularly rare. Therefore under Late Holocene conditions the hydraulic floodplain, i.e. the area of repeated flooding (cf. NANSON & CROKE 1992: 460), is limited to areas mapped in Figure 43 as having basal ages <1,500 cal BP. Before the large-scale anthropogenic modification of the rivers during the 20th century, fluvial dynamics during the Late Holocene were characterised by erosion and deposition along the rivers as well as in the central part of the Palpa Valley. The age of the valley margin debris aprons in terms of last fluvial impact by the rivers has to be considered high: minimum ages are 2,720–2,360 cal BP (basal age of irragric anthrosol in profile BJ) and 2,070–1,820 cal BP (near-basal age of profile RPF-2 on a lower terrace). However,

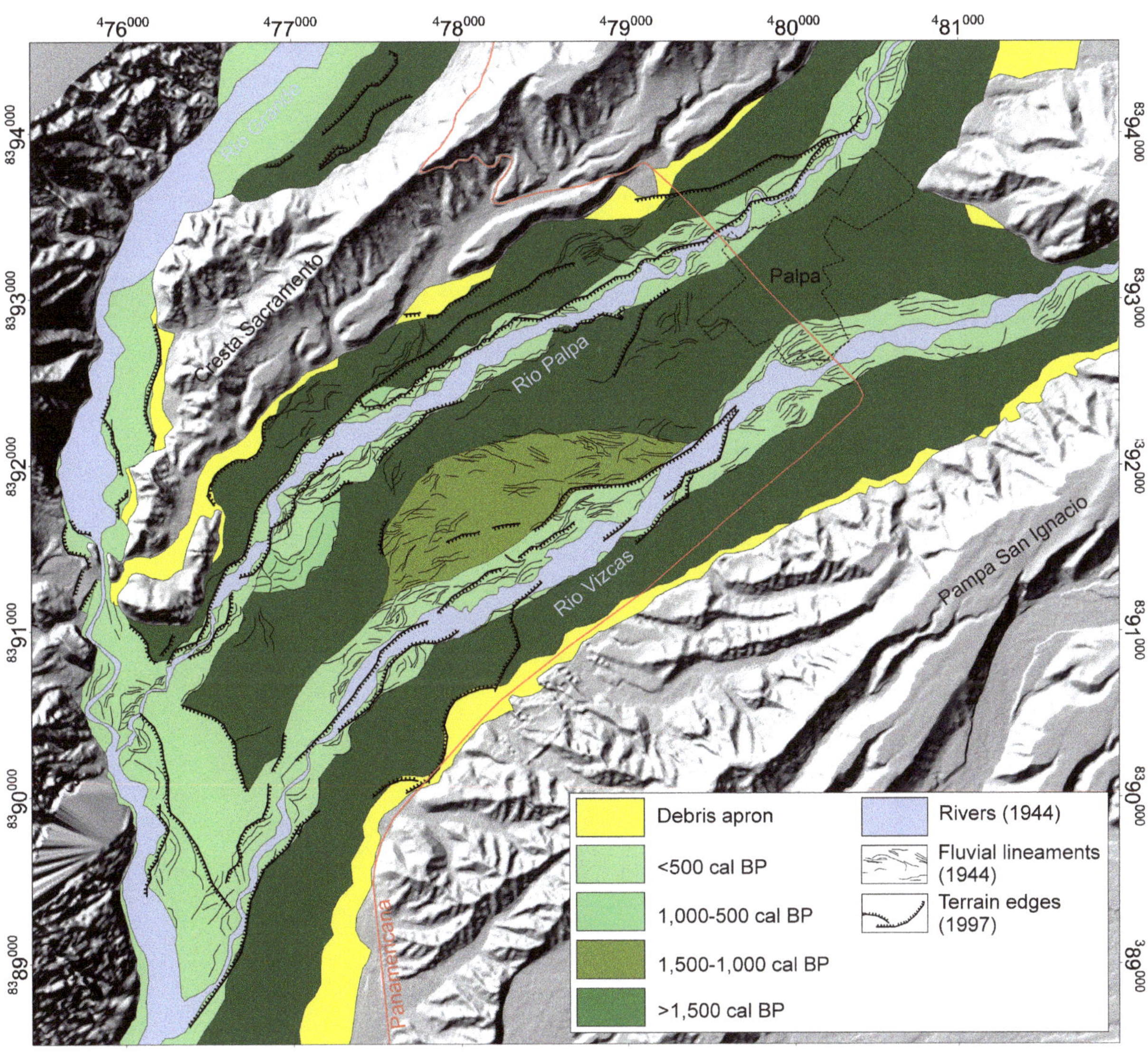

Fig. 43: Geomorphological-geochronological map of the Palpa Valley (valley bottom area classified by approximate minimum basal age above gravel).

Tab. 12: Quantitative results of the geomorphological-geochronological mapping of the Palpa Valley shown in Figure 43.

	Area [km^2]	Fraction of valley floor area [%]
Debris aprons	1.20	7.5
Rivers (1944)	1.54	9.7
Basal dates <500 cal BP	3.92	24.6
Basal dates 500 – 1,000 cal BP	0.15	0.9
Basal dates 1,000 – 1,500 cal BP	0.86	5.4
Basal dates >1,500 a	8.24	51.8
Total	**15.92**	**100.0**

taking into account height relationships relative to the riverbeds, they were last affected by the rivers before the mid-Holocene.

The generally ca. 50–150 m and locally up to ca. 300 m wide fringes of eroded and aggraded land along Rio Palpa and Rio Vizcas (see 4.1.2) have been interpreted by Unkel (2006: 120, 128) as a result of a regionally moist Little Ice Age. However, the analysis of geospatial data (see 4.1) shows that lateral migration and changes in width are characteristic features of the braided gravel-bed rivers Rio Palpa and Rio Vizcas. Recurrence intervals from profile DO do not support changes in the frequency of flood events before and after 490–320 cal BP (see 5.2.5). Figure 44 compares the summed probability distribution for samples from the erosional fringes of the Rio Palpa (8 radiocarbon ages from the present study and 16 radiocarbon ages taken from Unkel 2006: Anhang pages 38–39) with the summed probability distribution for 13 samples from sites with anthropogenic impact outside of the erosional fringes of the rivers (this study, limited to ages <1,500 ^{14}C BP). The general similarity of the two probability distributions shows that a large part of the variations in probability has to be attributed to variations in human impact rather than to variations in fluvial activity. Therefore, the fringes along the rivers likely represent areas within whose limits erosion and deposition have repeatedly taken place in the course of episodic lateral river channel adjustments; they do not necessarily record certain phases of fluvial activity.

The prominent area characterised by erosional features and fluvial lineaments in the central Palpa Valley (Figure 45) has a size of 1.21 km^2. On the basis of the available data it can not be determined whether this area represents long-term permanent channel migration or a singular or few erosional events before and after which the river followed its present-day course. In any case, an extreme event with high erosional and transport capacity has to be assumed judging from the entrance of a wide flow channel marked by a high concentration of large boulders (Figure 45 inset). High concentrations of boulders are also present along the driveway between the locations of well

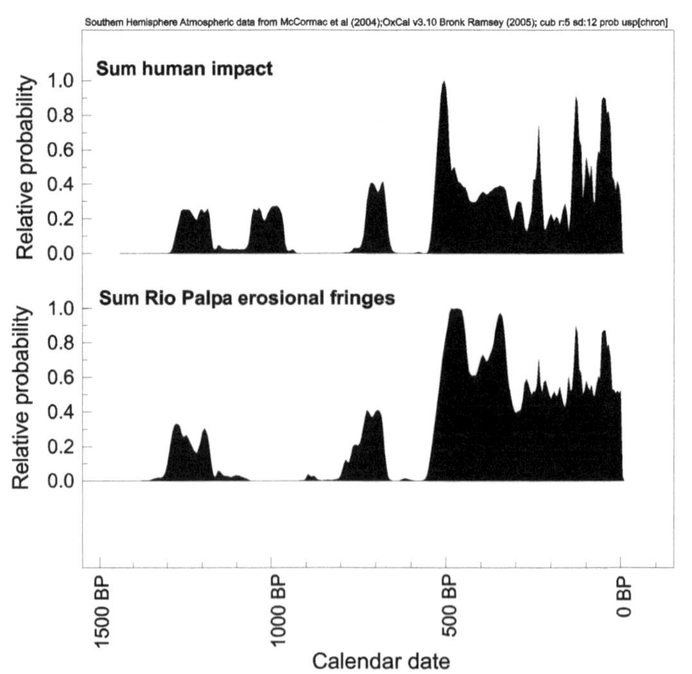

Fig. 44: Summed probability distributions for samples from unambiguous fluvial (bottom) and anthropogenic (top) contexts (data from this study and from Unkel 2006, see text).

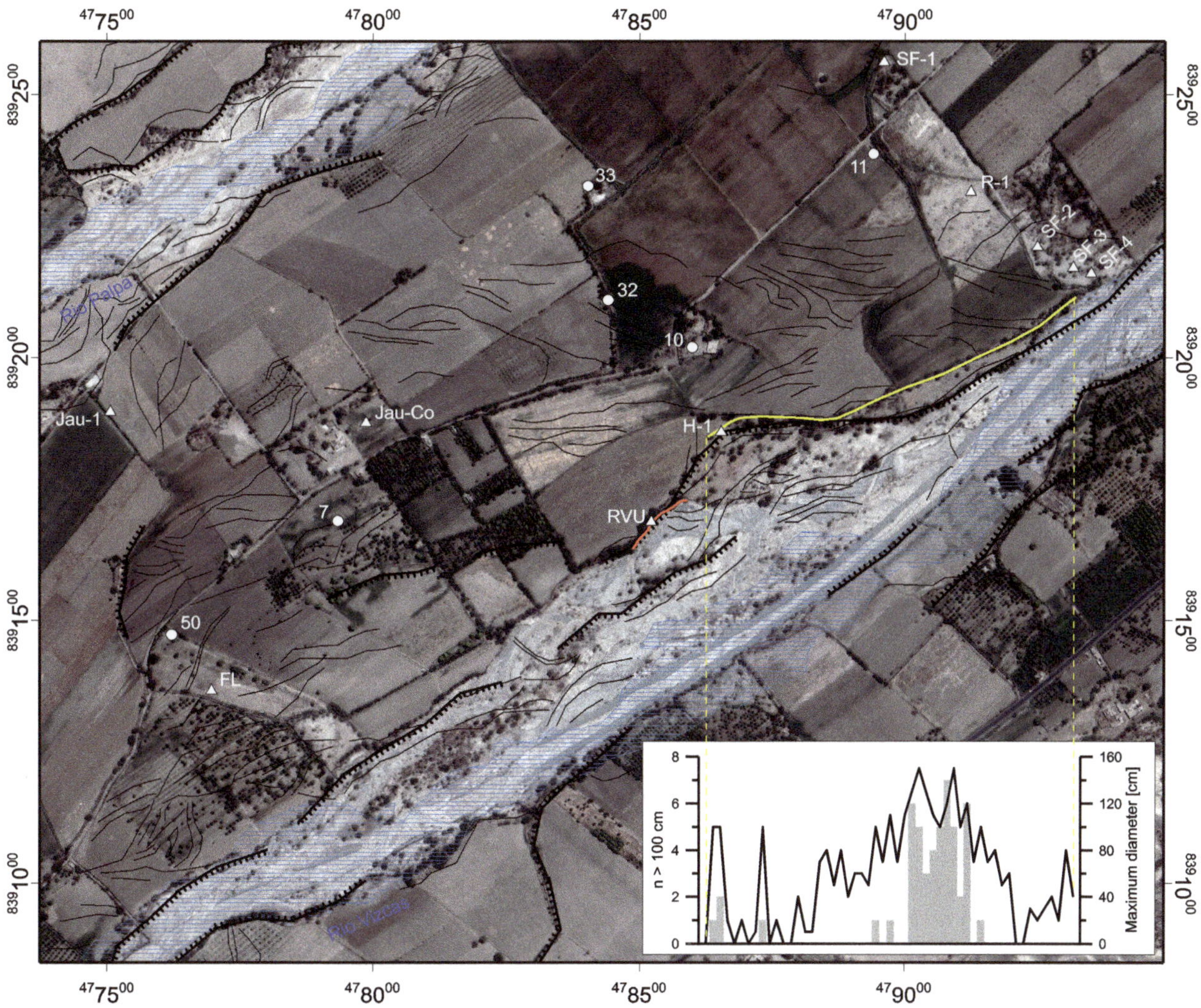

Fig. 45: Satellite image (Quickbird) of the central Palpa Valley showing terrain edges (from the 1997 DTM), fluvial lineaments (from the 1944 aerial photographs) as well as the positions of profiles (white triangles and red line) and wells (white circles). Inset shows the number of large clasts (>100 cm, grey bars) and maximum clast size (line graph) along the upper section of Canal San Fermin (yellow line in satellite image).

11 and profile Jau-Co (near the present-day settlement Jauranga), in particular in the vicinity of well 10 (the old Hacienda Jauranga). Piles of boulders are present on the field between the old Hacienda and the position of the deduced flood channel entrance.

Sedimentation of fine-grained fluvial overbank deposits and irragric anthrosol within this area has reached at most 1.8 m (profile FL) compared to >3.0 m in profiles SF-1 and R-2 situated up- and down-valley, respectively. It is constrained temporally to ≥1,260–1,060 cal BP (Poz-16827) for profile RVU and ≥1,060–920 cal BP (Poz-13547) for profile FL. In profile Jau-Co (see 5.2.5 and Appendix A), the fluvial activity seen as erosion followed by fluvial and irrigation sedimentation in profiles FL and RVU may be recorded by a flood deposit overlying sediment dated to 1,700–1,420 cal BP (Poz-16814). However, since sample Poz-16814 was charcoal from a unambiguous fluvial deposit, it may represent remobilised material. Sedimentation in the area probably began before 1,260–1,060 cal BP and possibly after 1,700–1,420 cal BP, thus placing the river channel migration roughly to its present course between the Early Nasca Period and the early Middle Horizon. This wide temporal range forbids interpretation in terms of event impact on pre-Columbian societies. Middle Horizon diagnostic archaeological finds in profile RVU (see Appendix A) and the presence of ceramic shards in wells 7 and 50, however, indicate subsequent use of the area by pre-Columbian cultures.

In the western part of the Palpa Valley (Figure 46), several profiles document erosion and sedimentation by the Rio Grande de Nazca. The Rio Grande, which is characterised by meandering behaviour in this stretch, has shifted ca. 150 m to the east from the mouth of Quebrada La Muña since 910–730 cal BP (Poz-8010) (see 5.2.3, transect La Muña). However, most of the lateral shift of the Rio Grande to its present course has taken place in historic times: The ca. 1.0–2.5 m thick fine-grained sedimentary body recorded in profiles LM-3, LM-5, LM-6 and LM-7 (see 5.2.2 and 5.3.1) has aggraded since ca. 320–150 cal BP (Poz-8014) mainly by fluvial processes, with the upper

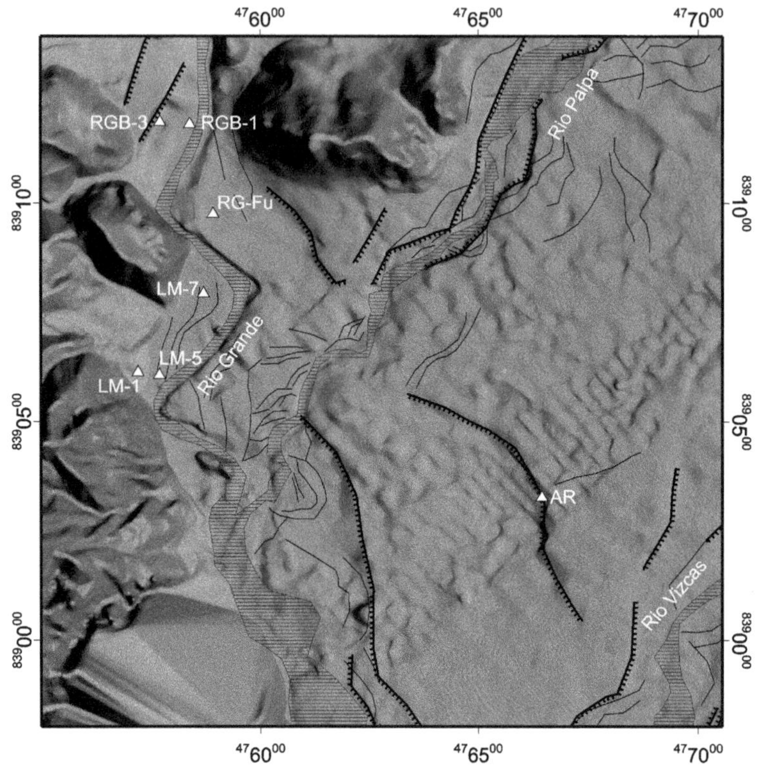

Fig. 46: DTM relief of the western segment of the Palpa Valley showing terrain edges (from 1997 DTM), fluvial lineaments (from the 1944 aerial photographs) as well as the positions of profiles (white triangles). DTM data: ETH Zurich.

ca. 0.4–0.8 m of the stratigraphy representing an agriculturally turbated mixture of fluvial overbank deposits and irrigation sediment. In a field ca. 500 m to the north, profiles RGB-1, RGB-2 and RGB-3 record a similar behaviour with near-basal dates of 910–680 cal BP (Poz-8023) in profile RGB-3 and ≤270 cal BP (Poz-8021) in profile RGB-1. Profile RG-Fu (albeit undated) shows a comparable stratigraphy of mainly fine-grained fluvial sediments. Fluvial lineaments (see 4.1.2) are clearly visible on the 1944 aerial photographs and confirm the relatively recent impact of fluvial events in this area. The data are compatible with the interpretations of (1) a unidirectional river channel shift from a more western position before 900–730 cal BP (combined calibration of Poz-8010 and Poz-8023) to near its present course by 320–150 cal BP (Poz-8014, profile LM-5) or (2) the repeated impact of river channel avulsions before 320–150 cal BP (Poz-8014).

The site of profile AR (Figure 46, see 5.2.5) is a large (670 m long, ca. 3 m high) erosional escarpment. Horizontal layers of fluvial and anthropogenically dominated sediments with an age of 530–330 cal BP (Poz-16829) at a depth of 1.37 m indicate that the escarpment must have been cut by the Rio Grande de Nazca (which is today situated at a distance of 610 m) after 530–330 cal BP. Taking into account the accumulation of the top 1.37 m of the profile at an assumed rate of 3.0–5.0 mm/a (cf. profiles ES and DO) constrains the age of this large erosional event to ≤260 cal BP. This interpretation is supported by the only thin fine-grained sediments overlying coarse fluvial gravel just west of profile AR, i.e. in the former riverbed. If the erosional activity at profile AR is attributable to the same event which caused river avulsion at transect La Muña (ca. 900 m to the north-west) and profile RGB-1 (ca. 1,200 m to the north-west), the available temporal constraints place the event between ≥320–150 cal BP (Poz-8014, profile LM-5) and ≤260 cal BP (profile AR), i.e. between 260 and 150 cal BP.

7.2 Irrigation systems and sediments

Besides indicating the time of the begin of fine-grained sedimentation after the last impact of erosional events, the ages mapped in Figure 43 in many cases represent the initiation of irrigation agriculture. Basal ages of irragric anthrosols along the rivers are low due to repeated erosion. A specific situation is present on the debris aprons along the valley margin which are the highest sectors of the valley bottom (Figure 43). Some of these areas record long-term agricultural use (e.g. profile BJ) while others have come under irrigation relatively late as indicated by the minor thickness of irragric anthrosols (e.g. profile RPF-3). This is due to their relatively high topographic positions which were only reached by the highest pre-Columbian irrigation canals, i.e. when the irrigation system reached its maximum extent. Despite their position at the foot of bajada slopes or at small quebrada mouths, debris flows are not recorded after 1,150–860 cal BP (profile RPF-3) and 2,720–2,360 cal BP (profile BJ). Contrary to the assertion that “the amount of land under cultivation would have been much less in the prehispanic era” (Silverman 1996: 102), the data show that a near-modern extent of the irrigation system was achived in the Nasca Period with minor additions in the Middle Horizon or Late Intermediate Period (profiles RPF-2 and RPF-3).

Based on the available geospatial data and the results of the sedimentological work, the total amount of irrigation sediment deposited in the Palpa Valley is estimated to be between 20 and 100 Mio. t, making human action an important geomorphic force. Early irrigation agriculture in the Palpa Valley may date back to the mid-Holocene / Archaic to Initial Period (see 6.2). However, no artefacts were recovered from the mid-Holocene (pre-ceramic, SILVERMAN 1996: 95) portions of the stratigraphy of profiles R-2, R-3 and SF-1 which are interpreted as irragric anthrosols.

Well-defined irragric anthrosols dated to the Paracas period are present in the central Palpa Valley (between Rio Palpa and Rio Vizcas; profiles RP-06, SF-2, R-2 and R-3) as well as south-east of the Rio Vizcas (profiles FV, LLPT and BJ). Irragric anthrosol in profile BJ with a basal date of 2,720–2,360 cal BP (Poz-13293) indicates the begin of irrigation at this site during the Early to Middle Paracas period. In profile BJ, the base of the irragric anthrosol lies ca. 4.4 m above the current riverbed of the Rio Vizcas (distance ca. 630 m). Irragric anthrosol of similar age (2,710–2,350 cal BP, Poz-13545) lies ca. 2.0 m above the current riverbed in profile FV (distance ca. 260 m). While a relatively short canal would have sufficed to supply the site of profile FV with irrigation water (today: Canal Valdivia, 1.7 km), a much longer canal is necessary to irrigate the higher-elevation site of profile BJ (today: Canal La Peña Baja, 3.8 km). The main field terrace north-west of the Rio Palpa (profile RPF-2; unfortunately no radiocarbon dates exist for profile YN-2) has come under irrigation significantly later, during the Initial Nasca or Early Nasca period.

The properties of fossil irragric anthrosols identified in the Palpa Valley are listed in table 13. The boundary between underlying fluvial or hyperconcentrated flow deposits and the base of irragric anthrosols commonly is diffuse or disturbed due to agricultural activities. However, the different facies are generally well-defined in terms of the properties presented in chapter 5.1, Table 6. Comparison of grain size distributions of recent A_p horizons and fossil irragric anthrosols dated mainly to the Paracas to Nasca periods (Figure 47) shows good agreement for most samples. One group of fossil irragric anthrosol samples (pertaining to profiles BJ, FV and R-2) is characterised by relatively low sand and high clay content (≤25% and ≥30%, respectively) and high CSR (0.59–0.84). While these properties characterise almost all samples of profile BJ between 2,720–2,360 and 1,610–1,410 cal BP, ages of samples with

Tab. 13: Properties of fossil irragric anthrosols in the Palpa Valley. Age constraints are based on radiocarbon dates with the exception of profiles RP-06 and RPF-2 (diagnostic ceramic shards). Ages in brackets are based on linear interpolation or extrapolation (* = identification as irragric anthrosol uncertain).

Profile	Depth (top) [m]	Thickness [m]	Age at bottom [cal BP]	Age at top [cal BP]	Upper boundary	Lower boundary
LLPT	2.36	0.94	2,720 – 2,360	1,830 – 1,610	diffuse, with silt lens	sharp, horizontal
BJ	1.40 – 2.15	0.55 – 1.20	2,720 – 2,360	1,610 – 1,410	sharp, erosional	distinct, slightly undulating
SF-2	2.00	0.70	after 3,220 – 2,950 (2,650 – 2,410)	before 1,080 – 930 (1,840 – 1,650)	sharp/diffuse	diffuse (over hyperconcentrated flow deposit)
RP-06	1.10	0.50	3,080 – 2,840	2,300 – 2,150	diffuse, anthropogenic	diffuse (over hyperconcentrated flow deposit)
R-2	1.00	0.80	1,720 – 1,540	(980 – 140)	diffuse	diffuse (over hyperconcentrated flow deposit)
R-2	1.90	1.48	(4,900 – 3,510)	after 2,700 – 2,340 (2,300 – 1,580)	sharp, undulating (hyperconcentrated flow deposit)	diffuse, hyperconcentrated flow deposit
Jau-1	2.88	0.52	1,490 – 1,300	before 550 – 505	turbated	sharp (over gravel)
R-3	2.45	1.15	(4,210 – 3,300)	2,160 – 1,950	undulating	slightly undulating, somewhat diffuse (hyperconcentrated flow deposit)
R-3	1.80	0.44	(2,160 – 1,950)	(1,355 – 960)	sharp, horizontal (fluvial layer)	horizontal, diffuse (hyperconcentrated flow deposit)
RPF-2	0.85 – 0.90	0.70 – 0.75	2,070 – 1,820	1,620 – 1,520	undulating (irragric anthrosol)	basal gravel
R-1	1.92	0.63	1,870 – 1,630	1,490 – 1,120	diffuse, broken-up	undulating, broken-up
FV	1.39	0.26	2,710 – 2,350	1,710 – 1,540	diffuse	plant holes in bedded sand
R-2*	3.38	0.38	(5,660 – 3,910)	(4,900 – 3,510)	diffuse	diffuse (over hyperconcentrated flow deposit)

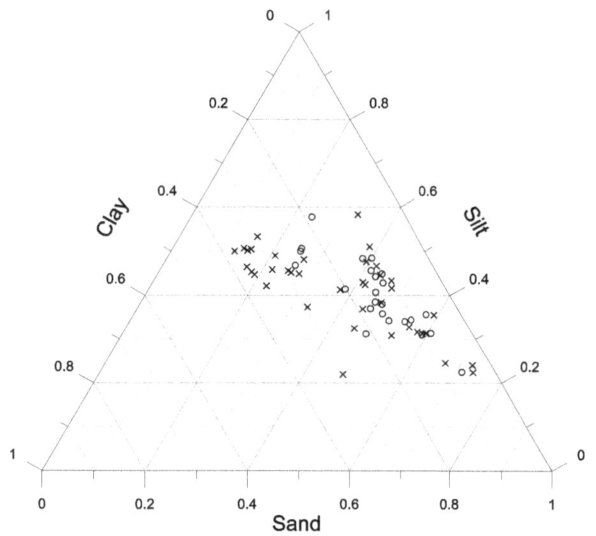

Fig. 47: Grain size distributions for the fine-grained fractions of facies F1 (recent irragric anthrosol, open circles) and (b) facies F2 (fossil irragric anthrosol, crosses).

these properties are linearly extrapolated to 1,900–1,640 cal BP in profile FV and interpolated to 1,510–1,140 cal BP in profile R-2 using the available radiocarbon dates from the respective profiles.

Despite the low number of profiles in which it occurs, this grain size fining is of considerable significance because several factors could explain the phenomenon:

(1) A change in grain size distribution of river (and hence irrigation) water towards higher clay and lower sand content could reflect climatic changes in the upper drainage basin. This effect should be observable in other samples of similar age from profiles throughout the Palpa Valley – this is not the case.

(2) Post-depositional changes of originally similar sediment (weathering or clay mobilisation) could lead to a change of CSR and thereby reflect changes e.g. in water availability (Schaetzl & Anderson 2006: 579, after Van Wambeke 1962; Noller 1993: 137). However, these processes would not significantly affect sand content.

(3) Grain size distribution may change during transport to the field due to sedimentation of larger grain sizes and preferential transport of smaller grain sizes. Prevailing irrigation technique (e.g. simultaneous or sequential watering of fields along a canal) also influences flow velocity and thus the amount and grain size of material transported and sedimented.

While the two former models (both implying climatic change) can be excluded, the latter model best explains the observed phenomenon because the site of profile BJ required the longest irrigation canal due to its elevation relative to the river.

While Paracas and Nasca period irragric anthrosols are generally well-defined in terms of radiocarbon dates and diagnostic ceramic shards, radiocarbon dates and diagnostic archaeological finds from later cultural phases are only present in a limited number of profiles. However, the issue is not one of lacking irragric anthrosols (earlier irragric anthrosols are generally overlain by later ones up to the present A_p / irrigation horizon) but one of lacking sufficient chronological information (i.e. due to financial constraints to date more samples) to place these sediments into one or another period. While the Paracas/Nasca period irragric anthrosols are generally characterised by great vertical homogeneity, land use changes (e.g. possibly between field and plantation) or hiatus (indicated by diffuse, turbated boundaries between similar or slightly differing irragric anthrosol layers within the stratigraphy) are more common in later periods. On the basis of the available data, a general chronological pattern of these changes beyond the statistical interpretation of available radiocarbon dates (see 7.3.1) can not be deduced.

7.3 *Pre-Columbian cultural landscape changes*

7.3.1 *Interpretation of the distribution of radiocarbon dates*

Figure 48 shows the summed relative probability distribution resulting from the calibration of all 63 radiocarbon dates of the present study which are younger than 4,000 ^{14}C BP. Due to the uneven probability distributions for individual samples (related to the wiggles in the calibration curve) and the relatively low number of samples spread over a long time span, a smooth summed probability distribution is not to be expected. Assuming that the probability of acquiring a sample of a given calendaric age decreases exponentially due to declining sample preservation within the stratigraphic column as well as erosion, a summed probability distribution was modelled using simulated radiocarbon dates based on an equal number (63) of calendaric ages following an exponential function fitted to the summed probability distribution of the actual samples. This modelled distribution is shown as a grey line in Figure 48. It serves (1) to indicate the general trend which would be expected for an idealised probability distribution and (2) to visualise the amount of variability induced by the wiggles of the calibration curve (note that the peaks and troughs may change if number and/or spacing of input samples are changed). Narrow spikes in the distribution curve are therefore of no interpretative value. Longer-term deviations of the actual from the modelled curve, however, indicate deviations from the assumption that charcoal was introduced into the sedimentary record at a constant rate. They may therefore represent changes in the intensity of human impact.

Despite the limitations posed by the low sample number, Figure 48 illustrates a number of issues. During most of the historic period, the actual probability distribution is

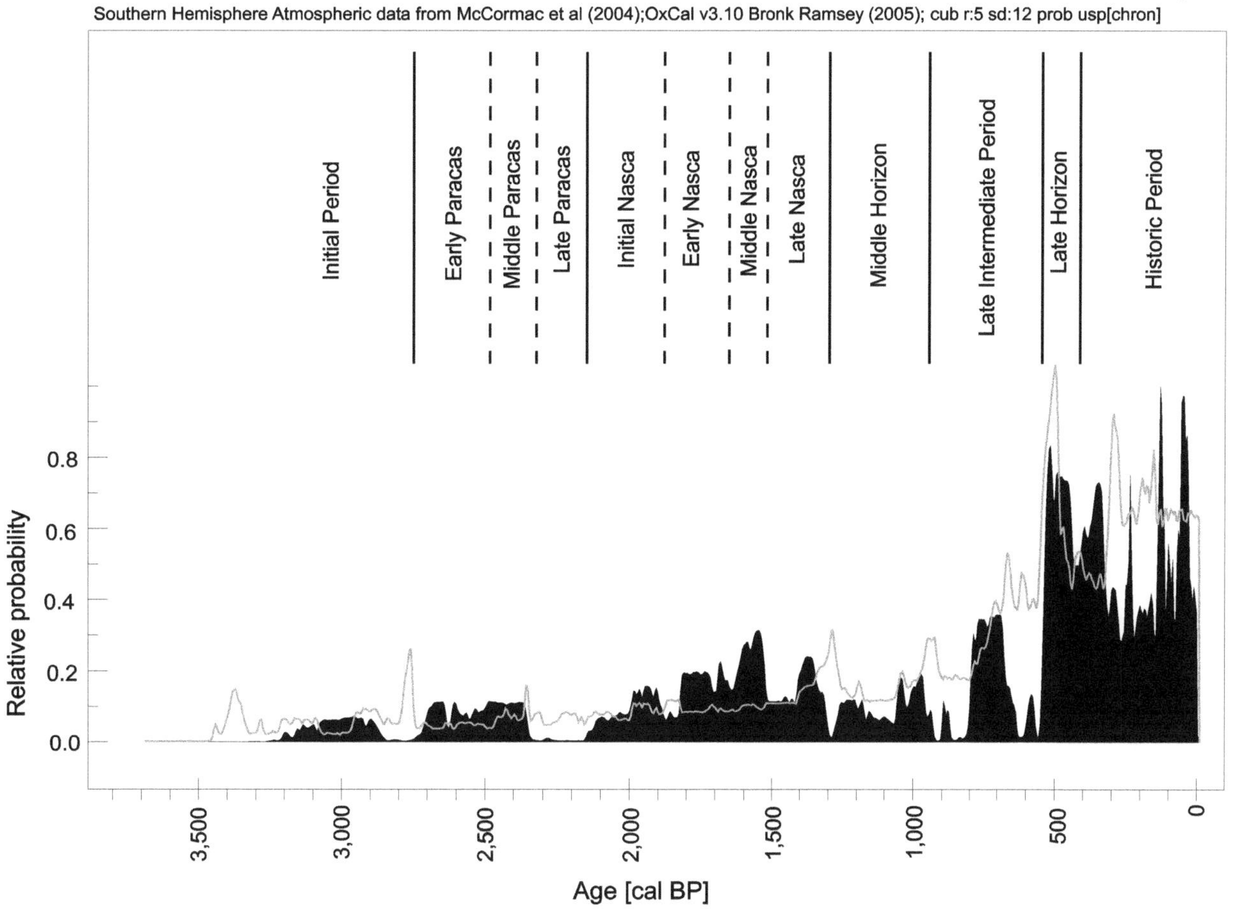

Fig. 48: Summed probability distributions for all samples younger than 4,000 ^{14}C BP (black; grey line: modelled distribution, see text).

below the modelled distribution. This is attributed to the early historic population decline (see Cook 1981) and may for the late Historic Period be due to a sampling bias because near-surface samples were largely excluded from dating. During the Late Horizon and early Historic Period, the two curves reach similar average values, albeit with very strong short-term variability which inhibits further interpretation. No Late Horizon (Inca) ceramic shards have been recovered in the course of this study despite the abundance of radiocarbon dates falling into this period. Domestic activities may therefore have been more separated from agricultural activities.

Between 1,350 and 550 cal BP (e.g. during the Middle Horizon and Late Intermediate Period), the actual distribution is below the modelled distribution for 800 years with the exception of a window from 800 to 680 cal BP (middle Late Intermediate Period). This may be showing a drop in human activity caused by reduced population density or by settlement pattern shifts away from the Palpa Valley during the Middle Horizon and high population densities during the Late Intermediate Period and is in accordance with archaeological studies in the region (e.g. Reindel et al. 2001: 95; Schreiber 1999: 168). Between 2,100 and 1,350 cal BP (Initial to Late Nasca period), the actual probability distribution is generally above the modelled distribution. This is interpreted as indicating extensive human activity and probably high population density and corresponds with the abundance of Nasca period archaeological sites in the Palpa Valley (Reindel et al. 2001; Reindel et al. 2002). The Paracas period is likewise characterised by higher-than-modelled relative probability from 2,700 to 2,350 cal BP. The apparent gap during the Late Paracas period is probably the result of sampling bias rather than a lack of human presence at this time – the Late Paracas findings in profile RP-06 have not been radiocarbon-dated since the consistent Ocucaje 9 ceramic shards were considered sufficient for dating the respective layer.

7.3.2 *The late Holocene cultural landscape: natural impact and anthropogenic changes*

Quantitative results regarding both the impact of fluvial events and the impact of human activities in the Palpa Valley have been attained above. Extreme discharge events did occur in the Palpa Valley during the Late Holocene. During the past >1,500 a, these events affected ca. 30% of the valley floor area (see 7.1). Human activities, on the other hand, have transformed the Palpa Valley into a fertile oasis by accumulating several metres of irrigation sediment upon a base of coarse gravel, sand and hyperconcentrated flow deposits. During the Late Holocene, the Palpa Valley has been developed as a cultural landscape in the sense of a anthropogenetically modified landscape. Human settlement in the Palpa Valley appears to have been continuous since the Initial Period; however, large fluctuations in population and shifts in settlement patterns have been deduced from the archaeological record (REINDEL et al. 2001). The long history of settlement in the hyperarid desert shows that irrigation as it has been practised here represents a sustainable technology, in particular in terms of the avoidance of soil salinisation. While some low-lying river-proximal areas are affected by salinisation today due to a shallow groundwater table, the low concentration of soluble salts in the river and irrigation water (1.0–4.8 ppm Cl^- and 3.6–14.4 ppm $[SO_4]^{2-}$ as measured in March 2006) and the application of abundant water to the fields has prevented salt accumulation.

Vulnerability to flood events and droughts

While the sedimentological, geomorphological and archaeological records allow quantitative conclusions regarding the physical impact of both fluvial events and human activity in the valley floor area, the potential impacts of fluvial events on pre-Columbian human societies are not easily assessed. Direct losses of life were probably small relative to the total population of the area (cf. HOFFMAN 1999: 305, 320) and impacts were probably highly localised (cf. DILLEHAY et al. 2004: 279). Extreme discharge events can cause the destruction of irrigation canal intakes and sedimentation within the canals. As exemplified by the 1999 flood event and discussed by CLEMENT & MOSELEY (1991: 427), damage to the simple and (in terms of necessary workforce) low-cost traditional irrigation infrastructure was quickly repaired. The same applies to the simple domestic architecture (insofar as settlements were present in the valley bottom) which was quantitatively dominant throughout the pre-Columbian period (e.g. SILVERMAN 2002; REINDEL et al. 2001). The only seasonal availability of water for irrigation implies the necessity of food storage or transhumance. Subterranean storage facilities (*collomas* with volumes of 0.2 to 1.2 m^3, VAUGHN 2005: 94) have been described as parts of Nasca domestic structures. However, to what extend would a destruction of at most one third of the agricultural area (if all areas mapped as having basal ages <1,500 cal BP in Figure 43 had been affected simultaneously) have impacted both survival and cultural development?

A number of questions related to the vulnerability and adaptability of these societies remain unanswered:

(1) Population sizes of pre-Columbian societies in the research area are unknown. Very rough estimates have been put forward by ENGEL (1981); however, settlement studies have since identified numerous additional sites (e.g. REINDEL et al. 2001). With unknown numbers of settlements in the valley bottoms possibly eroded or undetected due to burial under later sediments (e.g. SILVERMAN 1994: 361; see 7.3.3), without estimates regarding the numbers of inhabitants per settlement and without well-constrained settlement life-spans (cf. SILVERMAN 1994: 361 for the problem of temporally multicomponent sites), absolute population sizes have to be considered unknown.

(2) Carrying capacities (i.e. the number of people which could subsist on a given size of land using given technology) are unknown. The available estimates (one family of 5 to 7.5 persons per 1.5–4.5 hectares of farmland, ENGEL 1981: 12f.; 0.5 persons per hectare based on early maize yield, WILSON 1981: 106; 2.1–12.4 persons per hectare, COOK 1981: 20f., after Keith 1976 and Romero 1966) at best provide a rough approximation of an upper ceiling for plausible population sizes but can not be used to infer population pressure which would be an important component in assessing vulnerability.

(3) Cultural variables governing the adaptation and reaction of individuals and societies to internal and external influences are largely unknown for pre-Columbian cultures and can only be tentatively inferred from the material archaeological record and ethnographic analogies (cf. OLIVER-SMITH 1999: 74ff.; WISNER & LUCE 1993: 127). The distinction between cultural and natural variables thought to underlie cultural changes can thus only be speculative.
A limitation of the sedimentological work presented in this thesis is the impossibility of detecting highland drought events which would have affected discharge of the allogenic rivers in the coastal desert. This in turn may have impacted irrigation agriculture. The boundaries between the similar or slightly differing vertically stacked irragric anthrosols are generally diffuse or broken-up due to agricultural turbation. Even if the causes of changes detectable in the sedimentary record (land use change or hiatus, e.g. due to abandonment caused by drought and/or population decline or intentional fallow) could be determined, dating insecurity would inhibit the discrimination between e.g. a one-year and a thirty-year hiatus.

Moreover, as discussed above for flood events, definite qualitative or quantitative conclusions regarding the potential impact of droughts on populations with unknown size and vulnerability can not be made. However, irrigation water requirement for the entire valley floor area of 14.4 km^2 (excluding the rivers) can be estimated at 8.6 Mio m^3/a using an approximation of 600 mm/a necessary water input (cf. 4.2.2). While mean annual discharge of Rio Palpa and Rio Vizcas combined has been 47.8 Mio. m^3/a between 1975 and 2007, discharge has been below the minimum requirement for 4 years out of 29 (3 years without data). Assuming a 50% irrigation efficiency due to incomplete water capture on one hand and 100% water efficiency and lower (300 mm/a) water requirement of more primitive crops on the other hand increases or decreases the number of effective drought years to 8 and 3 per 29 years, respectively. Despite the short record, these data show that the natural discharge variability is indeed frequently a limiting factor for optimal irrigation agriculture. In comparison with northern coastal Peru where irrigable land and not available river discharge appears to have been the limiting factor for agricultural expansion (WELLS & NOLLER 1999: 781), water may have been the main constraint in the Palpa Valley.

Wells must have existed to provide potable water during the season without river discharge. No direct evidence for pre-Columbian water storage systems other than small reservoirs (*kochas*) at the lower end of *puquios* (SCHREIBER & LANCHO ROJAS 1995: 235) has been reported from the research area, although the utility of such systems under the given climatic and hydrologic conditions is obvious and water storage was practised in other parts of the Americas in pre-Columbian times (cf. SCARBOROUGH & GALLOPIN 1991: 6598ff. for Mexico). However, a few *kochas* (albeit of unknown age) tapping groundwater exist in the Palpa Valley (e.g. Laguna La Maquina). Since groundwater table lowering even during multi-year droughts was much less pronounced before the advent of mechanised pumping (see 4.3), groundwater remained within reach given the existing technology for large excavations (e.g. SCHREIBER & LANCHO ROJAS 1995; REINDEL & CUADRADO 2000: 85). Thus even during periods of drought, a curtailed continuation of agriculture was possible by bucket irrigation which has been shown to have been a feasible component of subsistence agriculture in coastal Peru at least until the 1970s (WEST 1981: 60ff.).

WITTFOGEL (1956: 153) has argued that "waters of rivers and large streams in arid or semiarid regions [...] defied the small-scale approach" and that "man had to create large-scale enterprises", thereby initiating a process which led to monopolised political power (cf. WITTFOGEL 1955; for criticism of the "hydraulic hypothesis" see LEES 1994: 364). While being problematic in terms of the postulated necessity of centralised control (as opposed to co-operative efforts) and the question of primacy of irrigation vs. social complexity, WITTFOGEL's (1955; 1956) key argument of correlating irrigation and social complexity may be applicable to the Palpa Valley: The large-scale expansion of the irrigation system is roughly coincident with the rise of the first regional culture (Paracas, cf. SILVERMAN 1996: 134). While small-scale irrigation may be managed by individuals or families, the main canals likely did require institutionalised collective labour beyond the household level, whether by voluntary co-operation or hierarchical control. Considering the present-day situation with average ratios of main canal length to irrigated area between 37.0 m/ha (between Rio Palpa and Rio Vizcas) and 72.5 m/ha (south-east of the Rio Vizcas and north-west of the Rio Palpa) (see 4.2.1, Table 5) and the simple construction of canals, co-ordination may be have been more important than the actual labour force mobilisation. The difference in canal length required per unit irrigated area (Table 5) suggests a potential preference for agricultural use of the central valley over the valley sides.

Climatic changes in the coastal desert

A model for climatic changes in the Palpa region has been presented by EITEL et al. (2005). As discussed in chapter 6.2.2 and HESSE & BAADE (2007), their argumentation regarding a monsoonally influenced semi-arid mid-Holocene in the Palpa area is at odds with the findings of this thesis and various paleoclimatological records (e.g. indicating a northward shift of the mean latitude of the ITCZ during this period: HAUG et al. 2001; TEDESCO & THUNELL 2003). According to EITEL et al. (2005: 148), there was a decrease in local precipitation from a semi-arid Initial Period to a peak in aridity during the Nasca period and the Middle Horizon. However, local precipitation in the Palpa Valley can not be satisfactorily inferred for any of these periods with the exception of rare and intense events likely related to El Niño. For the Late Intermediate Period, EITEL et al. deduce the occurrence of regular monsoonal precipitation at elevations between 250 and 700 m in the Palpa region from a population increase apparent from large archaeological sites outside of the valley bottoms and from features interpreted as water harvesting structures (EITEL et al. 2005: 153; MÄCHTLE 2007).

Further to the south, ZARO & ALVAREZ (2005: 717) find that the coastal spring site of Wawakiki (17°25'S) "was farmed most intensively between A.D. 1200 and 1400 [750–550 cal BP], a period characterized by prolonged highland drought and recurrent El Niño-induced floods." According to anecdotal evidence collected by CLEMENT & MOSELEY (1991: 430), coastal spring discharge is directly related to mountain rainfall 40 to 60 km inland. However, the groundwater flow velocities implied by this correlation (220–330 m/d under the assumption of a six month delay) are unrealistically high under the given geological conditions (fine-grained deposits according to SATTERLEE et al. 2000: 97; cf. DUNNE & LEOPOLD 1978: 204ff. for groundwater flow calculation and permeability values). Furthermore, if mountain rainfall was responsible for spring discharge, the alleged highland drought of A.D. 1040 to 1490 (910–460 cal BP) (ZARO & ALVAREZ 2005: 719, after Thompson et al. 1994 – see Figure 7 for discrepancies between the two Quelccaya ice cores) would have resulted in a reduction in spring flow during the time of most intensive use of the Wawakiki site.

On the other hand, ZARO & ALVAREZ (2005: 732, referring to Magilligan & Goldstein 2001 and Dillehay et al. 2004) acknowledge the possible influence of local precipitation (albeit limited to El Niño precipitation) on the coastal aquifer. However, an influence of El Niño on the precipitation on the far southern coast of Peru has not been verified except for the rare very strong events (see 2.3.1; ZARO & ALVAREZ 2005: 719f.; cf. RUNDEL et al. 1991: 28 for El Niño 1982; cf. MURPHY 1926 for El Niño 1925) which would not allow permanent agricultural use. Satellite images (Landsat, Quickbird) and digital elevation models (SRTM) indicate that the steeply rising coastal escarpment between the several spring-fed agricultural sites and the Clemesí desert is well-situated for the capture of fog precipitation and the development of lomas vegetation (cf. RUNDEL et al. 1991; KOEPCKE 1961). Anthropogenic structures like corrals and terraces can be identified in the elevation of the fog zone (ca. 400–800 m asl), well above the aquifer which intersects the land surface at an elevation of 100–150 m asl according to CLEMENT & MOSELEY (1991: 430). It is therefore likely that not rare El Niño events or mountain precipitation but fog intensity and duration control the pattern of moisture availability in the Wawakiki area (cf. 2.3.1 and 2.4). This interpretation is supported by COHEN (1975: 112) who states that "in the late Intermediate Period and Late Horizon, a number of superficial archaeological sites occur in the regions of lomas-vegetation, often including rectangular stone corrals which appear to suggest the use of this vegetation for grazing domestic flocks."

Like at Wawakiki, increased human activity occurs in the Palpa region during the Late Intermediate Period (see 7.3.1; REINDEL et al. 2001). Like at Wawakiki, settlements and possible agricultural infrastructure (*khadins* as water harvesting structures, MÄCHTLE 2007) appear outside of the allogenic river valleys (e.g. at Ciudad Perdida de Huayuri). Since regular monsoonal precipitation in the Palpa area has to be discounted (WEISCHET 1966: 1; HESSE & BAADE 2007; see 2.3), the only plausible local moisture sources are El Niño events or *garúa* precipitation. The Late Intermediate Period site of Ciudad Perdida de Huayuri lies ca. 1.2 km from the Rio Santa Cruz, a distance over which small quantities of water for domestic purposes could have easily been transported. However, even with precipitation of 150–200 mm/a (EITEL et al. 2005: 148, 153), local water supply would probably only be available seasonally due to the small size of the catchment and sedimentary bodies presumably used for agriculture and water storage, suggesting a reliance on wells tapping the main valley aquifer for most of the year.

If the terraced hilltop west of Ciudad Perdida de Huayuri as well as the filled-dam-like structures in the vicinity of the Ciudad interpreted as *khadins* (EITEL et al. 2005: 153; MÄCHTLE 2007) did indeed serve agricultural purposes, there must have been local precipitation. However, while terraces and *khadins* cover 4.6 and 8.8 hectares respectively, the dense settlement of Ciudad Perdida de Huayuri extends over 6.8 ha. Using the highest agricultural carrying capacity estimates of 7.5 persons per 1.5 ha (ENGEL 1981: 13) and 12.4 persons per hectare (COOK 1981: 20f., after Keith 1976 and Romero 1966), terraces and *khadins* combined could have sustained at most 67 to 166 persons – far less than the number of buildings in the Ciudad. By comparison, there are currently ca. 340 ha under cultivation in the Rio Santa Cruz valley within a distance of 2.5 km of the Ciudad (the *khadins* lie within this distance) resulting in maximum carrying capacity estimates of 1,700 to 4,200 persons on the basis of river valley irrigation. Terraces and *khadins* can therefore only have served for complementary or opportunistic use (cf. WEST 1981: 60ff. for opportunistic land use in historic times in Peru).

Distinction between opportunistic use of ephemeral resources and long-term productivity is a challenge which will have to be resolved before secure paleoclimatic conclusions can be drawn from the existence of the observed structures (cf. PARSONS 1970: 300ff. for the problems of interpreting vegetation extent and utilisation as ephemeral or long-term phenomena). However, El Niño events appear unlikely as the source of local moisture because many buildings of Ciudad Perdida de Huayuri lie at the bottom of a quebrada where they would have been threatened by runoff from intense precipitation. If such runoff was not a concern here during the Late Intermediate Period, precipitation probably was light.

At a distance of 5.7 km south-east of Ciudad Perdida de Huayuri, profile LM-1 records only one coarse-grained debris flow event attributable to intense local precipitation between 910–730 cal BP (Poz-8010) and 490–320 cal BP (Poz-5262), i.e. during the Late Intermediate Period (950–550 cal BP). This contrasts with six coarse-grained debris flow events which occurred after 490–320 cal BP. These findings correspond to 0.17–0.42 events per century between 910–730 and 490–320 cal and 1.10–1.60 events per century after 490–320 cal BP. The Cariaco Basin titanium record (HAUG et al. 2001) indicates a northward shift of the mean latitude of the ITCZ between 1,050 and 550 cal BP which would have reduced the frequency and strength / southward extent of El Niño events (see 2.3.2) and favoured southern hemisphere winter conditions which are characterised by coastal fog. This is compatible with the findings from profile LM-1, the evidence from Wawakiki as discussed above and the interpretation of Late Intermediate Period local moisture as being derived from fog precipitation.

A possible impact of El Niño events on the cultural sequence in the Rio Grande de Nazca drainage basin has been suggested by GRODZICKI (1990), SILVERMAN (2002: 24f.) and REINDEL et al. (2003: 127). The assumption that a mega-Niño event caused infilling of the Nazca Valley to a height of more than 30 m (GRODZICKI 1990: 89) can not be sustained because (1) this interpretation is not compatible with geomorphic processes of deposition and erosion (infilling with coarse gravel followed by dissection limited by morphologically soft previous structures such as adobe walls) and (2) the postulated total interruption of settlement activity is not confirmed by the archaeological record (SILVERMAN 2002: 24). Local intense precipitation and associated runoff and debris flows in the quebradas did occur in pre-Columbian times as exemplified by traces of runoff affecting the Early Nasca site of Cahuachi in the

Nazca valley after the site's period of maintenance (SILVERMAN & PROULX 2002: 51), damage to settlement infrastructure at La Muña and Los Molinos (REINDEL & ISLA CUADRADO 2001), debris flow deposits interfingering with archaeological layers of Nasca age in profile RG-Mo (see 5.2.3 and Appendix A) and debris flow deposits accumulated after 910–730 cal BP in profile LM-1 (see 5.2.2 and Appendix A). However, these events did not cause large-scale destruction throughout the valley (e.g. no debris flow deposits have been identified in profile BJ after 2,720–2,360 cal BP (Poz-13293) despite its vulnerable location at the mouth of a small quebrada). Thus, while intense precipitation events likely had some effect on domestic and agricultural infrastructure, they did not have the potential to physically destroy the subsistence base. However, contrary to the assumption that El Niño impacts were very rare and therefore ignored in the placing of settlements (REINDEL & ISLA CUADRADO 2001: 285), adaptive structures do exist (SILVERMAN 2002: 24).

7.3.3 Pre-Columbian settlement patterns and space use

Settlement pattern studies widen the scope of archaeological investigations beyond individual sites and have been used to infer cultural (e.g. BILLMAN 1999; BROWNE 1992) or environmental (e.g. DILLEHAY & KOLATA 2004: 4,329; REINDEL et al. 2001: 107) variables. Despite the explicit recognition that settlements may have existed on the valley floors and may since have been covered by sediment or eroded by the rivers (COHEN 1975: 106; WELLS 2001: 136; SILVERMAN 1994: 361, 365; SILVERMAN 1996: 130; SILVERMAN 2002: 48f., 142), until today, valley floors in the Rio Grande de Nazca basin have not been systematically studied by archaeologists with the exception of surface survey (SCHREIBER & LANCHO ROJAS 1995: 249). Rather, survey usually concentrated on the valley margins where traces of occupations are ubiquitous and well-preserved due to the prevailing climatic conditions (e.g. BROWNE 1992; SCHREIBER & LANCHO ROJAS 1995: 249; REINDEL et al. 2001; SILVERMAN 2002). Thus, while "[a]rchaeological settlement patterns are reflections of how humans have used the landscape through time" (WELLS 2001: 137), settlement pattern studies biased by ignoring the valley floor areas are at least in part reflections of how archaeologists *think* that humans have used the landscape (cf. FISH 1999: 205). However, the thus biased data have been used to infer environmental changes in the Rio Grande the Nazca area (e.g. SCHREIBER & LANCHO ROJAS 1995; REINDEL et al. 2001).

The findings of the present study help to elucidate temporal changes in space use within the valley floor area and thereby add to the extensive data on valley-margin sites published by BROWNE (1992), REINDEL et al. (1999), REINDEL et al. (2001) and ISLA et al. (2003). The fact that artefacts are ubiquitous in many profiles investigated in the Palpa Valley indicates long-term human presence and intensive space use throughout the valley. Particularly high concentrations of ceramic shards were recovered from profiles R-2 and RP-06 (Paracas), RPF-1, RPF-2 and RG-Mo (Middle and Late Nasca), SF-3 (Loro and Late Intermediate Period) and SF-4 (Loro). Specifically in these profiles, artefact concentrations are too high to be attributable to incidental loss or discard. Moreover, they occur in shallow, midden-like deposits (including marine mollusc shells and stone fragments; profiles R-2 and SF-4) or in layers associated with anthropogenic structures (wall foundations; profiles RP-06 and SF-3). Other contexts are paleosurfaces associated with graves (profile RPF-2 and possibly profiles RG-Mo and BJ) and a hearth (profile SF-2).

The sequence of the expansion of irrigation systems in the Palpa Valley can be reconstructed using the data presented in this thesis. As indicated by Table 5 for the present-day situation, the central valley (between Rio Palpa and Rio Vizcas) requires the least length of irrigation canal per hectare irrigated land due to its advantageous topographic position. The oldest archaeological finds (Paracas culture) in this study are from profiles in the central valley (RP-06 and R-2). Irragric anthrosols in the central valley date back to the Initial Period and possibly to the Archaic Period (profiles RP-06, SF-2, R-2 and R-3; see Table 13), suggesting that the ease of irrigation was a factor in the development of agricultural infrastructure. By the Early to Middle Paracas period, the irrigation system had been extended to the area south-east of the Rio Vizcas (profile FV) and reached the foot of the Pampa San Ignacio (profile BJ). While most areas in the central valley can be watered by relatively short canals, irrigation of the site of profile BJ required the construction of a canal of ca. 3.8 km length. The area between the Rio Palpa and the north-western valley margin came under irrigation somewhat later (Initial Nasca or Early Nasca period, profile RPF-2). Irrigation agriculture on the debris apron at the foot of the Cresta Sacramento is estimated to have commenced by the Middle Horizon or Late Intermediate Period (profile RPF-3), indicating the construction of the uppermost canal by this time. In the nearby section of the Rio Grande Valley, the uppermost irrigation canal serving the field on the debris apron was constructed during the Late Intermediate Period as shown by irragric anthrosol with a basal age of 745–655 cal BP in profile RG-Mo. Contrary to northern Peru where pre-Columbian irrigation systems have been more extensive than present systems (e.g. MOSELEY 1983: 778), in the Palpa Valley the present extent is equal to its maximum pre-Columbian extent with the exception of areas watered by canals built after 1944.

The sites which are characterised by evidence of particularly intense use during certain periods are interpreted as indicating within- or near settlement situations. While the Paracas culture settlement pattern had long been considered sparse (REINDEL et al. 1999: 339ff.; SILVERMAN 1994), a number of additional valley margin settlements pertaining to this period have since been identified (REINDEL et al. 2001; ISLA et al. 2003). In particular the discovery and excavation of a valley-floor settlement at Jauranga (ISLA et al. 2003) has shed new light on this period. Evidence for Paracas culture settlement activities in profiles R-2 and RP-06 (Figure 49) indicates that Jauranga is not a solitary site but part of a valley-floor settlement pattern.

During the Middle to Late Nasca period, particularly intensive use has occurred in the vicinity of profiles RPF-1, RPF-2, RPW and RPE (Figure 50). The large quantity of ceramic shards in profiles RPF-1 and RPF-2 and the presence of a grave associated with a Middle Nasca paleosurface (profile RPF-2) as well as high artefact concentrations in the surrounding area (profiles RPE and RPW; abundant Nasca period ceramic shards in the excavation spoil of the modern canal constructed in 2006) suggest the presence of a large Middle to Late Nasca settlement. In the 1944 aerial photograph, angular features possibly indicating buried anthropogenic structures are faintly visible to the north and south-east of profile RPF-2 (Figure 50) and may be related to the archaeological finds in this area. For the Middle Horizon (Loro), a near-settlement situation is suggested by the abundance of ceramic shards in profile SF-4 pertaining to this period. Architectural remains (a foundation-like arrangement of large stones, see Figure 30) pertaining to the Late Intermediate Period are present in profile SF-3. This departs from the pattern of settlement placement in locations at some distance from the valley floors during this period (REINDEL et al. 2001: 96).

Given the observed valley-floor presence of settlement contexts of all main pre-Columbian periods with the exception of the Late Horizon (Inca) despite the limited number and size of the excavations presented in this thesis, the need for systematic archaeological investigation of the valley floor and the necessity of including the valley floor in settlement pattern analyses have to be emphasised. Of particular relevance is the emerging pattern of river-proximal archaeological sites in the Palpa Valley (Figure 49). Abundant artefacts were found in profiles of high basal ages near past river channels. Given the strong lateral activity of the rivers (see 4.1.1 and 4.1.2), the probability of site destruction is high (e.g. profile RP-06 is presently being eroded by the Rio Palpa). However, the preservation of some sites for more than 2,000 years shows that this lateral activity was limited to relatively narrow fringes along the rivers.

Rather than being an exception, settlement in the valley bottom may have been common throughout pre-Columbian times. The growth of the known valley margin settlement pattern (e.g. REINDEL et al. 2001) generally at distances of ≥400 m from the rivers and ≥500 m from the main agricultural surface in the central valley may reflect the expansion of the irrigation system to its maximum, valley-margin extent (cf. COHEN 1975: 111). These distances are large in comparison to the distances between the closely spaced valley margin settlements (REINDEL et al. 1999; REINDEL et al. 2001) presumably placed close to the associated fields. Settlements may have been placed not only within short distance of associated fields throughout the valley floor but in particular near associated irrigation infrastructure such as canals (e.g. valley margins for the uppermost canals) and canal intakes (river-proximal). Notwithstanding the results of Chapter 7.3.1, in particular during the Early Horizon (Paracas) and Middle Horizon (Loro/Wari), the inferred low population size (REINDEL et al. 1999: 338ff.; REINDEL et al. 2001: 94ff.) may at least partially be an artefact of incomplete survey as intensive human impact is locally evident.

While "infrastructural elements such as subsistence patterns are generally invested with causal primacy over structural elements such as settlement patterns, not the reverse" (CLEMENT & MOSELEY 1991: 440, after Harris 1979), possible feedback mechanisms have to be acknowledged. For example, population growth after the initial agricultural development in the Palpa Valley may have been responded to by an up-valley expansion of the irrigation system, including labour-intensive terrace construction. The upstream tapping of river water may subsequently have led to water competition, in particular during periods of drought. Given the advantageous position of upstream communities regarding water access, their long-term sustainability would have been greater than for the downstream communities despite the less advantageous situation in terms of labour requirement for the construction and maintenance of agricultural infrastructure. This hypothetical scenario may explain the observed (albeit biased) up-valley settlement shifts (cf. Reindel et al. 1999, 2001 and 2002) without implying impacts of climatic changes on pre-Columbian societies. A settlement shift to agriculturally more efficient areas as a response to population increase as suggested by REINDEL et al. (1999: 348) for the Middle Nasca period appears implausible because population pressure generally leads to an expansion into more marginal areas.

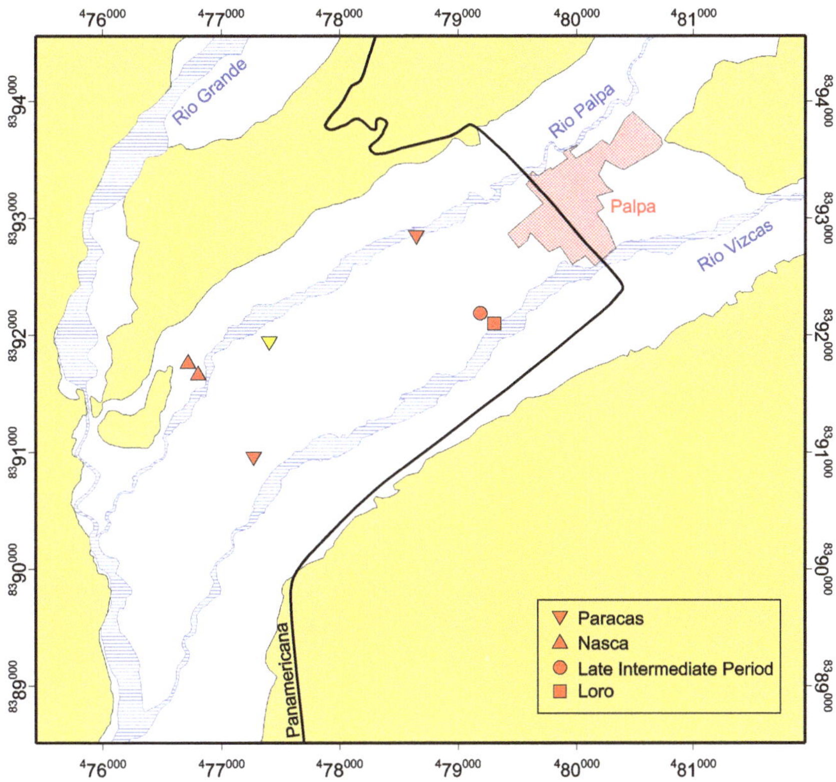

Fig. 49: Valley floor settlement locations inferred from abundant archaeological finds (red) and Paracas settlement Jauranga (yellow, ISLA et al. 2003).

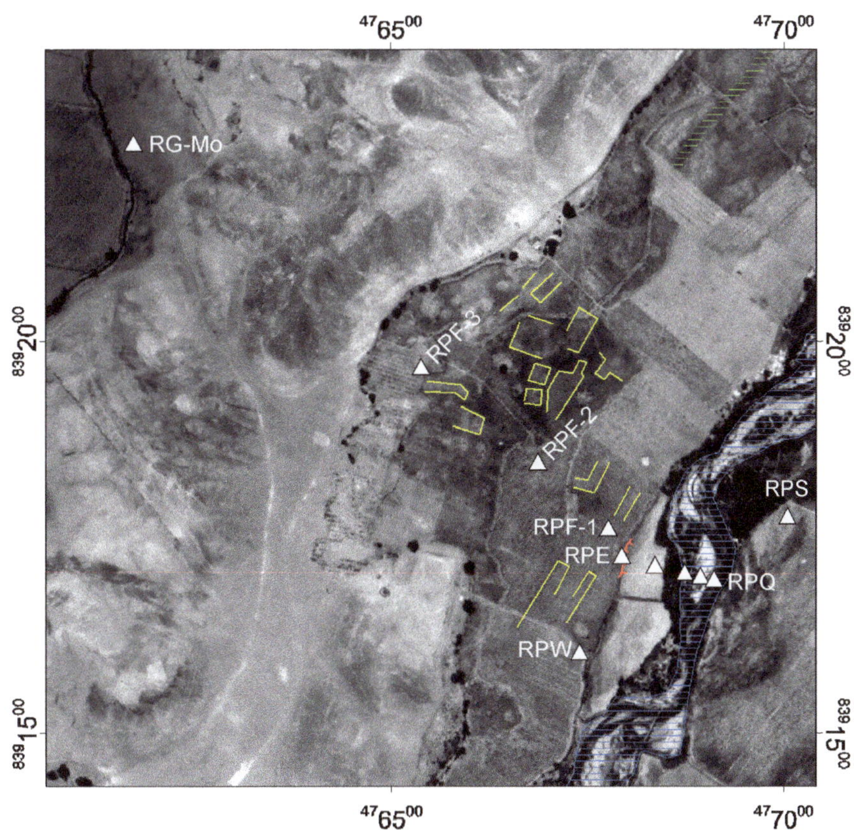

Fig. 50: Area interpreted to represent Middle to Late Nasca settlement activity (1944 aerial photograph with possible anthropogenic structures shown in yellow and possible former irrigation canal hatched in green).

8. Synopsis

The aim of the research presented in this thesis was to reconstruct the landscape evolution in the lower Rio Grande de Nazca drainage basin during the Holocene with an emphasis on human-environment interactions. While the work focused mainly on the Late Holocene, both the necessity to understand processes at work in a longer time frame and incidental findings have led to paleoenvironmental reconstructions spanning the past 30,000 years.

The long-term landscape evolution was reconstructed using a sequence of four stratigraphically consistent radiocarbon dates from a debris fan ca. 15 kilometres south-west of Palpa. During the Last Glacial, debris flows attributable to local intense precipitation led to aggradation of the fan. The Last Glacial Maximum was characterised by both autochthonous debris flow and allochthonous flood sedimentation. After the subsequent incision of the local quebrada, river incision to its present level took place. Morphological similarities with other quebradas suggest that this sequence is of regional relevance. The incision of the Rio Grande de Nazca has been ≤6 m since 30 ^{14}C ka BP, with most of the incision occurring after the LGM, likely during the Holocene.

The natural dynamics in the Palpa Valley before the inception of agriculture are recorded by coarse riverbed sediments, fine-grained fluvial sediments and hyperconcentrated flow deposits. These sediments indicate mid-Holocene dynamics similar to those represented by recent sediments in the Quebrada Yapona valley which is characterised by low discharge due to the small size of its discharge-relevant catchment area above 2,500 m asl. The mid-Holocene dry period in the high Andes with reduced discharge in the allogenic rivers interspersed by periods of strong precipitation events triggering erosional and depositional floods and hyperconcentrated flows is thus represented in the sedimentary record of the Palpa Valley.

For the Late Holocene, the quantitative analysis of the impact of fluvial dynamics on the landscape of the Palpa Valley shows that a large part of the valley floor (ca. 60%) has not been subject to fluvial erosion since at least 1,500 cal BP. While parts of the central valley have experienced fluvial erosion between and after 1,500 and 1,000 cal BP and the rivers are characterised by strong lateral activity, most of this lateral activity was restricted to relatively narrow fringes along the present-day riverbeds. These fringes have been subject to repeated erosion and sedimentation. The relatively young ages of sediments within these fringes therefore do not necessarily represent a specific phase of fluvial activity.

The impact of human action on the Late Holocene landscape evolution can be assessed by reconstructing the expansion of the irrigation system. In the Palpa Valley, irrigation agriculture can be confirmed for the Initial Period from the presence of irragric anthrosols. These are characterised by indicators for intense turbation (e.g. a lack of bedding or lamination), increased LoI and CSR values and the presence of charcoal and artefacts. While the age of possible irragric anthrosol (but lacking artefacts) can be extrapolated to the Archaic Period with maximum ages of 5,660–3,910 cal BP, the age of the base of the earliest securely identified irragric anthrosol is extrapolated to 4,900–3,500 cal BP. The begin of irrigation agriculture in the Palpa Valley is thus roughly coincident with the end of the mid-Holocene dry period in the Andes. The earliest irragric anthrosol occurs in the central valley (between Rio Palpa and Rio Vizcas).

By the Early to Middle Paracas period, the irrigation system had been expanded to include the area south-east of the Rio Vizcas. The uppermost pre-Columbian irrigation canal along the foot of the Pampa San Ignacio on the south-eastern valley margin had been constructed by 2,720–2,360 cal BP as indicated by the begin of irragric anthrosol deposition. The length of this canal was ca. 3.8 km as inferred from comparison with the present-day situation and indicates the construction of large-scale irrigation infrastructure. The area north-west of the Rio Palpa was the last large tract of land to be included in the irrigation system, with the main terrace under irrigation by the Initial Nasca or Early Nasca period. Along the north-western valley margin, the timing of the construction of the highest canal on this side of the valley (at the foot of the Cresta Sacramento) is reflected by the begin of irrigation by the Middle Horizon or Late Intermediate Period. In the neighbouring Rio Grande valley, the uppermost irrigation canal was constructed during the Late Intermediate Period.

Contrary to the assumption that irrigation systems were much smaller in pre-Columbian times than at present (SILVERMAN 1996: 102), by the Initial Nasca or Early Nasca period and slightly expanded by the Middle Horizon or Late Intermediate Period, the pre-Columbian irrigation system in the Palpa Valley had reached an extent similar to the area under cultivation shown by the 1944 aerial photographs. However, unlike in some coastal valleys in northern Peru (MOSELEY 1983: 778), irrigation in the Palpa Valley has never been expanded beyond its modern limits. With sedimentation rates of 0.4–3.0 mm/a, the long history of sedimentation from irrigation agriculture amounts to a raise in land surface elevation by one to more than three metres and an estimated accumulation of 25–100 Mio. t of irragric anthrosol. Thus, human action has to be considered an important geomorphic force in the Palpa Valley. The construction of a system of irrigation canals and the accumulation of fine-grained sediment constitute the long-term creation of landscape capital (cf. ERICKSON 1999, 2000) – the agricultural landscape as we see it today is an artefact created by more than three thousand years of human activity. Irrigation agriculture as it is practised in the Palpa Valley represents a sustainable adaptation to local environmental conditions.

However, while it is possible to quantify the impact of both natural dynamics and human action on the landscape, the available data do not allow a quantification of the vulnerability of pre-Columbian populations with respect to long-term environmental changes or short-term extreme events. Given the long-term persistence of human occupation in the Palpa Valley and the findings of this thesis regarding the history of irrigation agriculture, human–environment interactions have been characterised by adaptation to the prevailing conditions and their variability through large-scale (albeit gradual) modifications of the environment. Opportunistic and supplementary use of resources (i.e. expansion of resource use where and when possible and necessary) likely was an important factor throughout prehistory. It is exemplified by the maintenance of resource relationships with the coast (e.g. the presence of marine mollusc shells in the Palpa Valley; see SILVERMAN 2002: 152) and the presence of possible water-harvesting structures reported by MÄCHTLE (2007).

Relationships between climatic changes or extreme events and cultural change can not be deduced from the findings of this research. While destructive flood events have affected at most one third of the valley floor area since ≥1,500 a, the spatial extent of simultaneous destruction can not be quantified due to dating uncertainty. The identification of droughts from the sedimentological record is largely impossible; in particular they can not be quantified in terms of severity and duration. However, the quantification of recent groundwater dynamics suggests that before the advent of mechanised pumping, groundwater table lowering even during multi-year droughts may have been small, thus allowing a (albeit curtailed) continuation of agricultural production. El Niño events can be inferred from sediments recording debris flow events originating in local catchments (e.g. La Muña for the time since 910–730 cal BP); however, large-scale destruction of irrigation systems can not be demonstrated. To assess the potential impact of any of the described environmental perturbations on pre-Columbian societies, a quantification of physical and social vulnerability is necessary. However, this is impeded by the lack of reliable data on population sizes and carrying capacities.

The analyses of present-day fluvial dynamics as well as land use and irrigation systems show that in particular the hydraulic engineering measures of the past decades and the intense groundwater use may not be sustainable given the high inherent dynamics of the fluvial system. Anthropogenic narrowing and straightening has led to incision of the Rio Vizcas which has necessitated upstream displacement of irrigation canal intakes by between 70 and 310 m. Narrowing also impacts the recharge of the valley aquifer by drastically reducing the riverbed infiltration surface. Regarding flood protection, the artificial gravel embankments are largely inefficient given their high vulnerability to erosion even at moderate discharges. The economic needs for flood protection and reclamation of agricultural land (inefficient as the current measures may be) on one hand and optimal irrigation and domestic water supply on the other hand are not compatible. This exemplifies for the present what applies to pre-Columbian through historic times: Adaptation to environmental conditions is not a straightforward process including all relevant variables. Rather, it is based on the cumulative decision of individuals embedded in certain cultures and equipped with certain knowledge and technologies. Their respective interests are negotiated in the social and economic spheres – in any case the outcome is neither environmentally determined nor solely driven by cultural/social or technological factors.

Zusammenfassung

Das Ziel der in dieser Dissertation vorgestellten Forschungen war die Rekonstruktion der holozänen Landschaftsentwicklung am Rio Grande de Nazca mit dem Schwerpunkt auf Mensch-Umwelt-Interaktionen. Während die Arbeit in erster Linie auf die jungholozäne Entwicklung abzielte, führten die Notwendigkeit des Verständnisses von längerfristig ablaufenden Prozessen und Gelegenheitsfunde zu Rekonstruktionen der Paläo-Umwelt, welche die vergangenen 30 000 Jahre umfassen.

Die Rekonstruktion der langfristigen Landschaftsentwicklung basiert auf einer Sequenz von vier stratigraphisch konsistenten ^{14}C-Datierungen aus einem Murfächer etwa 15 Kilometer südwestlich von Palpa. Während des letzten Glazials führten Murereignisse, welche auf lokale Starkniederschlagsereignisse zurückzuführen sind, dort zur Aufschüttung. Das LGM war sowohl durch autochthone Mur- als auch durch allochthone Hochflutsedimentation gekennzeichnet. Erst nach der daran anschließenden Einschneidung des lokalen Trockentals kam es zur Eintiefung des Flussbettes auf das rezente Niveau. Morphologische Ähnlichkeiten mit anderen Trockentälern deuten an, dass diese Sequenz von regionaler Bedeutung ist. Die Ein-schneidung des Rio Grande de Nazca kann auf $\leq$6 m seit 30 ^{14}C ka BP eingegrenzt werden, wobei der größte Teil der Einschneidung nach dem LGM, vermutlich während des Holozäns, stattgefunden hat.

Die natürliche Dynamik im Palpa-Tal vor dem Beginn der Landwirtschaft war durch die Ablagerung von grobkörnigen Flussbettsedimenten, feinkörnigen fluvialen Sedimente sowie von *hyperconcentrated flow deposits* gekennzeichnet. Diese Sedimente deuten auf eine mittel-holzäne Dynamik im Palpa-Tal vergleichbar mit der durch oberflächennahe Sedimente repräsentierten rezenten Dynamik im Quebrada-Yapona-Tal hin. Dieses unterscheidet sich vom Palpa-Tal durch die geringen Abflussmengen aufgrund der geringen Ausdehnung des abflussrelevanten Anteils (>2 500 m) seines Einzugsgebietes. Die mittelholozäne Trockenperiode in den Hochanden, welche mit verringertem Abfluss in den Fremdlingsflüssen verbunden gewesen sein muss und durch Phasen mit Starkniederschlagsereignissen unterbrochen wurde, welche zu erosiven und akkumulativen Abflussereignissen sowie *hyperconcentrated flows* geführt haben, ist damit in den Sedimenten des Palpa-Tals belegt.

Für das späte Holozän zeigt die Analyse des Einflusses der fluvialen Dynamik auf die Landschaftsentwicklung des Palpa-Tals, dass weite Teile des Talbodens (ca. 60%) seit min-destens 1 500 cal BP nicht fluvial erodiert worden sind. Zwar sind die Flüsse durch starke la-terale Aktivität gekennzeichnet und es kam vor allem im mittleren Teil des Tales zwischen und seit 1 500 und 1 000 cal BP zu fluvialer Erosion, diese laterale Aktivität war jedoch im Wesentlichen auf relativ schmale Bereiche entlang der Flussläufe beschränkt. In diesen Bereichen kam es zu wiederholter Erosion und Sedimentation; die hier relativ geringen Alter der Sedimente repräsentieren nicht notwendigerweise eine spezifische Phase fluvialer Aktivität.

Der Einfluss des Menschen auf die jungholozäne Landschaftsentwicklung kann über die Rekonstruktion der Ausdehnung des Bewässerungssystems beurteilt werden. Bewässerungslandwirtschaft im Palpa-Tal wird bereits für die Initial-Zeit durch das Vorhandensein von *irragric anthrosol* belegt. Diese zeichnen sich durch Indikatoren für intensive Turbation (z.B. das Fehlen von Schichtung oder Laminationen), erhöhten Organikgehalt und erhöhtes Ton: Schluff-Verhältnis sowie das Vorhandensein von Holzkohle und Artefakten aus. Während das Alter möglicher *irragric anthrosols* (allerdings ohne Artefakte) mit maximal 5 660–3 910 cal BP bis ins Archaikum extrapoliert werden kann, ist das extrapolierte Alter der Basis des ältesten sicher identifizierten *irragric anthrosol* 4 900–3 500 cal BP. Der Beginn der Bewässerungslandwirtschaft fällt damit in etwa auf das Ende der mittelholozänen Trockenperiode in den Anden. Der früheste *irragric anthrosol* kommt im zentralen Talbereich (zwischen Rio Palpa und Rio Vizcas) vor.

Zur frühen bis mittleren Paracas-Zeit war das Bewässerungssystem bereits auf den Bereich südöstlich des Rio Vizcas ausgedehnt worden. Der höchstgelegene Bewässerungskanal entlang des südöstlichen Talrandes (am Fuß der Pampa San Ignacio) wurde etwa 2 720–2 360 cal BP angelegt, wie die Datierung des Beginns der Ablagerung von *irragric anthrosol* zeigt. In Anlehnung an die rezente Situation betrug die Länge dieses Kanals etwa 3,8 Kilometer, was die Schaffung eines Bewässerungssystems im großen Maßstab belegt. Die letzte große Fläche, welche in das Bewässerungssystem einbezogen wurde, befindet sich nordwestlich des Rio Palpa. Die Hauptterrasse auf dieser Talseite wird seit der Initial-Nasca- oder der frühen Nasca-Zeit bewässert. Entlang des nordwestlichen Talrandes fällt die Anlage des höchstgelegenen Kanals (am Fuß der Cresta Sacramento) in die Zeit des Mittleren Horizontes oder der Späten Zwischenperiode. Im benachbarten Tal des Rio Grande wurde der höchstgelegene Bewässerungskanal während der Späten Zwischenperiode angelegt. Im Gegensatz zu der Annahme, dass die präkolumbianischen Bewässerungssysteme im Einzugsgebiet des Rio Grande de Nazca eine wesentlich geringere Ausdehnung hatten als die heutigen (Silverman 1996: 102) zeigen die Befunde aus dem Palpa-Tal, dass zur Initial-Nasca- oder frühen Nasca-Zeit und geringfügig erweitert zum Mittleren Horizontes oder zur Späten Zwischenperiode das präkolumbianische Bewässerungssystem eine Ausdehnung ähnlich der in den Luftbildern aus dem Jahr 1944 erkennbaren erreicht hatte. Im Unterschied zum Norden Perus (Moseley 1983: 778) überschritt die Bewässerung hier allerdings nie die moderne Ausdehnung.

Mit Sedimentationsraten von 0,4–3,0 mm/a hat die langandauernde Bewässerungsland-wirtschaft zu einer Anhebung der Landoberfläche um einen bis über drei Meter und zu einer geschätzten Ablagerung von 25–100 Millionen Tonnen *irragric anthrosol* geführt. Damit gehört die Aktivität des Menschen zu den wichtigen geomorphischen Faktoren im Palpa-Tal. Die Schaffung eines Systems von Bewässerungskanälen und die Ablagerung von feinkörnigem Material sind als Schaffung von Landschaftskapital (*landsape capital*, cf. ERICKSON 1999, 2000) anzusehen – die heutige agrarische Landschaft ist ein durch über 3 000–jährige Aktivität des Menschen geschaffenes Artefakt. Bewässerungslandwirtschaft wie sie im Palpa-Tal praktiziert wird stellt eine nachhaltige Anpassung an die lokalen Umweltbedingungen dar.

Während es möglich ist, den Einfluss der natürlichen Dynamik sowie der Aktivität des Menschen auf die Landschaft zu quantifizieren, erlauben die vorhandenen Daten keine Quantifizierung der Vulnerabilität präkolumbianischer Gesellschaften bezüglich langfristiger Umweltveränderungen oder kurzfristiger Extremereignisse. Angesichts der langen Persistenz der Besiedlung des Palpa-Tals und der vorgestellten Befunde bezüglich der Rekonstruktion der Bewässerungslandwirtschaft ist festzustellen, dass die Mensch-Umwelt-Interaktionen durch Anpassung an die vorherrschenden Bedingungen und deren Variabilität mittels ausgedehnter, wenn auch allmählicher, Umgestaltungen der Umwelt charakterisiert sind. Opportunistische und ergänzende Ressourcennutzung (d.h. die Ausweitung der Ressourcennutzung wo und wann möglich und notwendig) waren vermutlich während der gesamten präkolumbianischen Zeit von Bedeutung. Dies wird durch die Unterhaltung von Ressourcenbeziehungen zur Küste (z.B. belegt durch die Präsenz von marinen Molluskenschalen im Palpa-Tal; siehe auch SILVERMAN 2002: 152) und die Existenz von möglichen Anlagen des *water harvesting* (MÄCHTLE 2007) angedeutet.

Beziehungen zwischen klimatischen Veränderungen oder Extremereignissen auf der einen und Kulturwandel auf der anderen Seite können aus den Ergebnissen der vorliegenden Arbeit nicht abgeleitet werden. Zwar haben destruktive Flutereignisse seit ≥1 500 cal BP bis zu einem Drittel des Talbodens betroffen, die räumliche Ausdehnung gleichzeitiger Auswirkungen kann allerdings aufgrund von Datierungsungenauigkeiten nicht quantifiziert werden. Die sedimentologische Identifikation von Dürreperioden ist generell problematisch; insbesondere betrifft dies quantitative Aussagen bezüglich Intensität und Dauer der Dürre. Allerdings kann auf-grund der Quantifizierung der rezenten Grundwasserdynamik davon ausgegangen werden, dass vor der Einführung motorgetriebener Pumpen die Grundwasserabsenkung im Palpa-Tal selbst während mehrjähriger Dürreperioden gering gewesen ist. Eine – wenn auch eingeschränkte – Fortsetzung der Landwirtschaft war somit vermutlich möglich. Das Auftreten von El Niño-Ereignissen kann aus Murablagerungen lokaler Einzugsgebiete abgeleitet werden (z.B. La Muña für die Zeit seit 910–730 cal BP). Allerdings sind umfangreiche Zerstörungen des Bewässerungssystems nicht nachweisbar. Um den potentiellen Einfluss der beschriebenen Umwelterscheinungen auf präkolumbianische Gesellschaften einschätzen zu können, ist eine Quantifizierung der physischen und sozialen Vulnerabilität erforderlich. Diese wird allerdings durch einen Mangel an belastbaren Daten bezüglich Bevölkerungsdichte und Tragfähigkeit verhindert.

Die Analyse der rezenten fluvialen Dynamik sowie der Landnutzung und des Bewässerungssystems zeigt, dass insbesondere die wasserbaulichen Maßnahmen der vergangenen Jahrzehnte und die intensive Grundwassernutzung angesichts der inhärenten Dynamik des fluvialen Systems möglicherweise nicht nachhaltig sind. Die anthropogene Einengung und Begradigung hat zu einer Einschneidung des Rio Vizcas geführt und damit die Verlegung von Bewässerungskanal-Einlässen um 70 bis 310 m flussaufwärts erforderlich gemacht. Die Einengung wirkt sich darüber hinaus durch die drastische Verringerung der Versickungsfläche in Flussbett negativ auf die Grundwassererneuerung aus. Bezüglich des Hochwasser- bzw. Erosionsschutzes ist festzustellen, dass die aus Schotter bestehenden Eindeichungen selbst moderaten Abflüssen kaum standhalten. Die wirtschaftlichen Notwendigkeiten von Hochwasser- und Erosionsschutz sowie Landgewinnung einerseits und optimaler Wasserversorgung für Landwirtschaft und Bevölkerung auf der anderen Seite sind nicht kompatibel. Dies zeigt exemplarisch für die Gegenwart, was für präkolumbianische wie auch historische Zeiten gilt: Die Anpassung an Umweltbedingungen ist kein geradliniger Prozess, in dem alle relevanten Variablen zur Geltung kommen. Vielmehr basiert sie auf den kumulativen Entscheidungen von Individuen, welche in bestimmte Kulturen eingebettet und mit bestimmten Technologien ausgestattet sind. Ihre jeweiligen Interessen werden auf sozialem und wirtschaftlichem Gebiet ausgeglichen – in jedem Fall ist das Ergebnis weder umweltdeterminiert noch ausschließlich durch kulturelle oder technologische Faktoren bestimmt.

Resumen

El objetivo de la investigación presentada en esta tesis es reconstruir la evolucion del paisaje cultural en la cuenca del Rio Grande de Nazca durante el Holoceno, teniendo como enfasis las interacciones poblaciones precolombinas y el ambiente. Mientras este trabajo se centra principalmente el Holoceno tardio, la necesidad de comprender los procesos de trabajo en un espacio de tiempo mas largo y los hallazgos casuales han conducido a reconstrucciones paleoambientales englobando los ultimos 30 mil años.

La evolución larga del ambiente es reconstruida usando una secuencia de cuatro fechas radiocarbonos del cono de deyección de una quebrada ubicado aproximadamente a 15 kilómetros al sudoeste de Palpa. Durante el último período glacial, flujos hiperconcentrados gruesos atribuibles a precipitaciones intensas locales han ocasionado la deposición del cono. El ultimo periodo glacial máximo fue caracterizado por ambas deposiciónes de flujos torrenciales locales y deposiciónes fluviales atribuibles a crecidas del Río Grande. Después de la incisión siguiente de la quebrada, el Río Grande se ha inciso hasta el nivel actual. Las semejanzas morfológicas con otras quebradas sugieren que esta secuencia tiene una relevancia regional. La incisión fluvial del Río Grande de Nazca fue ≤6 metros desde 30 mil ^{14}C BP. La mayoridad de la incisión ocurrió después del máximo del ultimo periodo glacial, probablemente durante el Holoceno.

La dinámica natural en el valle de Palpa antes del comienzo de la agricultura esta registrada por sedimentos fluviales gruesos (atribuidos a los cauces de los ríos), sedimentos fluviales finos (atribuidos a las crecidas de los ríos y a corrientes disminuidas) y sedimentos de flujos hiperconcentrados. Estos sedimentos indican dinámicas en el valle de Palpa durante el Holoceno medio siendo similares a las presentadas en la quebrada Yapona. La Quebrada Yapona se caracteriza por presentar caudales muy escasos debido a la menor dimensión de su área sobrepasa 2 500 metros donde hay precipitaciones relevantes. Así, la fase árida den Holoceno medio en las cordilleras altas es representada en el valle de Palpa. Esta fase árida a ocasionadó caudales disminuidos en los ríos alógenos del desierto costero, siendo interrumpido por periodos cortos de precipitaciónes fuertes provocando tanto eventos de erosión y deposición fluvial como de flujos hiperconcentrados.

Para el Holoceno tardío, los análisis cuantitativos concerniento al impacto de la dinámica fluvial en el valle de Palpa demuestran que una gran parte del fondo del valle (60% approx.) no ha sido afectado por la erosión fluvial desde ≥1 500 cal BP. Mientras partes del valle han sido afectadas por la erosión fluvial entre y despues 1 500 y 1 000 cal BP y los ríos estan caracterizados por una actividad lateral fuerte, la mayoría de esta actividad fue limitada a franjas relativamente estrechas a lo largo de los ríos. Estas franjas estaban sujetas a repeticiones de erosión y sedimentación. Los fechoados relativamente recientes de los sedimentos en estas franjas no necesariamente señalan fases específicas de la actividad fluvial.

El impacto de la actividad humana en la evolución de paisaje durante el Holoceno tardío se puede determinar mediante la reconstrucción de la expansión del sistema de la irrigación. En el valle de Palpa, la agricultura con irrigación manifesta en el Periodo Inicial de la presencia de antrosoles irrágricos. Estos son caracterizados por indicadores de turbació intensa (por ejemplo la falta de estratificación), valores aumentados de carbón organico y de la proporción de arcilla: limo y la presencia de carbón y artefactos. Mientras la edad de antrosoles irrágricos probables (pero sin artefactos) se pueden extrapolar al Periodo Arcaico con edades máximas de 5 660–3 910 cal BP, la edad máxima del mas antiguo antrosol irrágrico seguro se pueden extrapolar a 4 900–3 500 cal BP. Este antrosol se encuentran en el valle central (entre los ríos Palpa y Vizcas). En el Periodo Paracas temprano o medio, el sistema de irrigación se ha expandido acoplando terrenos ubicados al sudeste del Río Vizcas. El mas alto canal de irrigación precolombiano a lo largo del borde al sudeste del valle de Palpa (al pie de la Pampa de San Ignacio) se ha construido aproximadamente 2 720–2 360 cal BP como indica el comienzo de la deposición de antrosol irrágrico. La extensión de este canal alcanzó aproximadamente de 3.8 kilómetros como sugiere una comparación con la situación actual e indica la construcción a gran escala de infraestructura para irrigación durante el Periodo Paracas. La ultima gran área incluida en el sistema de irrigación era el terreno al noroeste del Río Palpa donde en la terraza principal se ha comenzado la agricultura de irrigación en el Periodo Nasca Inicial o Nasca temprano.
A lo largo del borde noroeste del valle la construcción del canal mas alto al pie de la Cresta Sacramento esta reflejada en el comienzo de la irrigación en el Horizonte Medio o el Periodo Intermedio Tardío. Contrario a la suposición de que los sistemas de irrigación precolombianos en la cuenca del Río Grande de Nazca estaban menos extendidos que en el presente (Silverman 1996: 102), en el Periodo Nasca Inicial o Nasca temprano y mínimamente aumentado en el Horizonte Medio o el Periodo Intermedio Tardío el sistema precolombiano de la irrigación en el valle de Palpa llegó a una dimensión semejante al área cultivada mostrada en las imágenes aéreas del año 1944. Sin embargo, a diferencia de algunos valles en el norte del Perú (Moseley 1983: 778) la irrigación en el valle de Palpa nunca se ha expandido más allá de los limites actuales.

Con tasas entre 0.4 y 3.0 milímetros por año, la sedimentación por irrigación se asciende a un aumento en la elevación de la superficie de uno a mas que tres metros y una acumulación de antrosol irrágrico estimado a 25–100 millones toneladas. Así, actividades humanas deben ser consideradas una importante fuerza geomórfica en el valle de Palpa. La construcción de un sistema de canales para irrigación y la deposición de sedimento fino constituyen la creación de capital del paisaje (*landscape capital*, cf. Erickson 1999, 2000) – el paisaje agrícola como lo vemos hoy día es un artefacto creado en mas que tres mil años de actividades humanas. La agricultura de irrigación en el valle de Palpa representa una adaptación sostenible a las condiciones ambientales.

No obstante, mientras el impacto de las dinámicas naturales y la acción humana sobre el paisaje puede ser cuantificado, los datos disponible no permiten la cuantificación de la vulnerabilidad de poblaciones precolombianas con respecto a cambios largos de las condiciones ambientales o eventos extremos. Considerando la larga persistencia de la ocupación del valle de Palpa y los resultados de esta tesis concerniente a la historia de la agricultura de irrigación, las interacciones entre sociedades precolombianas y el medio ambiente estaban caracterizados por la adaptación a las condiciones prevalecientes y su variabilidad mediante modificaciones a gran escala del medio ambiente. Uso oportunista y suplementario de recursos (por ejemplo expansión del uso de recursos donde y cuando posible y necesario) probablemente fué un factor importante a través la prehistoria. Eso es ejemplificado por las relaciones concernientes a los recursos del mar (indicado por ejemplo por las conchas marinas encontradas en el valle de Palpa; ver Silverman 2002: 152) y por la presencia de estructuras posiblemente relacionadas a la agricultura reportadas por Mächtle (2007).

Relaciones entre cambios climáticos o eventos extremos y cambios culturales no se pueden deducir de los resultados de esta investigación. Mientras eventos fluviales destructivos han afectado a lo máximo un tercio del fondo del valle desde ≥1 500 cal BP, la extensión espacial de la destrucción contemporánea no se pueden cuantificar a causa de inseguridad de fechado. La identificación en el registro sedimentario de sequías es casi imposible, particularmente la cuantificación concerniente a la severidad y duración. Sin embargo, la cuantificación de la dinámica del agua subterránea sugiere que antes de la venida de bombas mecánicas, el bajar del nivel friático estaba escaso aun durante sequías plurianuales. Por tanto una continuación de la producción agraria (aunque reducido) puede haber sido posible. Eventos El Niño se pueden deducir de sedimentos de flujos hiperconcentrados originado de cuencas locales (por ejemplo La Muña para el tiempo desde 910–730 cal BP). Sin embargo, la destrucción a gran escala de sistemas de la irrigación no se pueden demostrar. Para evaluar el impacto potencial a las poblaciones precolombianas de cualquiera de las perturbaciones ambientales mencionados, es necesario una cualificación de la vulnerabilidad física y social. Sin embargo, esto es impedido por la falta de datos seguros concerniente a los tamaños de las poblaciones y las capacidades sustentadoras.

Los análisis de la recién dinámica fluvial y de los sistemas de irrigación y la utilización de la tierra, indican que particularmente las medidas de la ingeniería hidráulica durante las décadas pasadas y el uso intenso del agua subterránea no puede ser sostenible considerando la dinámica fuerte inherente en el sistema fluvial. Medidas de la ingeniería hidráulica para estrechar y poner derechos los ríos ha resultado de la incisión del Río Vizcas que ha necesitado la dislocación de las entradas de cuatro canales de irrigación por 70 a 310 metros río arriba. Estrechando los ríos también influye la recarga del acuífero del valle mediante la reducción drástica de la superficie de infiltración en los cauces de los ríos. Concerniente a la protección de crecidas, las contenciones construidas de grava gruesa en gran parte no son efectivas como indicado por su gran vulnerabilidad a la erosión, aún con caudales moderados. Las necesidades económicas para la protección de crecidas y para la utilización de la tierra (poco efectivos como las medidas pueden ser) por un lado y por el aprovisionamiento optimo de agua para irrigación y uso domestico por el otro lado no son compatibles. Eso ejemplifica para el presente que aplica a tiempos precolombianos hasta históricos: La adaptación a condiciones ambientales no es un proceso simple o rectilíneo incluyendo todas los variables relevantes. Más sin embargo, es un proceso basado en decisiones cumulativas de individuales empotrado en su cultura específica y equipado con tecnologías específicas. Sus interés son negociados en las esferas sociales y económicas – de igual manera el resultado de los procesos de adaptación no es determinado por al ambiental ni exclusivamente motivado por factores culturales/sociales o tecnológicos.

References

Abbott, M.B., Binford, M.W., Brenner, M. & Kelts, K. (1997): A 3500 ^{14}C yr high resolution record of lake level changes in Lake Titicaca, South America. ***Quaternary Research***, 47, 169–180.

Abbott, M.B., Wolfe, B.B., Aravena, R., Wolfe, A.P. & Seltzer, G.O. (2000): Holocene hydrological reconstructions from stable isotopes and paleolimnology, Cordillera Real, Bolivia. ***Quaternary Science Reviews***, 19, 1,801–1,820.

Abele, G. (1992): Landforms and climate on the western slope of the Andes. ***Zeitschrift für Geomorphologie***, Supplement-Band 84, 1–11.

Aceituno, P. (1988): On the functioning of the Southern Oscillation in the South American sector. ***Monthly Weather Review***, 116, 505–524.

AG Boden (2005): Bodenkundliche Kartieranleitung. 5. Auflage. Schweizerbart: Stuttgart.

Alabyan, A.M. & Chalov, R.S. (1998): Types of river channel patterns and their natural controls. ***Earth Surface Processes and Landforms***, 23, 467–474.

Alpers, C.N. & Brimhall, G.H. (1988): Middle Miocene climatic change in the Atacama desert, northern Chile: evidence for supergene mineralization at La Escondida. ***Bulletin of the Geological Society of America***, 100(10), 1,640–1,656.

Ammann, C., Jenny, B., Kammer, K. & Messerli, B. (2001): Late Quaternary glacier response to humidity changes in the arid Andes of Chile (18–29°S). ***Palaeogeography, Palaeoclimatology, Palaeoecology***, 172, 313–326.

Anonymous (2007a): Los huaicos llegaron a escasos dos metros de trapecios Nasca. ***El Comercio***, Lunes, 19 de febrero de 2007. Edición Impresa – Economia. http://www.elcomercio.com.pe/EdicionImpresa/Html/2007-02-19/ImEcNacional0673893.html (download: 02.05.2007).

Anonymous (2007b): Vientos desplazaron nubes hacia la costa. ***El Comercio***, Lunes, 19 de febrero de 2007. Edición Impresa – Perú. http://www.elcomercio.com.pe/EdicionImpresa/Html/2007-02-19/ImEcNacional0674040.html (download: 02.05.2007).

Anonymous (2007c): Envían 27 toneladas a afectados por huaicos. ***El Comercio***, Martes, 20 de febrero 2007. Edición Impresa – Perú. http://www.elcomercio.com.pe/edicionimpresa/html/2007-02-20/imecnacional0674673.html (download: 02.05.2007)

Arancibia, G., Matthews, S. & de Arce, C.P. (2005): K–Ar and $^{40}Ar/^{39}Ar$ ages from supergene minerals in northern Chile: Prevalence of humid climate and tectonic uplift until the Upper Miocene in the Atacama Desert. In: 6th International Symposium on Andean Geodynamics (ISAG 2005, Barcelona), Extended Abstracts, 50–52.

Aveni, A.F. (1990): An assessment of previous studies of the Nazca Lines. In: Aveni, A.F.: The Lines of Nasca. Memoirs of the American Philosophical Society, 183. Philadelphia. 1–40.

Baade, J. (2001): Jungholozäne fluviale Dynamik und Kulturlandschaftsentwicklung am Rio Grande de Nazca, Süd-Peru (unpublished project grant application).

Baied, C.A. & Wheeler, J.C. (1993): Evolution of high Andean Puna ecosystems: Environment, climate, and culture change over the last 12,000 years in the central Andes. ***Mountain Research and Development***, 13(2), 45–156.

Baker, P.A., Seltzer, G.O., Fritz, S.C., Dunbar, R.B., Grove, M.J., Tapia, P.M., Cross, S.L., Rowe, H.D. & Broda, J.P. (2001): The history of South American tropical precipitation for the past 25,000 years. ***Science***, 291, 640–643.

Barber, R.T. & Chavez, F.P. (1983): Biological consequences of El Niño. ***Science***, 222, 1,203–1,210.

Baucom, P.C. & Rigsby, C.A. (1999): Climate and lake-level history of the northern Altiplano, Bolivia, as recorded in Holocene sediments of the Rio Desaguadero. ***Journal of Sedimentary Research***, 69(3), 597–611.

Beaumont, P. (1998): An ill wind: child of destruction. ***The Observer*** (London). March 8, 1998, 11.

Beresford-Jones, D.G. (2004): Pre-hispanic Prosopis-human relationships on the south coast of Peru: riparian forests in the context of environmental and cultural trajectories of the lower Ica valley. Dissertation, University of Cambridge.

Berninger, O. (1925): Extreme Ausbildung einer Nebelvegetation in der nordchilenischen Wüste. ***Zeitschrift der Gesellschaft für Erdkunde***, 383–384.

Betancourt, J.L., Latorre, C., Rech, J.A., Quade, J. & Rylander, K.A. (2000): A 22,000-year record of monsoonal precipitation from nothern Chile's Atacama desert. ***Science***, 289, 1,542–1,546.

Billman, B.R. (1999): Reconstructing prehistoric political economies and cycles of political power in the Moche valley, Peru. In: Billman, B.R. & Feinman, G.M.: Settlement pattern studies in the Americas. Fifty years since Virú. Smithsonian Institution Press: Washington. 131–159.

Blum, M.D. & Törnqvist, T.E. (2000): Fluvial respones to climate and sea-level change: a review and look forward. ***Sedimentology***, 47(Suppl. 1), 2-48.

Bobst, A.L., Lowenstein, T.K., Jordan, T.E., Godfrey, L.V., Ku, T.-L. & Luo, S. (2001): A 106 ka paleoclimate record from drill core of the Salar de Atacama, northern Chile. ***Palaeogeography, Palaeoclimatology, Palaeoecology***, 173, 21–42.

Bonavia, D. & Chauchat, C. (1990): Presencia del Paijanense en el desierto de Ica. ***Bulletin de l'Institute Français d'Études Andines***, 19(2), 399–412.

BOND, G., SHOWERS, W., CHESEBY, M., LOTTI, R., ALMASI, P., DEMENOCAL, P., PRIORE, P., CULLEN, H., HAJDAS, I. & BONANI, G. (1997): A pervasive millennial-scale cycle in North Atlantic Holocene and Glacial climates. ***Science***, 278, 1,257–1,266.

BROGGI, J.A. (1946): Las terrazas marinas de la Bahia de San Juan en Ica. ***Boletin de la Sociedad Geológica del Perú***, 19, 21–33.

BRONK RAMSEY, C. (1995): Radiocarbon calibration and analysis of stratigraphy: the OxCal program. ***Radiocarbon***, 37(2), 425–430.

BRONK RAMSEY, C. (2001): Development of the radiocarbon calibration program. ***Radiocarbon***, 43(2A), 355–363.

BROOKS, N. (2004): Beyond collapse: the role of climatic desiccation in the emergence of complex societies in the middle Holocene. In: LEROY, S. & COSTA, P.: Environmental catastrophes in Mauritania, the desert and the coast. Abstract volume and field guide. Mauritania, 4–18 January 2004. First joint meeting of ICSU dark nature and IGCP 490. 26–30.

BROOKS, N. (2006): Cultural responses to aridity in the Middle Holocene and increased social complexity. ***Quaternary International***, 151(1), 29–49.

BROWNE, D.M. (1992): Further archaeological reconnaissance in the Province of Palpa, Department Ica, Peru. In: SAUNDERS, N.J.: Ancient America: contributions to New World Archaeology. Oxbow Monograph, 24. Oxbow: Oxford. 77–116.

CANE, M.A. (1983): Oceanographic events during El Niño. ***Science***, 222, 1,189–1,195.

CANE, M.A., CLEMENT, A., GAGAN, M.K., AYLIFFE, L.K. & TUDHOPE, S. (2000): ENSO through the Holocene, depicted in corals and a model simulation. ***PAGES News***, 8(1), 3–7.

CARNEIRO, R.L. (1977): A theory of the origin of the state. Studies in Social Theory No. 3. Institute for Humane Studies: Menlo Park. 3–21.

CAVIEDES, C. (1973): A climatic profile of the North Chilean desert at latitude 20°S. In: AMIRAN, D.H.K. & WILSON, A.W.: Coastal deserts, their natural and human environments. University of Arizona Press: Tuscon. 115–121.

CAVIEDES, C.N. (1975): El Niño 1972: its climatic, ecological, human, and economic implications. ***Geographical Review***, 65(4), 493–509.

CAVIEDES, C.N. (1982): On the genetic linkages of precipitation in South America. ***Freiburger Geographische Hefte*** (18), 55–77.

CAVIEDES, C.N. (1984a): Geography and the lessons from El Niño. ***Professional Geographer***, 36(4), 428–436.

CAVIEDES, C.N. (1984b): El Niño 1982–83. ***Geographical Review***, 74, 267–290.

CLAPPERTON, C.M. (1993): Nature of environmental changes in South America at the Last Glacial Maximum. ***Palaeogeography, Palaeoclimatology, Palaeoecology***, 101(3–4), 189–208.

CLARKE, J.D.A. (2006): Antiquity of aridity in the Chilean Atacama Desert. ***Geomorphology***, 73, 101–114.

CLEMENT, C.O. & MOSELEY, M.E. (1991): The spring-fed irrigation system of Carrizal, Peru: A case study of the hypothesis of agrarian collapse. ***Journal of Field Archaeology***, 18, 425–443.

COHEN, M.N. (1975): Population pressure and the origins of agriculture: an archaeological example from the coast of Peru. In: POLGAR, S.: Population, ecology, and social evolution. Mouton Publishers: The Hague. 79–121.

COOK, N.D. (1981): Demographic collapse: Indian Peru, 1520–1620. Cambridge Latin American Studies 41. Cambridge University Press: Cambridge.

CRAIG, A.K. (1992): Archaeological occurrences of Andean land snails. ***Andean Past***, 3, 127–136.

CRAIG, A.K. & PSUTY, N.P. (1968): The Paracas Papers: Studies in marine desert ecology, Vol. 1. Occasional Papers, Department of Geography. Florida Atlantic University: Boca Raton.

CRAIG, A.K. & SHIMADA, I. (1986): El Niño flood deposits at Batán Grande, northern Peru. ***Geoarchaeology***, 1, 29–38.

CROSS, S.L., BAKER, P.A., SELTZER, G.O., FRITZ, S.C. & DUNBAR, R.B. (2000): A new estimate of the Holocene lowstand level of Lake Titicaca, central Andes, and implications for tropical palaeohydrology. ***The Holocene***, 10(1), 21–32.

DE ABREU, M.L. & BANNON, P.R. (1993): Dynamics of the South American coastal desert. ***Journal of the Atmospheric Sciences***, 50(17), 2,952–2,964.

DEFRANCE, S.D. & ÁLVAREZ, A.U. (2004): Quebrada Tacahuay: un sitio marítimo del Pleistoceno Tardío en la costa sur del Perú. ***Revista de Antropología Chilena***, 36(2), 257–278.

DEMENOCAL, P.B. (2001): Cultural responses to climate change during the late Holocene. ***Science***, 292, 667–673.

DENTON, G.H., HEUSSER, C.J., LOWEL, T.V., MORENO, P.I., ANDERSEN, B.G., HEUSSER, L.E., SCHLUHTER, C. & MARCHANT, D.R. (1999): Interhemispheric Linkage of Paleoclimate During the Last Glaciation. ***Geografiska Annaler: Series A, Physical Geography***, 81(2), 107–153.

DEUTSCHES INSTITUT FÜR NORMUNG (1973): DIN 19683, Blatt 2: Bestimmung der Korngrößen-zusammensetzung nach Vorbehandlung mit Natriumpyrophosphat. Berlin.

DEUTSCHES INSTITUT FÜR NORMUNG (1977): Bodenuntersuchungensverfahren im landwirtschaftlichen Wasserbau. Chemische Laboruntersuchungen. DIN 19684, Teil 3: Bestimmung des Glühverlustes und des Glührückstandes. Berlin.

DEUTSCHES INSTITUT FÜR NORMUNG (1993): Bodenbeschaffenheit. DIN ISO 10693: Bestimmung des Carbonatgehaltes. Volumetrisches Verfahren. Berlin.

DEVRIES, T.J. & WELLS, L.E. (1990): Thermally-anomalous Holocene molluscan assemblages from coastal Peru: evidence for paleogeographic, not climate change.

Palaeogeography, Palaeoclimatology, Palaeoecology, 81, 11–32.

DILLEHAY, T.D., ELING JR., H.H. & ROSSEN, J. (2005): Preceramic irrigation canals in the Peruvian Andes. ***PNAS***, 107(47), 17,241–17,244.

DILLEHAY, T.D. & KOLATA, A.L. (2004): Long-term human response to uncertain environmental conditions in the Andes. ***PNAS***, 101(12), 4,325–4,330.

DILLEHAY, T., KOLATA, A.L. & PINO Q., M. (2004): Pre-industrial human and environment interactions in northern Peru during the late Holocene. ***The Holocene***, 14(2), 272–281.

DILLON, M.O. (2005): The solanaceae of the lomas formations of coastal Peru and Chile. In: KEATING, R.C., HOLLOWELL, V.C. & CROAT, T.B.: A Festschrift for William G. D'Arcy, The legacy of a taxonomist. Monographs in systematic botany from the Missouri Botanical Garden Press, Vol. 104. 131–156.

DORNBUSCH, U. (1997): Geomorphologische Untersuchungen zur jungquartären Vergletscherung der Westanden Südperus zwischen. 14°25'S und 15°30'S, daraus ableitbare Bedingungen und deren Vergleich mit neu erarbeiteten Angaben zur rezenten Verteilung von Temperatur, Niederschlag und Schneegrenzhöhen in Südperu südlich 12°S. Dissertation, Fachbereich Geowissenschaften der FU Berlin.

DORNBUSCH, U. (1998): Current large-scale climatic conditions in southern Peru and their influence on snowline altitudes. ***Erdkunde***, 52, 41–54.

DUNNE, T. & LEOPOLD, L.B. (1978): Water in environmental planning. Freeman: New York.

EITEL, B., HECHT, S., MÄCHTLE, B., SCHUKRAFT, G., KADEREIT, A., WAGNER, G.A., KROMER, B., UNKEL, I. & REINDEL, M. (2005): Geoarchaeological evidence from desert loess in the Nazca-Palpa region, southern Peru: palaeoenvironmental changes and their impact on precolumbian cultures. ***Archaeometry***, 47(1), 137–158.

ELLENBERG, H. (1959): Über den Wasserhaushalt tropischer Nebeloasen in der Küstenwüste Perus. ***Bericht über das geobotanische Forschungsinstitut Ruebel in Zürich*** für das Jahr 1958. Vol. 12, 47–74.

ENGEL, F. (1963): Notes relatives a des explorations archaeologiques a Paracas et sur la cote sud du Perou. ***Travaux de l'Institut Français d'Etudes Andines***, Vol. IX.

ENGEL, F. (1973): New facts about pre-Columbian life in the Andean lomas. ***Current Anthropology***, 14(3), 271–280.

ENGEL, F.A. (1981): Prehistoric Andean ecology. Man, settlement and environment in the Andes: 2. The deep south. Humanities Press for Hunter College: New York.

ENGEL, F.A. (1987): De las begonias al maíz, vida y produccion en el Peru antiguo. Universidad Nacional Agraria La Molina: Lima.

ENGEL, F.A. (1991): Un desierto en tiempos prehispanicos – Rio Pisco, Paracas, Rio Ica. Universidad Nacional Agraria La Molina: Lima.

ERICKSON, C.L. (1999): Neo-environmental determinism and agrarian "collapse" in Andean prehistory. ***Antiquity***, 73(281), 634–642.

ERICKSON, C.L. (2000): The Lake Titicaca basin: a precolumbian built landscape. In: LENTZ, D.L.: Imperfect balance. Landscape transformations in the precolumbian Andes. Columbia University Press: New York. 311–356.

EVENSTAR, L., HARTLEY, A., RICE, C., STUART, F., MATHER, A. & CHONG, G. (2005): Miocene-Pliocene climate change in the Peru–Chile Desert. 6th International Symposium on Andean Geodynamics (ISAG 2005, Barcelona), Extended Abstracts. 258–260.

FAO (FOOD AND AGRICULTURE ORGANIZATION OF THE UNITED NATIONS) (1998): World reference base for soil resources. FAO: Rome.

FELDMAN, R.A. (1992): Preceramic architectural and subsistence traditions. ***Andean Past***, 3, 67–86.

FERREYRA, R. (1983): Los tipos de vegetación de la costa Peruana. ***Anales Jardín Botánico de Madrid***, 40(1), 241–256.

FISCHER, K. (1985): Grundzüge der jungtertiären Geomorphogenese der mittleren Anden. ***Erdkunde***, 39, 248–259.

FISH, S.K. (1999): Conclusions: the settlement pattern concept from an Americanist perspective. In: BILLMAN, B.R. & FEINMAN, G.M.: Settlement pattern studies in the Americas. Fifty years since Virú. Smithsonian Institution Press: Washington. 203–208.

FONTUGNE, M., USSELMANN, P., LAVALLÉE, D., JULIEN, M. & HATTÉ, C. (1999): El Niño variability in the coastal desert of southern Peru during the mid-Holocene. ***Quaternary Research***, 52, 171–179.

FÜCHTBAUER, H. (1988): Sandsteine. In: FÜCHTBAUER, H.: Sedimente und Sedimentgesteine. Sediment-Petrologie II. 4. Auflage. Schweizerbart: Stuttgart. 97–184.

GARLEFF, K., SCHÄBITZ, F., STINGL, H. & VEIT, H. (1991): Jungquartäre Landschaftsentwicklung und Klimageschichte beiderseits der ariden Diagonale Südamerikas. ***Bamberger Geographische Schriften***, 11, 359–394.

GARREAUD, R., VUILLE, M. & CLEMENT, A.C. (2003): The climate of the Altiplano: observed current conditions and mechanisms of past changes. ***Palaeogeography, Palaeoclimatology, Palaeoecology***, 194, 5–22.

GAY, P. (1962): Origen, distribución y movimiento de las arenas eólicas en al área de Yauca a Palpa. ***Boletín de la Sociedad Geológica del Perú***, 37, 37–58.

GAY JR., S.P. (1999): Observations regarding the movement of barchan sand dunes in the Nazca to Tanaca area of southern Peru. ***Geomorphology***, 27, 279–293.

GAY JR., S.P. (2005): Blowing sand and surface winds in the Pisco to Chala area, southern Peru. ***Journal of Arid Environments***, 61(1), 101–117.

GOUDIE, A. (2002): Great warm deserts of the world: landscapes and evolution. Oxford University Press: Oxford.

GRODZICKI, J. (1990): Las catastrofes ecologicas en la pampa de Nazca a fines del Holoceno y el fenomeno "El Niño". In: GRODZICKI, J.: El fenómeno El Niño a través de las fuentes arqueológicas y geológicas. 64–102.

GROSJEAN, M. (2001): Mid-Holocene climate in the south-central Andes: humid or dry? Technical comment. ***Science***, 292, 2391a.

GROSJEAN, M., CARTAJENA, I., GEYH, M.A. & NUÑEZ, L. (2003): From proxy data to paleoclimate interpretation: the mid-Holocene paradox of the Atacama Desert, northern Chile. ***Palaeogeography, Palaeoclimatology, Palaeoecology***, 194, 247–258.

GROSJEAN, M., MESSERLI, B., AMMANN, C., GEYH, M.A., GRAF, K., JENNY, B., KAMMER, K., NUÑEZ A., L., SCHREIER, H., SCHOTTERER, U., SCHWALB, A., VALERO-GARCÉS, B. & VUILLE, M. (1995): Holocene environmental changes in the Atacama altiplano and paleoclimatic implications. ***Bulletin de l'Institut Français d'Études Andines***, 24(3), 585–594.

GROSJEAN, M., MESSERLI, B. & SCHREIER, H. (1991): Seenhochstände, Bodenbildung und Vergletscherung im Altiplano Nordchiles: Ein interdisziplinärer Forschungsbeitrag zur Klimageschichte der Atacama. Erste Resultate. ***Bamberger Geographische Schriften***, 11, 99–108.

GROSJEAN, M., NÚÑEZ, L., CARTAJENA, I. & MESSERLI, B. (1997): Mid-Holocene climate and culture change in the Atacama desert, northern Chile. ***Quaternary Research***, 48, 239–246.

GROSJEAN, M., NÚÑEZ, L. & CARTAJENA, I. (2005): Cultural response to climate change in the Atacama Desert. In: SMITH, M. & HESSE, P.: 23° South: The archaeology and environmental history of the southern deserts. National Museum of Australia. 156–171.

GROSJEAN, M., VAN LEEUWEN, J.F.N., VAN DER KNAAP, W.O., GEYH, M.A., AMMANN, B., TANNER, W., MESSERLI, B., NÚÑEZ, L.A., VALERO-GARCÉS, B.L. & VEIT, H. (2001): A 22,000 ^{14}C year BP sediment and pollen record of climate change from Laguna Miscanti (23°S), northern Chile. ***Global and Planetary Change***, 28, 35–51.

GU, D. & PHILANDER, S.G.H. (1997): Interdecadal climate fluctuations that depend on exchange between the tropics and extratropics. ***Science***, 275, 805–807.

HABERLE, S.G. & CHEPSOW LUSTY, A. (2000): Can climate influence cultural development? A view through time. ***Environment and History***, 6(3), 349–369.

HAMPEL, A. (2002): The migration history of the Nazca Ridge along the Peruvian active margin: a re-evaluation. ***Earth and Planetary Science Letters***, 203, 665–679.

HAMPEL, A., KUKOWSKI, N., BIALAS, J. & BARCKHAUSEN, U. (2002): The obliquely subducting Nazca Ridge at the Peruvian active margin: the present collision zone and a re-evaluation of its migration history. 5th International Symposium on Andean Geodynamics, 2002. 291–293.

HANSEN, B.C.S., RODBELL, D.T., SELTZER, G.O., LEÓN, B. YOUNG, K.R. & ABBOTT, M. (2003): Late-glacial and Holocene vegetational history from two sites in the western Cordillera of southwestern Ecuador. ***Palaeogeography, Palaeoclimatology, Palaeoecology***, 194, 79–108.

HAUG, G.H., GÜNTHER, D., PETERSON, L.C., SIGMAN, D.M., HUGHEN, K.A. & AESCHLIMANN, B. (2003): Climate and the collapse of Maya civilization. ***Science***, 299, 1,731–1,735.

HAUG, G.H., HUGHEN, K.A., SIGMAN, D.M., PETERSON, L.C. & RÖHL, U. (2001): Southward migration of the Intertropical Convergence Zone through the Holocene. ***Science***, 293, 1,304–1,308.

HEALY, R.W. & COOK, P.G. (2002): Using groundwater levels to estimate recharge. ***Hydrogeology Journal***, 10, 91–109.

HEINE, K. (2000): Tropical South America during the Last Glacial Maximum: evidence from glacial, periglacial and fluvial records. ***Quaternary International***, 72, 7–21.

HESSE, R. (2004): Landschaftsevolution, Klima und fluviale Dynamik im Gebiet des Rio Grande de Nazca, Süd-Peru. unpublished diploma thesis, Friedrich Schiller University Jena.

HESSE, R. & BAADE, J. (2007): Palaeoenvironmental changes in the Nazca-Palpa region, Southern Peru – alternative interpretations of geoarchaeological evidence: A comment on Eitel et al. (2005) in Archaeometry, Vol. 47(1). ***Archaeometry*** (OnlineEarly Articles) doi:10.1111/j.1475-4754.2007.00322.x.

HODELL, D.A., BRENNER, M., CURTIS, J.H. & GUILDERSON, T. (2001): Solar forcing of drought frequency in the Maya lowlands. ***Science***, 292, 1,367–1,370.

HOFFMAN, S.M. (1999): After Atlas shrugs: cultural change or persistence after a disaster. In: OLIVER-SMITH, A. & HOFFMAN, S.M.: The angry earth: disaster in anthropological perspective. Routledge: New York, London. 302–325.

HOLMGREN, C.A., BETANCOURT, J.L., RYLANDER, K.A., ROQUE, J., TOVAR, O., ZEBALLOS, H., LINARES, E. & QUADE, J. (2001): Holocene vegetation history from fossil rodent middens near Arequipa, Peru. ***Quaternary Research***, 56, 242–251.

HSU, J.T. (1992): Quaternary uplift of the Peruvian coast related to the subduction of the Nazca Ridge: 13.5 to 15.6 degrees south latitude. ***Quaternary International***, 15/16, 87–97.

HSU, J.T., LEONARD, E.M. & WEHMILLER, J.F. (1989): Aminostratigraphy of Peruvian and Chilean Quaternary terraces. ***Quaternary Science Reviews***, 8, 255–262.

ISLA C., J., REINDEL, M. & DE LA TORRE Z., J.C. (2003): Jauranga: un sitio Paracas en el valle de Palpa, costa sur del Perú. ***Beiträge zur Allgemeinen und Vergleichenden Archäologie***, 23, 227–274.

KASTNER, S. (2005): Jungholozäne Landschaftsgenese am archäologischen Grabungsplatz La Muña, Palpa, Süd–

Peru. unpublished diploma thesis, Friedrich Schiller University Jena.

KEEFER, D.K., DEFRANCE, S.D., MOSELEY, M.E., RICHARDSON III, J.B., SATTERLEE, D.R. & DAY-LEWIS, A. (1998): Early maritime economy and El Niño events at Quebrada Tacahuay, Peru. ***Science***, 281, 1,833–1,835.

KEEFER, D.K., MOSELEY, M.E. & DE FRANCE, S.D. (2003): A 38 000–year record of floods and debris flows in the Ilo region of southern Peru and its relation to El Niño events and great earthquakes. ***Palaeogeography, Palaeoclimatology, Palaeoecology***, 194, 41–77.

KESSLER, A. (1990): Das El Niño-Phänomen und der Titicacaseespiegel. ***Mainzer Geographische Studien***, 34, 91–100.

KLOHN CRIPPEN SVS S.A. INGENIEROS CONSULTORES (1998): Evaluación ambiental territorial grupo de cuencas de la Costa Sur: Palpa–Nasca–Marcona–Acarí–Ocoña y Cerro Verde–Yarabamba–Puquina. MEM (Ministerio de Energia y Minas). www.mem.gob.pe/wmem/publica/aa/evats/costa sur/costa.zip (download: 24.05.2003).

KOEPCKE, H.-W. (1961): Synökologische Studien an der Westseite der peruanischen Anden. Bonner Geographische Abhandlungen, 29.

KOUTAVAS, A., DEMENOCAL, P.B., OLIVE, G.C. & LYNCH-STIEGLITZ, J. (2006): Mid-Holocene El Niño–Southern Oscillation (ENSO) attenuation revealed by individual foraminifera in eastern tropical Pacific sediments. ***Geology***, 34(12), 993–996.

KOUTAVAS, A. & LYNCH-STIEGLITZ, J. (2004): Variability of the marine ITCZ over the eastern Pacific during the past 30,000 years. Regional perspective and global context. In: BRADLEY, R.S. & DIAZ, H.F.: The Hadley Circulation: present, past and future. Kluwer Academic Publishers: Dordrecht. 347–369.

KULL, C., GROSJEAN, M. & VEIT, H. (2002): Modelling modern and late Pleistocene glacio-climatological conditions in the north Chilean Andes (19–30°S). ***Climatic Change***, 52, 359–381.

KULL, C., HÄNNI, F., GROSJEAN, M. & VEIT, H. (2003): Evidence of an LGM cooling in NW-Argentina (22°S) derived from a glacier climate model. ***Quaternary International***, 108, 3–11.

LAMBERS, K. (2004): The geoglyphs of Palpa (Peru): documentation, analysis, and interpretation. PhD Thesis, University of Zurich.

LANNING, E.P. (1963): A pre-agricultural occupation on the central coast of Peru. ***American Antiquity***, 28, 360–371.

LANNING, E.P. (1967): Peru before the Incas. Prentice Hall: New York.

LATORRE, C., BETANCOURT, J.L. & ARROYO, M.T.K. (2006): Late Quaternary vegetation and climate history of a perennial river canyon in the Río Salado basin (22°S) of Northern Chile. ***Quaternary Research***, 65, 450–466.

LATORRE, C., BETANCOURT, J.L., RECH, J.A. QUADE, J., HOLMGREN, C., PLACZEK, C., MALDONADO, J.C., VUILLE, M. & RYLANDER, K. (2005): Late Quaternary history of the Atacama Desert. In: SMITH, M. & HESSE, P.: 23° South: Archaeology and environmental history of the southern deserts. National Museum of Australia Press: Canberra. 73–90.

Lees, S.H. (1994): Irrigation and Society. ***Journal of Archaeological Research***, 2(4), 361–378.

LEOPOLD, L.B. & WOLMAN, M. (1957): River channel patterns: baided, meandering and straight. Geological Survey Professional Paper 282-B. U.S. Government Printing Office: Washington.

LLAGOSTERA, A. (1992): Early occupations and the emergence of fishermen on the Pacific coast of South America. ***Andean Past***, 3, 87–109.

MACHARÉ, J., VELIZ, Y., ORTLIEB, L. & DUMONT, J.-F. (1990): A review of recent paleoclimatic studies in Peru. ***Quaternary of South America and Antarctic Peninsula***, 8, 157–176.

MÄCHTLE, B. (2007): Hygrische Fluktuationen am Ostrand der nördlichen Atacama (Region Palpa – Südperu) und Anpassungsstrategien präkolumbischer Kulturen. In: LEOPOLD, M. & VÖLKEL, J.: Tagungsband. Arbeitskreis für Geoarchäologie im Verband der Deutschen Gesellschaft für geographie (DGfG) und der Gesellschaft für Naturwissenschaftliche Archäologie (GNAA). 3. Jahrestagung, 10. bis 12. Mai 2007 in Regensburg. 22.

MÄCHTLE, B., KADEREIT, A., EITEL, B. & UNKEL, I. (2006): Holocene environmental changes in the northern Atacama desert, southern Peru (14°30'S) and their impact on the rise and fall of the pre-Columbian cultures. ***Zeitschrift für Geomorphologie*** N.F. Suppl., 142, 47–62.

MAGILLIGAN, F.J. & GOLDSTEIN, P.S. (2001): El Niño floods and culture change: A late Holocene flood history for the Rio Moquegua, southern Peru. ***Geology***, 29, 431–434.

MANN, C. (2005): Oldest Civilization in the Americas Revealed. ***Science***, 307, 34–35.

MCCORMAC, F.G., HOGG, A.G., BLACKWELL, P.G., BUCK, C.E., HIGHAM, T.F.G. & REIMER, P.J. (2004): ShCal04 southern hemisphere calibration, 0–11.0 cal kyr BP. ***Radiocarbon***, 46(3), 1,087–1,092.

MCFADDEN, L.D., MCDONALD, E.V., WELLS, S.G., ANDERSON, K., QUADE, J. & FORMAN, S.L. (1998): The vesicular layer and carbonate collars of desert soild and pavements: formation, age and relation to climate change. ***Geomorphology***, 24(2–3), 101–145.

MERCIER, J.L., SEBRIER, M., LAVENU, A., CABRERA, J., BELLIER, O., DUMONT, J.-F. & MACHARÉ, J. (1992): Changes in the tectonic regime above a subduction zone of the Andean type: the Andes of Peru and Bolivia during the Pliocene-Pleistocene. ***Journal of Geophysical Research***, 97(B8), 11,945–11,982.

MESSERLI, B., GROSJEAN, M., BONANI, G., BÜRGI, A., GEYH, M.A., GRAF, K., RAMSEYER, K., ROMERO, H., SCHOTTERER, U., SCHREIER, H. & VUILLE, M. (1993): Climate change and natural resource dynamics of the Atacama

Altiplano during the last 18 000 years: a preliminary synthesis. ***Mountain Research and Development***, 13(2), 117–127.

MESSERLI, B., GROSJEAN, M., GRAF, K., SCHOTTERER, U., SCHREIER, H. & VUILLE, M. (1992): Die Veränderungen von Klima und Umwelt in der Region Atacama (Nordchile) seit der letzten Kaltzeit. ***Erdkunde***, 46, 257–272.

MIALL, A.D. (1996): The geology of fluvial deposits: sedimentary facies, basin analysis and petroleum geology. Springer: New York.

MOHTADI, M. (2004): Reconstruction of upwelling and productivity in the southern part of the Peru-Chile Current: A multi parameter approach. PhD dissertation, Universität Bremen.

MOSELEY, M.E. (1983): The good old days were better: agrarian collapse and tectonics. ***American Anthropologist***, 85, 773–799.

MOSELEY, M.E. (1992): Maritime foundations and multilinear evolution: retrospect and prospect. ***Andean Past***, 3, 5–42.

MOSELEY, M. (1999): Convergent catastrophe: past patterns and future implications of collateral natural disasters in the Andes. In: OLIVER-SMITH, A. & HOFFMAN, S.M.: The angry earth: disaster in anthropological perspective. Routledge: New York. 59–88.

MURPHY, R.C. (1926): Oceanic and climatic phenomena along the west coast of South America during 1925. ***Geographical Review***, 16, 26–54.

NANSON, G.C. & CROKE, J.C. (1992): A genetic classification of floodplains. In: BRAKENRIDGE, G.R. & HAGEDORN, J.: Floodplain evolution. Geomorphology, 4. 459–486.

NOLLER, J.S. (1993): Late cenozoic stratigraphy and soil geomorphology of the Peruvian desert, 3° to 18°S: A long-term record of hyperaridity and El Niño. PhD thesis, University of Colorado: Boulder.

NORABUENA, E., LEFFLER-GRIFFIN, L., MAO, A., DIXON, T., STEIN, S., SACKS, I.S., OCOLA, L. & ELLIS, M. (1998): Space geodetic observations of Nazca–South America convergence across the central Andes. ***Science***, 279, 358–362.

NORDT, L., HAYASHIDA, F., HALLMARK, T. & CRAWFORD, C. (2004): Late prehistoric soil fertility, irrigation management, and agricultural production in northwest coastal Peru. ***Geoarchaeology***, 19(1), 21–46.

OBERLANDER, T.M. (1994): Global deserts: a geomorphic comparison. In: ABRAHAMS, A.D. & PARSONS, A.J.: Geomorphology of desert environments. Chapman and Hall: London. 13–36.

OLIVER-SMITH, A. (1999): Peru's five-hundred-year earthquake: vulnerability in historical context. In: OLIVER-SMITH, A. & HOFFMAN, S.M.: The angry earth: disaster in anthropological perspective. Routledge: New York, London. 74–88.

ONERN (OFICINA NACIONAL DE EVALUACIÓN DE RECURSOS NATURALES) (1971a): Inventario, evaluación y uso racional de los recursos naturales de la costa: cuenca de Río Grande (Nazca), Vol. I.

ONERN (OFICINA NACIONAL DE EVALUACIÓN DE RECURSOS NATURALES) (1971b): Inventario, evaluación y uso racional de los recursos naturales de la costa: cuenca de Río Grande (Nazca), Vol. II.

ORTLIEB, L. & MACHARÉ, J. (1990): Geocronología y morfostratigrafía de terrazas del Pleistoceno superior: El caso de San Juan–Marcona, Peru. ***Boletín de la Sociedad Geológica del Perú***, 81, 87–106.

ORTLOFF, C.R. & KOLATA, A.L. (1993): Climate and collapse: agroecological perspectives on the decline of the Tiwanaku state. ***Journal of Archaeological Science***, 20, 195–221.

PARSONS, M.H. (1970): Preceramic subsistence on the Peruvian coast. ***American Antiquity***, 35, 292–304.

PASKOFF, R. (1978): Sobre la evolucion geomorfologica del gran acantilado costero del Norte Grande de Chile. ***Norte Grande*** (Instituto de Geografía, Universidad Católica de Chile), 1978–79(6), 7–22.

PASKOFF, R. & NARANJO, J.A. (1983): Formation et évolution du piémont andin dans le désert du nord du Chili (18–21° latitude sud) pendant le Cénozoique supérieur. ***Comptes Rendus Académie des Sciences Paris***, Serie 2, 297, 743–748.

PAULSEN, A.C. (1976): Environment and empire: climatic factors in prehistoric Andean culture change. ***World Archaeology***, 8, 121–132.

PESSENDA, L.C.R., LEDRU, M.P., GOUVEIA, S.E.M., ARAVENA, R., RIBEIRO, A.S., BENDASSOLLI, J.A. & BOULET, R. (2005): Holocene palaeoenvironmental reconstruction in northeastern Brazil inferred from pollen, charcoal and carbon isotope records. ***The Holocene***, 15(6), 812–820.

PINTO, R., BARRÍA, I. & MARQUET, P.A. (2006): Geographical distribution of Tillandsia lomas in the Atacama desert, northern Chile. ***Journal of Arid Environments***, 65, 543–552.

PLACZEK, C., QUADE, J. & BETANCOURT, J.L. (2001): Holocene lake-level fluctuations of Lake Aricota, southern Peru. ***Quaternary Research***, 56, 181–190.

POORE, R.Z., DOWSETT, H.J., VERARDO, S. & QUINN, T.M. (2003): Millennial- to century-scale variability in Gulf of Mexico Holocene climate records. ***Paleoceanography***, 18(2), 1048.

POZNAN RADIOCARBON LABORATORY (2007): Sample preparation. http://www.radiocarbon.pl/sampprep.htm (download: 21.01.2007).

POZORSKI, S. & POZORSKI, T. (1990): Reexamining the critical preceramic/ceramic period transition: new data from coastal Peru. ***American Anthropologist***, 92, 481–491.

PROHASKA, F.J. (1973): New evidence on the climatic controls along the Peruvian coast. In: AMIRAN, D.H.K. & WILSON, A.W.: Coastal deserts, their natural and human environments. University of Arizona Press: Tuscon. 91–107.

Pulwarty, R.S., Barry, R.G. & Riehl, H. (1992): Annual and seasonal patterns of rainfall variability over Venezuela. ***Erdkunde***, 46, 273–289.

Quilter, J. (1992): To fish in the afternoon: beyond subsistence economies in the study of early Andean civilization. ***Andean Past***, 3, 111–125.

Quinn, W.H. & Neal, V.T. (1992): The historical record of El Niño events. In: Bradley, R.S. & Jones, P.D.: Climate since A.D. 1500. Routledge: London. 623–648.

Rech, J.A., Currie, B.S., Cowan, A. & Michalski, G. (2006): Mid-Miocene nitrate paleosols from the Atacama Desert: Implications for the antiquity of the Atacama Desert. 18th World Congress of Soil Science, July 9–15, 2006 – Philadelphia, Pennsylvania, USA.

Rech, J.A., Pigati, J.S., Quade, J. & Betancourt, J.L. (2003): Re-evaluation of mid-Holocene deposits at Quebrada Puripica, northern Chile. ***Palaeogeography, Palaeoclimatology, Palaeoecology***, 194, 207–222.

Rech, J.A., Quade, J. & Betancourt, J.L. (2002): Late Quaternary paleohydrology of the central Atacama Desert (lat 22°–24°S), Chile. ***Geological Society of America Bulletin***, 114(3), 334–348.

Reindel, M. & Cuadrado, J.I. (2000): Ausgrabungen in Los Molinos und La Muna: Ergebnisse der Grabungskampagne 1999 des Archäologischen Projektes Nasca–Palpa, Süd-Peru. ***Jahresbericht der Schweizerisch-Liechtensteinischen Stiftung für Archäologische Forschungen im Ausland*** (1999), 67–95.

Reindel, M. & Isla Cuadrado, J. (2001): Los Molinos und La Muna. Zwei Siedlungszentren der Nasca-Kultur in Palpa, Südperu / Los Molinos y La Muna. Dos centros administrativos de la cultura Nasca en Palpa, Costa Sur del Peru. ***Beiträge zur Allgemeinen und Vergleichenden Archäologie***, 21, 241–319.

Reindel, M. & Isla Cuadrado, J. (2004): Archäologisches Projekt "Paracas in Palpa", Peru. Bericht über die Grabungskampagne 2003. ***Jahresbericht der Schweizerisch-Liechtensteinischen Stiftung für Archäologische Forschungen im Ausland*** (2004), 137–156.

Reindel, M., Isla Cuadrado, J., Grün, A. & Lambers, K. (2001): Neue Erkenntnisse zu Siedlungen, Bodenzeichnungen und Kultplätzen in Palpa, Süd-Peru. Ergebnisse der Feldkampagne 2000 des Archäologischen Projektes Nasca–Palpa. ***Jahresbericht der Schweizerisch-Liechtensteinischen Stiftung für Archäologische Forschungen im Ausland*** (2000), 81–104.

Reindel, M., Isla Cuadrado, J. & Koschmieder, K. (1999): Vorspanische Siedlungen und Bodenzeichnungen in Palpa, Südperu / Asentamientos prehispánicos y geoglifos en Palpa, Costa Sur del Peru. ***Beiträge zur Allgemeinen und Vergleichenden Archäologie***, 19, 313–381.

Reindel, M., Isla Cuadrado, J. & Lambers, K. (2002): Abschließende Untersuchungen zu Geoglyphen und Siedlungen in Palpa, Südperu. Ergebnisse der Feldkampagne 2001 des Archäologischen Projektes Nasca–Palpa. ***Jahresbericht der Schweizerisch-Liechtensteinischen Stiftung für Archäologische Forschungen im Ausland*** (2001), 57–74.

Reindel, M., Isla, J. & Lambers, K. (2003): Die Arbeiten des Archäologischen Projektes Nasca–Palpa, Peru, im Jahr 2002. ***Jahresbericht der Schweizerisch-Liechtensteinischen Stiftung für Archäologische Forschungen im Ausland*** (2002), 119–132.

Reitz, E.J. (2001): Fishing in Peru between 10000 and 3750 BP. ***International Journal of Osteoarchaeology***, 11, 163–171.

Richter, M. (1981): Klimagegensätze in Südperu und ihre Auswirkungen auf die Vegetation. ***Erdkunde***, 35, 12–30.

Riedinger, M.A., Steinitz-Kannan, M., Last, W.M. & Brenner, M. (2002): A ~6100 ^{14}C yr record of El Niño activity from the Galapagos Islands. ***Journal of Paleolimnology***, 27, 1–7.

Rigsby, C.A., Baker, P.A. & Aldenfelder, M.S. (2003): Fluvial history of the Rio Ilave valley, Peru, and its relationship to climate and human history. ***Palaeogeography, Palaeoclimatology, Palaeoecology***, 194, 165–185.

Roberts, N. (1998): The Holocene. An environmental history. 2nd edition. Blackwell: Oxford.

Rodbell, D.T., Seltzer, G.O., Anderson, D.M., Abbott, M.B., Enfield, D.B. & Newman, J.H. (1999): An ~15,000-year record of El Niño-driven alluviation in southwestern Ecuador. ***Science***, 283, 516–520.

Rollins, H.B., Richardson III, J.B. & Sandweiss, D.H. (1986): The birth of El Niño: geoarchaeological evidence and implications. ***Geoarchaeology***, 1, 3–15.

Roosevelt, A.C., Housley, R.A., da Silveira, I., Maranca, S. & Johnson, R. (1991): Eighth millennium pottery from a prehistoric shell midden in the Brazilian Amazon. ***Science***, 254, 1621–1624.

Rosgen, D.L. (1994): A classification of natural rivers. ***Catena***, 22, 169–199.

Rowe, H.D., Dunbar, R.B., Mucciarone, D.A., Seltzer, G.O., Baker, P.A. & Fritz, S. (2002): Insolation, moisture balance and climate change on the South American Altiplano since the Last Glacial Maximum. ***Climatic Change***, 52, 175–199.

Rundel, P.W., Dillon, M.O., Palma, B., Mooney, H.A., Gulmon, S.L. & Ehleringer, J.R. (1991): The phytogeography and ecology of the coastal Atacama and Peruvian deserts. ***Aliso***, 13, 1–50.

Rust, B.R. (1978): A classification of alluvial channel systems. In: Miall, A.D.: Fluvial sedimentology. Canadian Society of Petroleum Geologists Memoir 5. 187–198.

Sandweiss, D.H. (2003): Terminal Pleistocene through Mid-Holocene archaeological sites as paleoclimatic archives for the Peruvian Coast. ***Palaeogeography, Palaeoclimatology, Palaeoecology***, 194, 23–40.

SANDWEISS, D.H., MAASCH, K.A., BURGER, R.L., RICHARDSON III, J.B., ROLLINS, H.B. & CLEMENT, A. (2001): Variation in Holocene El Niño frequencies: Climate records and cultural consequences in ancient Peru. ***Geology***, 29, 603–606.

SANDWEISS, D.H., MCINNIS, H., BURGER, R.L., CANO, A., OJEDA, B., PAREDES, R., SANDWEISS, M.D.C. & GLASCOCK, M.D. (1998): Quebrada Jaguay: early South American maritime adaptations. ***Science***, 281, 1,830–1,832.

SANDWEISS, D.H., RICHARDSON III, J.B., REITZ, E.J., ROLLINS, H.B. & MAASCH, K.A. (1997): Determining the early history of El Niño – response. ***Science***, 276, 966–967.

SANTORO, C.M., ARRIAZA, B.T., STANDEN, V.G. & MARQUET, P.A. (2005): People of the coastal Atacama Desert. Living between sand dunes and waves of the Pacific Ocean. In: VETH, P., SMITH, M. & HISCOCK, P.: Desert peoples. Blackwell: Malden. 243–260.

SATTERLEE, D.R., MOSELEY, M.E., KEEFER, D.K. & TAPIA A., J.E. (2000): The Miraflores El Niño disaster: convergent catastrophes and prehistoric agrarian change in southern Peru. ***Andean Past***, 6, 95–116.

SCARBOROUGH, V.L. & GALLOPIN, G.G. (1991): A water storage adaptation in the Maya lowlands. ***Science***, 251, 658–662.

SCHAETZL, R.J. & ANDERSON, S. (2006): Soils: genesis and geomorphology. Cambridge University Press: Cambridge.

SCHEMENAUER, R.S. & CERECEDA, P. (1993): Meteorological conditions at a coastal fog collection site in Peru. ***Atmosfera***, 6, 175–188.

SCHRADER, H. & SORKNES, R. (1991): Peruvial coastal upwelling: late Quaternary productivity changes revealed by diatoms. ***Marine Geology***, 97(3–4), 233–249.

SCHREIBER, K.J. (1999): Regional approaches to the study of prehistoric empires. Examples from Ayacucho and Nasca, Peru. In: BILLMAN, B.R. & FEINMAN, G.M.: Settlement pattern studies in the Americas. Fifty years since Virú. Smithsonian Institution Press: Washington. 160–171.

SCHREIBER, K.J. & LANCHO ROJAS, J. (1995): The puquios of Nasca. ***Latin American Antiquity***, 6(3), 229–254.

SÉBRIER, M. & MACHARÉ, J. (1980): Observaciones acerca del Cuaternario de la costa del Perú central. ***Bulletin de l'Institut Français d'Études Andines***, 9(1–2), 5–22.

SELLEY, R.S. (1988): Applied Sedimentology. Academic Press: London.

SELTZER, G.O. (1990): Recent glacial history and paleoclimate of the Peruvian–Bolivian Andes. ***Quaternary Science Reviews***, 9(2/3), 137–152.

SELTZER, G.O. (1993): Late-Quarternary glaciation as a proxy for climate change in the central Andes. ***Mountain Research and Development***, 13(2), 129–138.

SELTZER, G.O., RODBELL, D.T., BAKER, P.A., FRITZ, S.C., TAPIA, P.M., ROWE, H.D. & DUNBAR, R.B. (2002): Early warming of tropical South America at the Last Glacial–Interglacial transition. ***Science***, 296, 1,685–1,686.

SELTZER, G., RODBELL, D. & BURNS, S. (2000): Isotopic evidence for late Quaternary climatic change in tropical South America. ***Geology***, 28(1), 35–38.

SERVANT, M. & FONTES, J.-C. (1984): Les basses terrasses fluviatiles du Quaternaire récent des Andes Boliviennes. ***Cahiers ORSTOM, Série Géologie***,14(1), 15–28.

SILVERMAN, H. (1994): Paracas in Nazca: new data on the Early Horizon occupation of the Rio Grande de Nazca drainage. ***Latin American Antiquity***, 5(4), 359–382.

SILVERMAN, H. (1996): The formative period on the south coast of Peru: a critical review. ***Journal of World Prehistory***, 10(2), 95–146.

SILVERMAN, H. (2002): Ancient Nasca settlement and society. University of Iowa Press: Iowa City.

SILVERMAN, H. & PROULX, D. (2002): The Nasca. The peoples of America. Blackwell: Malden, Oxford.

SMITH, J.A., SELTZER, G.O., FARBER, D.L., RODBELL, D.T. & FINKEL, R.C. (2005): Early local Last Glacial Maximum in the tropical Andes. ***Science***, 308, 678–681.

SMITH, W.H.F. & SANDWELL, D.T. (1997): Global seafloor topography from satellite altimetry and ship depth soundings. ***Science***, 277, 1,956–1,962.

SOLIS, R.S, HAAS, J. & CREAMER, W. (2001): Dating Caral, a Preceramic site in the Supe Valley on the central coast of Peru. ***Science***, 292, 723–726.

SPENCE, W., MENDOZA, C., ENGDAHL, E.R., CHOY, G.L. & NORABUENDA, E. (1999): Seismic subduction of the Nazca Ridge as shown by the 1996–97 Peru earthquakes. ***Pure and Applied Geophysics***, 154, 753–776.

STANISH, C. (1999): Settlement patern shifts and political ranking in the Lake Titicaca basin, Peru. In: BILLMAN, B.R. & FEINMAN, G.M.: Settlement pattern studies in the Americas. Fifty years since Virú. Smithsonian Institution Press: Washington. 116–128.

STANISH, C. (2001): The origin of sate societies in South America. ***Annual Reviews of Anthropology***, 30, 41–64.

STEFFEN, D., SCHLUNEGGER, F. & PREUSSER, F. (2007): Correlating sediment aggradation and climate by means of luminescence dating, Valley de Pisco, Peru. ***Geophysical Research Abstracts***, 9, 03347.

STOTHERT, K.E. (1992): Early economies of coastal Ecuador and the foundations of Andean civilization. ***Andean Past***, 3, 43–54.

SWENSON, J.L. & BECK, S.L. (1999): Source characteristics of the 12 November 1996 Mw 7.7 Peru subduction zone earthquake. ***Pure and Applied Geophysics***, 154, 731–751.

TAPIA, P.M., FRITZ, S.C., BAKER, P.A., SELTZER, G.O. & DUNBAR, R.B. (2003): A late Quaternary diatom record of tropical climatic history from Lake Titicaca (Peru and Bolivia). ***Palaeogeography, Palaeoclimatology, Palaeoecology***, 194, 139–164.

Tedesco, K. & Thunell, R. (2003): High resolution tropical climate record for the last 6,000 years. ***Geophysical Research Letters***, 30(17 CLM), 1891.

Thompson, L.G. (1992): Quelccaya Ice Core Database. IGBP PAGES/World Data Center-A for Paleoclimatology Data Contribution Series # 92-008. NOAA/NGDC Paleoclimatology Program, Boulder CO, USA. ftp.ncdc.noaa.gov/pub/data/paleo/icecore/trop/quelccaya (download: 25.07.2006).

Thompson, L.G. (2001a): Huascarán Ice Core Data. IGBP PAGES/World Data Center A for Paleoclimatology Data Contribution Series #2001-008. NOAA/NGDC Paleoclimatology Program, Boulder CO, USA. ftp.ncdc.noaa.gov/pub/data/paleo/icecore/trop/huascaran (download: 25.07.2006).

Thompson, L.G. (2001b): Sajama Ice Core Data. IGBP PAGES/World Data Center A for Paleoclimatology Data Contribution Series #2001-009. NOAA/NGDC Paleoclimatology Program, Boulder CO, USA. ftp.ncdc.noaa.gov/pub/data/paleo/icecore/trop/sajama (download: 25.07.2006).

Thompson, L.G., Davis, M.E., Mosley-Thompson, E. & Liu, K.-B. (1988): Pre-Incan agricultural activity recorded in dust layers in two tropical ice cores. ***Nature***, 336, 763–765.

Thompson, L.G., Davis, M.E., Mosley-Thompson, E., Sowers, T.A., Henerson, K.A., Zagorodnov, V.S., Lin, P.-N., Mikhalenko, V.N., Campen, R.K., Bolzan, J.F., Cole-Dai, J. & Francou, B. (1998): A 25,000 year tropical climate history from Bolivial ice cores. ***Science***, 282, 1,858–1,864.

Thompson, L.G., Mosley-Thompson, E., Davis, M.E., Lin, P.-N., Henderson, K.A., Cole-Dai, J., Bolzan, J.F. & Liu, K.-B. (1995): Late glacial stage and Holocene tropical ice core records from Huascarán, Peru. ***Science***, 269, 46–50.

Thornburg, T. & Kulm, L.D. (1981): Sedimentary basins of the Peru continental margin: Structure, stratigraphy, and Cenozoic tectonics from 6°S to 16°S latitude. In: Kulm, L.D., Dymond, J., Dasch, E.J., Hussong, D.M. & Roderick, R.: Nazca Plate: crustal formation and Andean convergence. ***Geological Society of America Memoir*** 154. 393–422.

Troll, C. (1968): The cordilleras of the tropical Americas. Aspects of climatic, phytogeographical and agrarian ecology. In: Troll, C.: Geo-ecology of the mountainous regions of the tropical Americas. Colloquium Geographicum, Band 9. Geographisches Institut, Universität Bonn: Bonn. 15–56.

Tucker, G.E. & Slingerland, R. (1997): Drainage basin responses to climate change. ***Water Resources Research***, 33(8), 2,031–2,047.

Tucker, M. (1982): The field description of sedimentary rocks. Geological Society of London handbook. Wiley: Chichester.

Unkel, I. (2006): AMS-^{14}C-Analysen zur Rekonstruktion der Landschafts- und Kulturgeschichte in der Region Palpa (S-Peru). Dissertation, Universität Heidelberg.

Usselmann, P., Fontugne, M., Lavallee, D., Julien, M. & Hatté, C. (1999): Estabilidad y rupturas dinámicas en el Holoceno de la costa surperuana: el valle de Quebrada de los Burros (departamento de Tacna). ***Bulletin de l'Institut Français d'Études Andines***, 28, 1–11.

Vargas, G. & Ortlieb, L. (1998): Patrones de variaciones climáticas durante el Cuaternario tardío en la costa de la región de Antofagasta, Chile. ***Bulletin d'Institut Français d'Études Andines***, 27(3), 385–394.

Vaughn, K.J. (2005): Household approaches to ethnicity on the south coast of Peru: The domestic architecture of Early Nasca society. In: Reycraft, R.: Us and Them: The assignation of ethnicity in the Andean region, methodological approaches. UCLA Institute of Archaeology Press: Los Angeles. 86–103.

Veit, H. (2000): Klima- und Landschaftswandel in der Atacama. ***Geographische Rundschau***, 52(9), 4–9.

Wang, Y., Cheng, H., Edwards, R.L., He, Y., Kong, X., An, Z., Wu, J., Kelly, M.J., Dyoski, C.A. & Li, X. (2005): The Holocene Asian Monsoon: Links to solar changes and North Atlantic climate. ***Science***, 308, 854–857.

Weischet, W. (1966): Zur Klimatologie der Nordchilenischen Wüste. ***Meteorologische Rundschau***, 19(1), 1–7.

Weischet, W. (1996): Regionale Klimatololgie. Teil 1: Die Neue Welt. B.G. Teubner: Stuttgart.

Wells, L.E. (1987): An alluvial record of El Niño events from northern coastal Peru. ***Journal of Geophysical Research***, 92(C13), 14,464–14,470.

Wells, L.E. (1990): Holocene history of the El Niño phenomenon. ***Geology***, 18, 1,134–1,137.

Wells, L.E. (2001): A geomorphological approach to reconstructing archaeological settlement patterns based on surficial artifact distribution. In: Goldberg, P., Holliday, V. & Ferring, R.: Earth sciences and archaeology. Plenum: New York. 107–141.

Wells, L.E. & Noller, J.S. (1997): Determining the early history of El Niño. Technical comment. ***Science***, 276, 966.

Wells, L.E. & Noller, J.S. (1999): Holocene coevolution of the physical landscape and human settlement in the northern coastal Peru. ***Geoarchaeology***, 14, 755–789.

West, M. (1981): Agricultural resource use in an Andean coastal ecosystem. ***Human Ecology***, 9(1), 47–78.

Wille, M., Hooghiemstra, H., Behling, H., van der Borg, L. & Negret, A.J. (2001): Environmental change in the Colombian subandean forest belt from 8 pollen records: the last 50 kyr. ***Vegetation History and Archaeobotany***, 10, 61–77.

Wilson, D. (1981): Of maize and men: a critique of the maritime hypothesis of state origins on the coast of Peru. ***American Antiquity***, 83, 93–120.

Wisner, B. & Luce, H.R. (1993): Disaster vulnerability: scale, power and daily life. ***GeoJournal***, 30(2), 127–140.

Wittfogel, K. (1955): Developmental aspects of hydraulic societies. In: Steward, J.H.: Irrigation civilizations: a comparative study. Pan American Union: Washington, D.C. 43–52.

Wittfogel, K.A. (1956): The hydraulic civilizations. In: Thomas, W.L.: Man's role in changing the face of the earth. University of Chicago Press: Chicago. 152–164.

Yoffee, N. (1988): The collapse of ancient Mesopotamian states and civilization. In: Yoffee, N. & Cowgill, G.L.: The collapse of ancient states and civilizations. University of Arizona Press: Tucson. 44–68.

Yokoyama, Y., Lambeck, K., Decker, P.D., Johnston, P. & Fifield, L.K. (2000): Timing of the Last Glacial Maximum from observed sea-level minima. ***Nature***, 406, 713–716.

Zaro, G. & Alvarez, A.U. (2005): Late Chiribaya agriculture and risk management along the arid Andean coast of southern Peru, 1200–1400. ***Geoarchaeology***, 20(7), 717–737.

Zeil, W. (1979): The Andes, a geological review. Beiträge zur regionalen Geologie der Erde, Band 13. Gebrüder Bornträger: Berlin.

Zhou, J. & Lau, K.-M. (1998): Does a monsoon climate exist over South America? ***Journal of Climate***, 11, 1,020–1,040.

Maps and data sources

Maps and geospatial data

Topographic map of Peru 1: 100,000
(US Army Topographic Command, Washington, ca. 1970)
– Ed. 3–IGN, Serie J631, sheet1842 (29–m), Cordova
– Ed. 3–IGN, Serie J631, sheet 1942 (29–n), Laramate
– Ed. 3–IGN, Serie J631, sheet 1841 (30–m), Palpa
– Ed. 3–IGN, Serie J631, sheet 1941 (30–n), Nasca
Topographic Map of Peru 1: 50,000
(Insituto Geográfico Nacional, Lima, ca. 1985)
– Ed. 1–TPC, Serie J731, sheet 1841 I, Palpa
Geological Map of Peru 1: 100,000
– sheet 29–m, Cordova (INGEMMET 1979)
– sheet 29–n, Laramate (INGEMMET 1973)
– sheet 30–m, Palpa (INGEMMET 1999)
– sheet 30–n, Nasca (INGEMMET 1999)
– sheet 31–m, San Juan (INGEMMET 1998)
Aerial photographs
– aerial photographs of the Palpa Valley from the years 1944, 1970 and 1997 were provided by M. Reindel in 2002.
Digital geospatial data
– high-resolution DTM of the Palpa Valley: provided by A. Grün and M. Sauerbier (Institute of Geodesy and Photogrammetry, ETH Zurich
– Quickbird images: Google Earth, http://earth.google.com (download: 20.06.2006)
– Global Seafloor Topography (Smith & Sandwell 1997), ftp://topex.ucsd.edu/pub/global_topo_2min (download: 29.11.2003)
– GTOPO30: US Geological Survey, ftp://edcftp.cr.usgs.gov/pub/data/gtopo30/global/w020n90.tar.gz (download: 30.01.2001)
– Landsat TM ca. 1990 and Landsat ETM ca. 2000 composites: NASA Earth Science Applications Directorate, https://zulu.ssc.nasa.gov/mrsid/mrsid.pl (download: 26.01.2005)
– Landsat 7 ETM scene (26.04.2000): Global Land Cover Facility, Institute for Advanced Computer Studies, University of Maryland, http://glcf.umiacs.umd.edu (download: 05.03.2007)
– SRTM version 2: NASA, ftp://e0srp01u.ecs.nasa.gov/srtm/version2 (download: 13.07.2006)
– MODIS Vegetation Continuous Fields (01.11.2000–01.11.2001): Global Land Cover Facility, http://glcfapp.umiacs.umd.edu (download: 09.01.2006)
– GOES weather satellite data 1998–2007: NOAA Comprehensive Large Array–Data Stewardship System, http://www.class.noaa.gov (download: 10.–19.03 2007)
– near-shore bathymetric data: National Geophysical Data Center, Boulder, http://www.ngdc.noaa.gov/mgg/gdas/gd_sys.html (download: 06.05.2004)

Other data

– radiocarbon calibration: OxCal Version 3.10 (Bronk Ramsey 1995, 2001) with southern hemisphere correction (shcal04.14c, McCormac et al. 2004), http://c14.arch.ox.ac.uk/embed.php?File=oxcal.html (download: 14.08.2006)
– climatic data: ONERN 1971a, 1971b
– daily discharge data (Rio Grande, Rio Palpa, Rio Vizcas) 1975–2007: data provided by Junta de Usarios Palpa
– irrigation system data for the Palpa Valley: data provided by Junta de Usarios Palpa
– Quelccaya ice core data: Thompson (1992), ftp.ncdc.noaa.gov/pub/data/paleo/icecore/trop/quelccaya (download: 25.07.2006)
– Sajama ice core data: Thompson (2001b), ftp.ncdc.noaa.gov/pub/data/paleo/icecore/trop/sajama (download: 25.07.2006)
– Huascarán ice core data: Thompson (2001a), ftp.ncdc.noaa.gov/pub/data/paleo/icecore/trop/huascaran (download: 25.07.2006).
– Cariacon basin Ti data: Haug et al. (2001), ftp://wdc.cricyt.edu.ar/paleo/contributions_by_author/haug2001/cariaco_ti.txt (download: 25.06.2007)
– Palpa Valley radiocarbon dates: Unkel (2006: Anhang page 38–39)

Appendix

Contents of Appendix A and B

List of figures

List of tables

Appendix A – Sediment profiles

Appendix A contains sketches and detailed descriptions of the recorded sediment profiles as well as graphs showing the results of laboratory analyses. Profile locations are shown in Figure 22 (page 37) The layer-by-layer description is top-down for all profiles irrespective of layer numbering. The subsuming interpretations (which similarly appear in chapter 5.2 of this thesis) are generally bottom-up, i.e. in the order of deposition. Final sketches were produced by R. Mendler after hand drawings prepared in the field and/or drawings made from photographs combined with information collected in the field by R. Hesse, J. Baade, S. Kastner and S. Angelstein. The legend for all sketches is shown in Figure A-109 (page A-129). To ensure consistency between and within profiles even though not all profiles were sampled and sampling was omitted for some layers in most sampled profiles, grain size depicted in the sketches is shown as recorded in the field and might differ somewhat from the results of the laboratory analyses which are presented as graphs where applicable. In particular the percentage of clasts >2 mm may be subject to sampling error due to limited sample sizes. The graphs show the following results of laboratory analyses: cumulative weight percent of clay, silt and sand; median grain size; sorting according to Füchtbauer (1988: 136); clay: silt ratio (CSR, Schaetzl & Anderson 2006: 579, after Van Wambeke 1962; cf. Noller 1993: 137); weight percent $CaCO_3$; weight percent loss on ignition (LoI); AAS results for the elements Cu, Cr, Fe, Mn, Ca, K, Mg and Na. In the genetic interpretation of sedimentary characteristics I follow Miall (1996: 105–127) and FAO (1998) (cf. 5.1). Where not otherwise noted or discussed, all chronological results are stratigraphically consistent, i.e. younger ages occur higher up in the stratigraphy than older ages.

Rio Grande narrows

Tab. A-1: Position of profile RGE at Rio Grande narrows.

Profile	Depth [m]	Northing [m]	Easting [m]	Altitude [m]
RGE	5.00	8,381,293.11	475,027.81	219.79

Profile RGE

Profile RGE was recorded at a road-cut and quebrada exposure just downstream of the Rio Grande narrows between Cerro Corpante and Cerro Piedra Gorda and approximately 700 m upstream of the confluence of the Rio Grande and Rio Ingenio (14°38'S, 75°13'W, 220 m asl, Figure A-1). There, a local quebrada has deposited a debris fan. The quebrada drains a watershed of approximately 1.2 km² with elevations of up to 550 m asl which is completely devoid of vegetation. It is characterised by steep relief in igneous rock of Cretaceous age that is at least partially covered with a thin layer of loose, angular rock fragments. The debris fan extends between 50 and 160 m from the mountain front and has been dissected by the quebrada mouth.

On its surface, looted graves of unknown age (including human bones and fragments of plainware ceramics) have been found. At a distance of ca. 50 m from the current river bank, up to 5 metres of fine- and coarse-grained sediments are exposed in a near-vertical road cut. The base of this exposure is at the same level as the incised quebrada mouth, six metres above the current river bed. High water marks under a bridge across the Rio Grande about 100 m upstream (cf. Figure A-1) indicate that the highest water level in recent years, likely the 1999 flood event (see 4.1.4; peak discharge of Rio Grande, Rio Palpa, Rio Vizcas combined approximately 700 m³/s on 02.03.1999, discharge data from Junta de Usarios in Palpa) was approximately 2.7 m.

The sediments in the exposure record a sequence of deposition and erosion. Figure A-2 shows segment A. This 5 m high segment is next to the mouth of the quebrada and faces south-west. In the left part of the segment, at least 14 layers (unit 1) can be distinguished, all but one of them consisting of poorly sorted, coarse-grained, angular material from coarse sand to blocks of up to 25 cm length. Bedding is horizontal and erosional features can be discerned at some layer boundaries. Between 7.2 and 7.4 m above the present-day riverbed (a.r.), there is a thin fine-grained layer (silty sand). At 7.6 and 7.8 m a.r. there are two small sand lenses. An erosional discontinuity and associated sediments (unit 2) separate the left part of segment A from the right part over the entire height of the exposure. The erosional surface is steeper in the lower part (56°) and less steep in the upper part (32°). In the upper part, unit 2 consists of thin (5 to 30 cm) slope debris. Before a new sequence of sediments (units 3 to 5) was deposited, the oversteepened lower part appears to have been adjusted by slope processes to an angle of 21°, resulting in a wedge of sediment 1.4 m wide. In the right part of the exposure, three stratigraphic units can be discerned.

Unit 3 (6.0–7.8/8.0 m a.r.) consists of at least 8 layers of mainly horizontally bedded, poorly sorted, angular material (from sand to blocks of up to 16 cm). Unit 4 (7.8/8.0–9.4/10.7 m a.r.) is a sequence of fine- and coarse-grained layers not all of which are laterally continuous. The lower boundary of unit 4 is erosional and is followed by 3 fine-

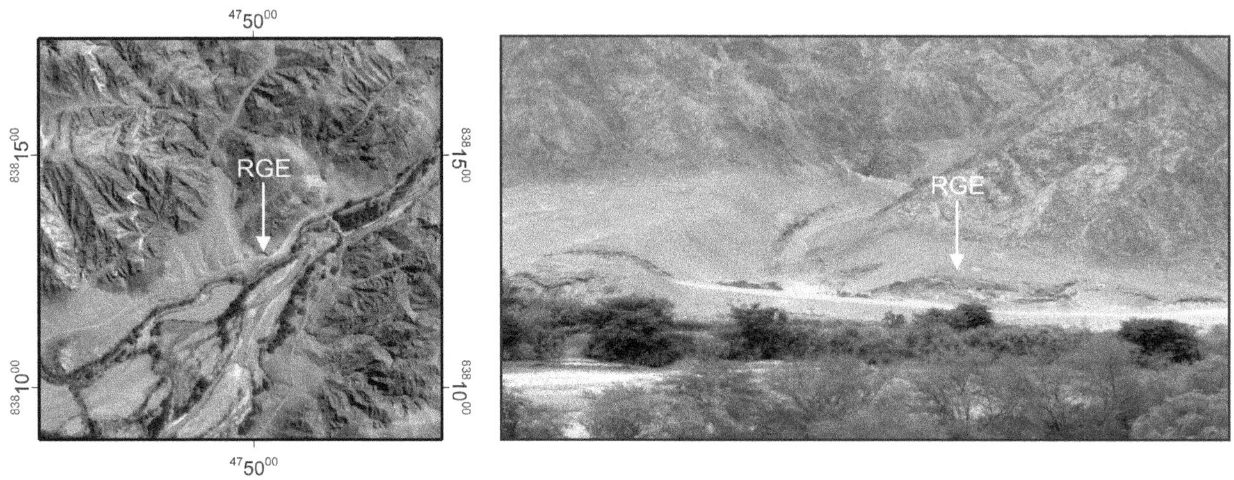

Fig. A-1: Satellite image (Quickbird) and photograph of dissected debris fan. Arrows show the position of segment A of profile RGE.

grained layers 5 to 30 cm strong. A 10 to 20 cm thick coarse-grained, angular layer is followed by a package of at least 4 fine-grained layers (together 50 to 60 cm), the surface of which is erosional. Not all layers of this package are present over the whole width of the segment. These layers are overlain by 1 coarse (12–30 cm), 1 fine (5 cm), 1 coarse (20 cm), 1 fine (7 cm), 1 coarse (<5 cm) and 1 fine (15 cm) layer.

Unit 5 (9.4/10.7–11.0 m a.r.) is similar to units 1 and 3 and consists of at least 8 layers of horizontally bedded, poorly sorted, angular material (from sand to blocks up to 17 cm). The layers in unit 5 are laterally less continuous with some of them appearing lens-like. Segment B (not shown) is the continuation of the right-hand side of the exposure roughly parallel to the Rio Grande in the upriver direction, facing south-east. Only the top of unit 3 is exposed in some parts of segment B. The lower part of unit 4 can be followed along the entire segment and consists here of one to two discernible fine-grained layers overlain by a thin coarse-grained layer (layer M). This layer was used as a marker layer to connect segments A and B. It is overlain by a package of at least 6 discernible fine-grained layers. However, due to large amounts of slope debris the top layers of unit 4 as well as the entire unit 5 in the right part of segment B could not directly be related to their counterparts in segment A. The fine-grained package is partially overlain by a coarse-grained layer and a lens of coarse material, followed by another fine-grained layer. In segment B, unit 5 consists of at least 5 coarse layers.

Table A-2 gives an overview over the stratigraphy of segments A and B. Some of the fine-grained layers of unit 3 contain dispersed charcoal. Sediment samples containing dispersed charcoal were taken from layers below and above the marker layer M. Charcoal was sampled from a charcoal layer near the top of unit 4 in segment B approximately 38 m north-east of Segment A. In addition, one charcoal sample was taken from a hearth close to the surface. The coarse-grained layers of angular clasts in units 1 through 5 are clearly of local origin since rounded clasts (which would indicate long-distance transport by the Rio Grande) are not present. These layers are interpreted as hyperconcentrated (debris) flow deposits resulting from local runoff in the quebrada or from the adjacent slopes of nearby Cerro Corpante. Similar deposits have been described by Sébrier & Macharé (1980: 12) for the coastal desert of central Peru and have been termed Colorada Alloformation by Noller (1993: 51ff.). The fine-grained layers of unit 4 are interpreted as overbank deposits of the Rio Grande due to their grain size, sorting and the presence of bedding.

The results of radiocarbon dating are shown in Table A-3. Since all but one of the dates are outside the limit for calibration (McCormac et al. 2004), all results for profile

Tab. A-2: Stratigraphy of segments A and B of profile RGE.

	Segment A	Segment B
Unit 5	≥8 coarse	≥5 coarse
Unit 4	1 fine	
	1 coarse	
	1 fine	
	1 coarse	
	1 fine	1 fine
	1 coarse	1 coarse
	≥4 fine	≥6 fine
	1 coarse (marker)	1 coarse (marker)
	≥ 3 fine	≥2 fine
Unit 3	≥8 coarse	≥1 coarse
Unit 2	slope debris	–
erosional hiatus		
Unit 1	≥14 coarse	–

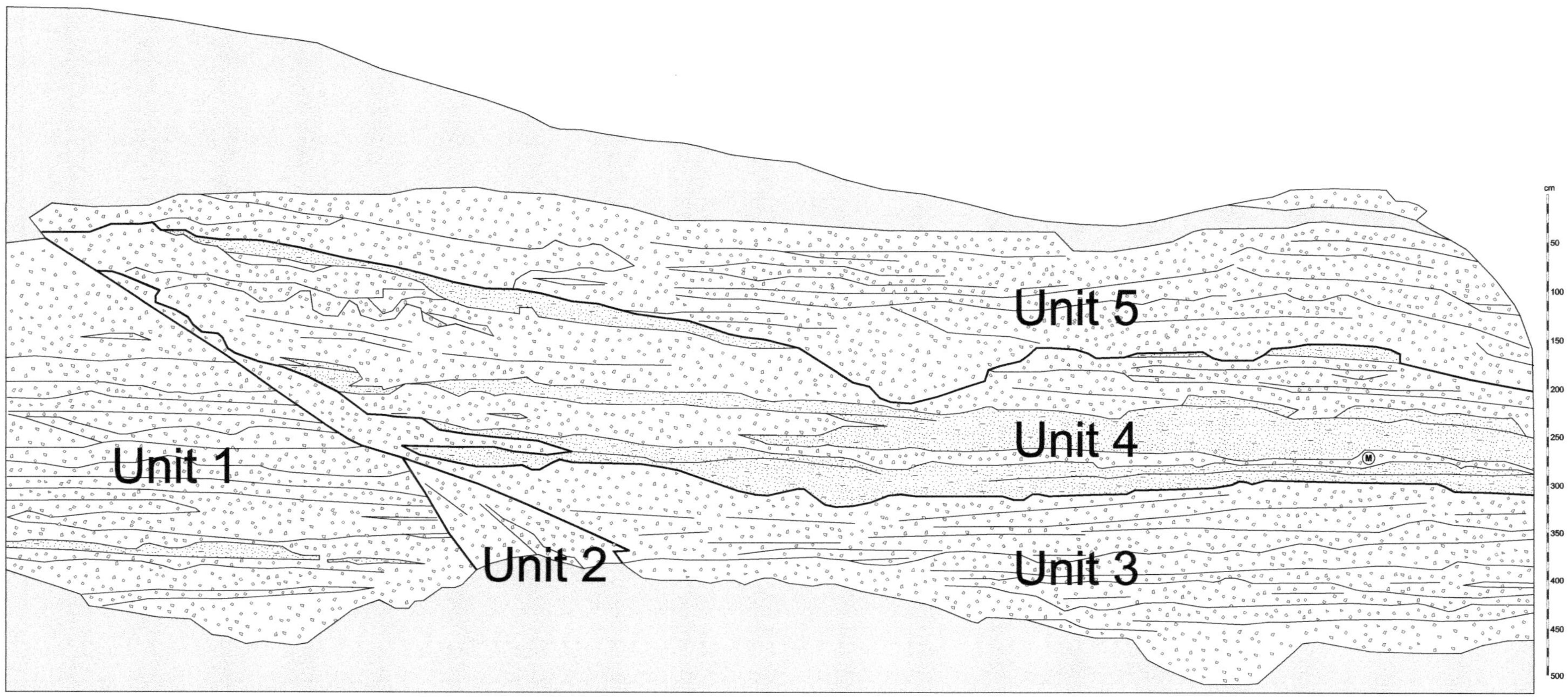

Fig. A-2: Profile RGE, segment A. Grey denotes slope debris; **M** - marker layer. (drawing: R.Mendler, scale 1: 67).

Tab. A-3: Sequence of radiocarbon dates (not calibrated) from profile RGE.

Sample ID	Stratigraphic position	Sample age (^{14}C BP)
Poz-13541	segment B, unit 5, near top surface	2,040 ±35
Poz-16830	segment B, near top of unit 4	21,800 ±150
Poz-13606	segment B, unit 4, 6th layer above marker layer M	22,060 ±120
Poz-13292	segment A, unit 4, directly below marker layer M	22,650 ±140
Poz-13491		22,610 ±230

RGE are given as uncalibrated ^{14}C ages. From one sample, two subsamples (Poz-13292 and Poz-13491) were measured; the close correspondence of the results indicates a high precision of the measurements. The results are stratigraphically consistent and reveal a high age for the sedimentary package exposed here. Even the youngest sample, taken from a hearth near the top surface, gives a ^{14}C age of 2,040 ±35 BP (Poz-13541). This date for the hearth shows that the site has been used by humans during the Inital Nasca period (cf. UNKEL 2006: 111). Since the hearth was a pit dug into the surface and did not appear to have been covered by another layer of sediment except for the collapsed walls of the pit, deposition on the landform must have ended before 2,040 ±35 ^{14}C BP.

The other three samples fall into the time of the Last Glacial Maximum (LGM) in South America (22.8–21.4 ^{14}C ka BP for mid-latitude Chile, DENTON et al. 1999; >34 to ~21 ka BP (^{10}Be) for moraines in Peru and Bolivia, SMITH et al. 2005; 30–18 ^{14}C ka BP from pollen studies in Colombia, WILLE et al. 2001). While they are somewhat older than the LGM as defined globally by the sea-level lowstand at 22–19 ka cal BP (YOKOYAMA et al. 2000) and by other records from South America (e.g. 21.0 ±1.0 ka cal BP for offshore Chile, MOHTADI 2004; 21.0 cal ka BP for the end of the LGM from magnetic susceptibility in Lake Titicaca sediments, BAKER et al. 2001), they do represent the time of full Glacial conditions (cf. 2.3.2). The dates for the two statistically contemporary samples Poz-13606 (22,060 ±120 BP) and Poz-16830 (21,800 ±150) and the two dates for the samples Poz-13292 (22,650 ±140 BP) and Poz-13491 (22,610 ±230) bracket one coarse-grained debris-flow layer followed by 6 fine-grained fluvial layers. In the following analysis and interpretation, the dates for the samples Poz-13606 and Poz-13292 are used. The fluvial layers total approximately 1.3 m. Given the interval of 590 ^{14}C years in which these layers were deposited, the average recurrence interval for deposition events was ca. 100 ^{14}C years. The average sedimentation rate from fluvial activity was ca. 22 cm/100 a. The one debris flow event at the beginning of this 590 ^{14}C year interval is not sufficient for a reliable extrapolation in terms of recurrence intervals. However, assuming a plausible range of recurrence intervals from half to twice the recurrence interval of one event per 590 ^{14}C years for periods during which accumulative debris flows occurred allows an estimation of the timing of erosional and depositional phases. Under this assumption, the deposition of the at least 8 debris-flow layers of unit 3 began before 32.3–25.2 ka BP and ended between 24.1 and 23.1 ka BP. The incision into unit 1 is thus constrained to have occurred before 25.2 ka BP. The deposition of unit 5 began after 20.6–17.0 ka BP and ended between 18.9 and 9.9 ka BP. This is consistent with the finding that the surface of the entire body of sediments is older than 2,040 ^{14}C BP.

Profile RGE records a sequence of erosion and autochthonous as well as allochthonous deposition ca. 15 km south-west of the Palpa Valley. A series of stratigraphically consistent radiocarbon dates constrains river valley incision to ≤6 m since ca. 30 ^{14}C ka BP and indicates a phase of fine-grained fluvial overbank sedimentation during the LGM.

Profiles La Muña

Tab. A-4: Positions of profiles: La Muna.

Profile	Depth [m]	Northing [m]	Easting [m]	Altitude [m]
LM-1	4.10	8,390,599.94	475,712.24	285.59
LM-2	3.50	8,390,590.13	475,721.67	284.86
LM-3	1.30	8,390,576.03	475,762.06	283.26
LM-4	2.25	8,390,569.15	475,773.28	281.35
LM-5	2.20	8,390,579.56	475,735.95	283.07
LM-6	1.35	8,390,572.60	475,754.70	283.29
LM-7	1.50	8,390,795.28	475,873.68	284.94
LM-8	2.90	8,390,584.85	475,727.39	284.35

Profile LM-1

Profile LM-1 (Figures A-3 and A-4; see also Figure 23) was a pit in the depositional fan at the mouth of Quebrada La Muña, a dry valley with a catchment area of 34.7 ha. The thalweg of the quebrada cuts the southern edge of the archaeological site of La Muña. Profile LM-1 is situated 14 m north-west (upslope) of profile LM-2.

Layer 0 consists of angular to subangular stones (>80%, up to 15 cm) in a matrix of sand. The surface is covered by sand. Layer 1 is a thin (up to 5 cm) band of fine sand with a sharp, undulating upper boundary. Layer 2 is fine sand with abundant organic debris. Its upper boundary is diffuse. Layer 3 consists of angular to subangular stones (>80%, up to 12 cm) in a sandy matrix. A piece of charcoal from the base of layer 3 (depth 0.70 m) was radiocarbon-dated at 450–150 cal BP (Poz-5259, 285 ±30 BP). Layer 4 is charcoal-rich silty sand with a horizontal upper boundary. Layer 5 is structureless silty fine sand with abundant small silty nodules and an undulating upper boundary. Layer 6 consists of slightly imbricated angular stones (80%, up to 6 cm) in a sandy matrix. Layer 7 is a downward-fining sequence from laminated fine sand (top) through structureless silty fine sand with small silt nodules to structureless fine sandy silt with silt clods (bottom). Layer 8 consists of angular stones (>80%, up to 12 cm) in a sandy matrix with an undulating upper boundary. Layer 9 consists of slightly imbricated angular stones (ca. 90%, up to 5 cm) in a sandy matrix. Its undulating upper boundary is marked by a 2 cm thick band of laminated fine sand.

Layer 10 is structureless silty sand with ca. 20% angular clasts (up to 6 cm) which are concentrated in lenses or bands. A radiocarbon assay on charcoal from layer 10 (depth 1.60 m) gave an age of 545–495 cal BP (Poz-5260, 525 ±30 BP). This age is stratigraphically not consistent with a sample age from layer 14 (too high age); hence it is assumed that this sample represents the deposition of remobilised older charcoal. The sample Poz-5260 is therefore excluded from further interpretation. Layer 11 is blackish silty fine sand with high content of disperse charcoal or organic matter. Root molds are present. Its upper boundary is sharp and horizontal. Layer 12 is silty fine sand with faint bedding. The diffuse upper boundary is marked by a locally broken-up band of laminated silt (ca. 5 mm thick). Root molds from the overlying layer 11 penetrate layer 12. Some charcoal is present. Layer 13 is compacted silty fine sand with ca. 15% coarse sand and fine angular gravel (up to 3 mm). Faint bedding is recognisable in the top of the layer. Layer 14 is a 2–5 cm thick band of blackish silty fine sand with a sharp, slightly undulating upper boundary. Root molds and some charcoal are present. A radiocarbon date of charcoal from layer 14 (depth 2.00 m) is 490–320 cal BP (Poz-5262, 385 ±30 BP).

Layer 15 is sand with undulating, broken-up lamination. Silt clods are present locally along its diffuse, undulating upper boundary. Root molds are present. Layer 16 is dark grey silty fine sand with some angular stones (up to 8 cm). Disperse charcoal and/or organic matter as well as root molds are present. The upper boundary is undulating. Layer 17 is structureless silty sand with some clods of cohesive silt. Its upper boundary is undulating and disturbed by root molds. Layer 18 consists of angular stones (>80%, up to 15 cm) some of which are strongly weathered. The matrix is fine sand. At its lower boundary, a band of structureless coarse sand up to 5 cm thick is locally present. Layer 19 is silty fine sand with disturbed bedding to lamination (locally broken up). Near the top of the layer, a disturbed charcoal-rich band (2 mm thick) is present. Layer 20 is a thin (2 cm), undulating band of blackish silty fine sand with faint lamination. Root molds are present. Layer 21 is silty fine sand with faint bedding and locally broken-up lamination. Charcoal from layer 21 (depth 2.93 m) was radiocarbon-dated at 910–730 cal BP (Poz-8010, 925 ±25 BP). Layer 22 is silt to fine sand with an undulating upper boundary. The top 10 cm of the layer are characterised by lamination, in the middle part there are lenses of cross-laminated medium to coarse sand (up to 10 cm thick), the bottom part is structureless. Oxidation stains and nodules (up to 5 cm) are present; however, given the very low carbonate content of layer 22, the nature of these nodules is unknown. Layer 23 is well-rounded fine gravel (>50%) with ca. 20% well-rounded coarse gravel (up to 10 cm). Imbrication and faint bedding are present.

As far as covered by sampling, the sediment in profile LM-1 is generally well sorted with the exception of layer 10. CSR is generally low (≤0.28), and LoI is low (≤1.0%) to intermediate (1.0–2.0%) except in layers 2 (3.4%) and 4 (4.3%). The blackish layer 14 has an LoI of 1.7%; the blackish layers 11, 16 and 20 which have not been sampled likely also have intermediate to high carbon contents. Some carbonate (≥2.1%) is present in all layers above a depth of 3.20 m, with small peaks occurring in layers 6 (4.3%, depth 0.95–1.05 m) and 18 (4.5%, depth 2.45–2.70 m). The bottom part of the profile (3.20–4.10 m) has a very low carbonate content (0.1–0.8%).

Profile LM-1 records a sequence of allochthonous and autochthonous sedimentation at the mouth of Quebrada La Muña. Fluvial gravel attributable to a former riverbed situation of the Rio Grande are present at a depth of 3.80 m. Fine-grained fluvial (allochthonous) sediment related to flood events is present between depths of 3.80 to 2.70 m. Evidence for a period of plant growth (indicated by root molds) and accumulation of disperse charcoal or organic matter on top of fine-grained fluvial sediment after 910–730 cal BP is overlain by the last secure allochthonous fluvial deposit and the first autochthonous debris flow deposit of coarse-grained, angular stones. Three further couples of fine sand overlain by thin blackish bands

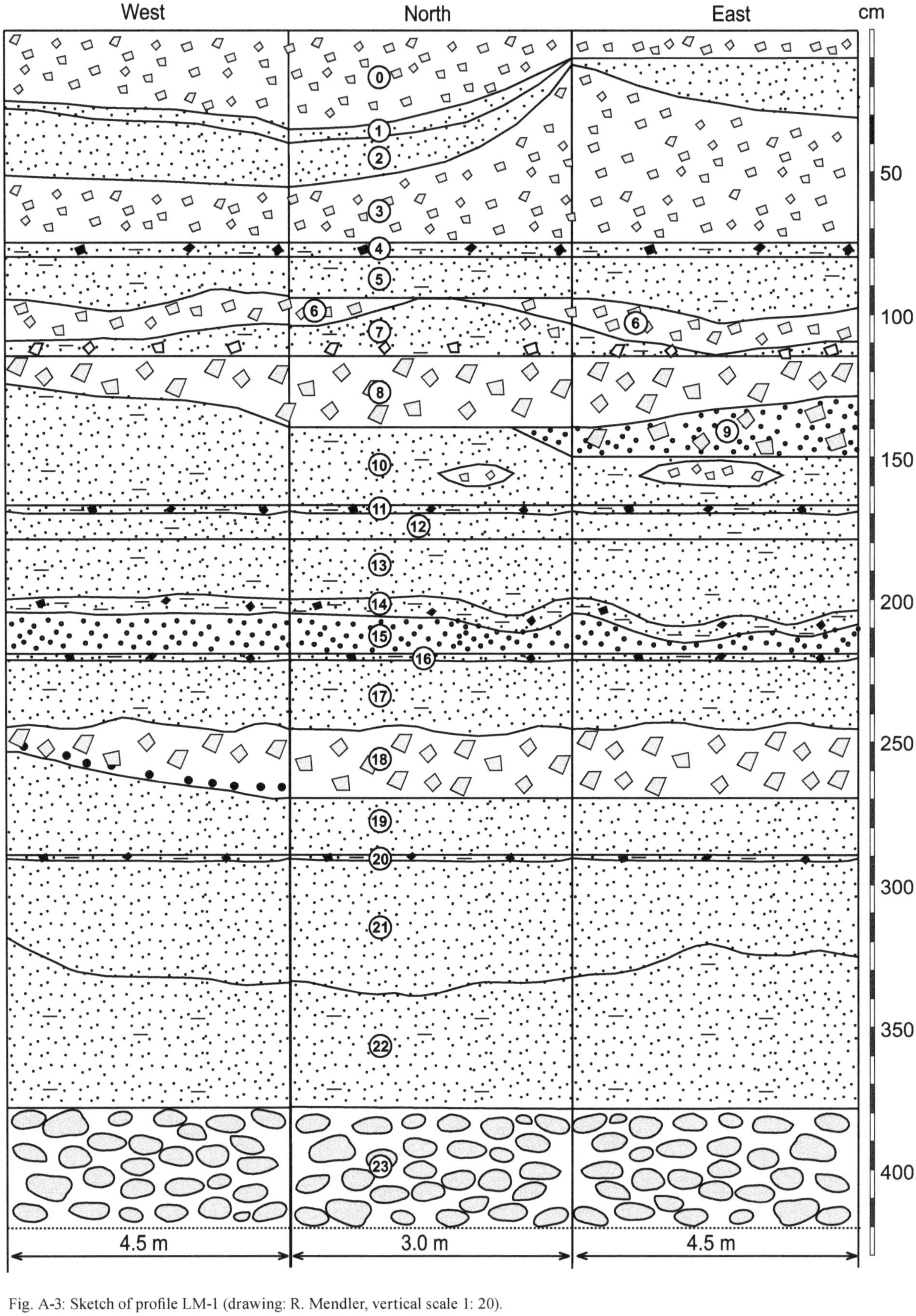

Fig. A-3: Sketch of profile LM-1 (drawing: R. Mendler, vertical scale 1: 20).

Fig. A-4 (following page): Laboratory results of samples from profile LM-1 (vertical scale 1: 30). Cumulative weight percent graph shows clay, silt and sand from left to right. Note that the lines connecting data points do not imply values between data points but serve solely to aid visualisation.

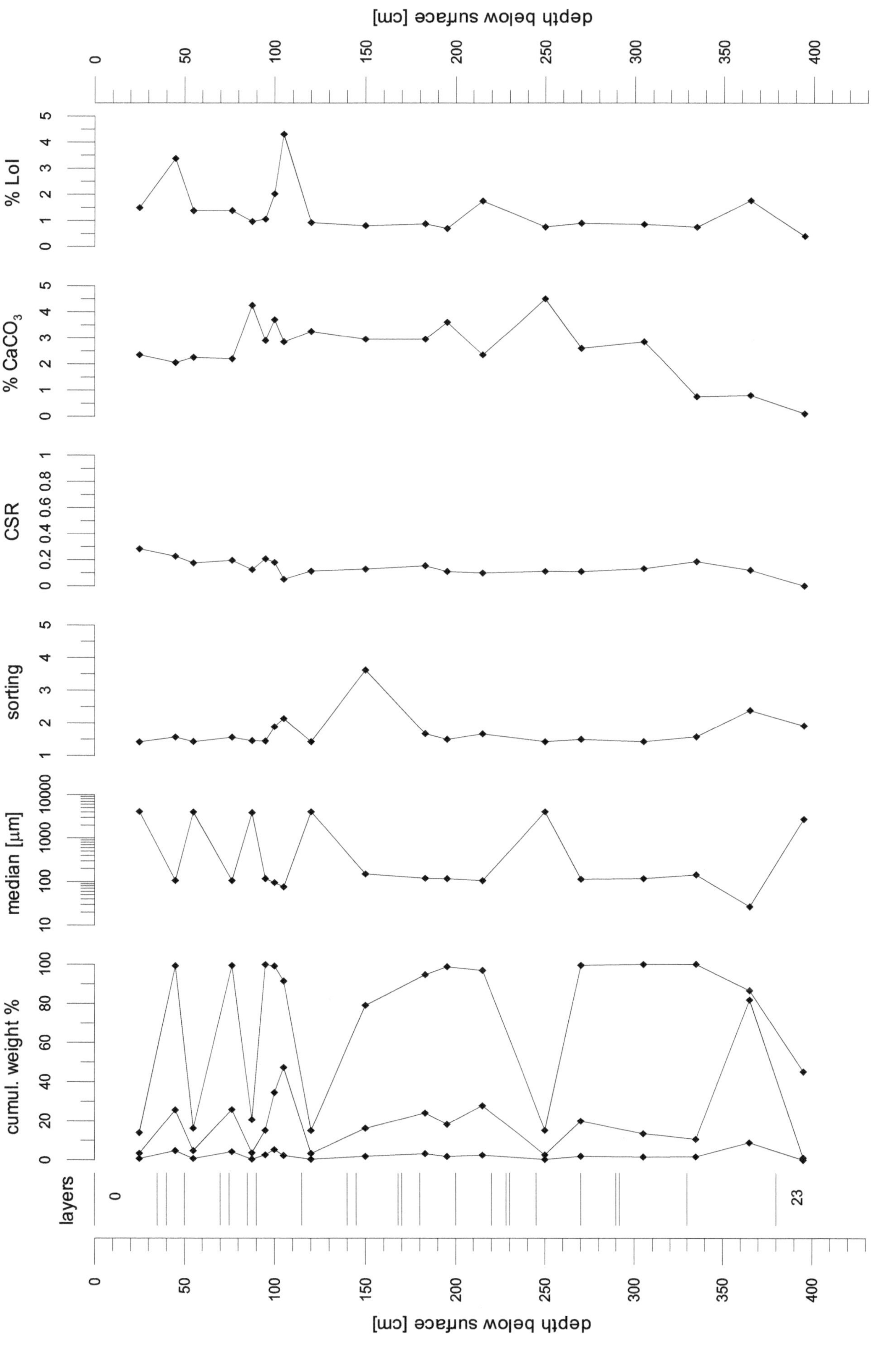
depth below surface [cm]
% LoI
% CaCO3
CSR
sorting
median [µm]
cumul. weight %
layers
0
23
depth below surface [cm]

are present between depths of 2.45 and 1.68 m. They are characterised by thin blackish top layers with high concentrations of disperse charcoal or organic matter and root molds extending down into the underlying sediment. While the fine-grained sediments can not be securely attributed to either allochthonous overbank flow of the Rio Grande or autochthonous fine-grained sedimentation from Quebrada La Muña, the diminution of layer thickness (30–25–20–12 cm) with increasing vertical accretion conforms with the pattern which would be expected from flood events (i.e. a decrease in flood sediment thickness with increasing height above the riverbed). A relatively fine-grained hyperconcentrated debris flow deposit (layer 13) is present between the uppermost couple at a depth of 2.10–1.80 m. Above a depth of 1.68, all layers are attributed to coarse- and fine-grained debris flow or fluvial sedimentation from Quebrada La Muña, with 6 coarse-grained events recorded since 490–320 cal BP (depth 2.00 m). Fine-grained fluvial sediment from Quebrada La Muña has to be attributed to fluvially mobilised aeolian material. Human activity in the vicinity of the site is indicated by the presence of charcoal.

Profile LM-2

Profile LM-2 (Figures A-5 and A-6; see also Figure 23) was a pit on the depositional fan at the mouth of Quebrada La Muña, a dry valley with a catchment area of 34.7 ha. The thalweg of the quebrada cuts the southern edge of the archaeological site of La Muña. Profile LM-2 is situated 14 m south-east (downslope) of profile LM-1 and 8 m north-west of profile LM-8.

Layer 0 is silty fine sand with abundant organic debris, angular stones (up to 5 cm) and charcoal. Layer 1 is blackish silty fine sand with an undulating upper boundary. It contains abundant disperse charcoal. Layer 2 is silty fine sand with a diffuse, undulating upper boundary which is marked by a band of disturbed, laminated sandy silt. A lens of coarse sand, a thin band of silt and clods of cohesive silt as well as charcoal are present. Layer 3 is a filled-in pit of ca. 60 cm diameter. The fill consists of silty fine sand with distinct colour changes (reddish, probably heat-induced oxidation, white ash and dark grey charcoal-rich material). Layer 4 is structureless silty fine sand with a diffuse upper boundary. Silt nodules are common in the top and bottom 5 cm of the layer. Some oxidation stains are present; root molds penetrate the lower boundary. A charcoal sample from layer 4 (depth 0.68 m) was used for radiocarbon dating, resulting in an age of ≤300 cal BP (Poz-5263, 195 ±30 BP). Layer 5 is a thin (2 cm), undulating, dark grey band of sandy silt with disperse organic matter. A radiocarbon date of charcoal from this layer (depth 0.83 m) is 450–300 cal BP (Poz-8013, 335 ±25 BP).

Layer 6 is silty fine sand with an undulating upper boundary. Upper and lower boundary are disturbed by root molds. Broken-up laminated silt and clods of cohesive silt are present. Layer 7 is a 4 cm thick, dark grey band of sandy silt with an undulating upper boundary. Charcoal from layer 7 (depth 0.97 m) was radiocarbon-dated at 520–330 cal BP (Poz-8012, 450 ±25 BP). Layer 8 is structureless silty fine sand with a few small silt clods. Oxidation stains are present. The undulating upper and lower boundaries are penetrated by root molds. Layer 9 is an undulating, disturbed, black, organic-rich band of 2–4 cm thickness. It is penetrated by root molds. A radiocarbon assay on charcoal from layer 9 (depth 1.20 m) resulted in an age of 545–495 cal BP (Poz-8011, 520 ±30 BP).

Layer 10 is structureless loamy silt with an undulating upper boundary which is disturbed by root molds. Oxidation stains and abundant clods of cohesive silt are present. Layer 11 is structureless silty fine sand with a diffuse, undulating upper boundary. Oxidation stains and silt clods are present. Layer 12 is structureless fine sandy silt with a slightly undulating upper boundary. Laminated cohesive clods of silt (up to 8 cm) as well as strong red and black oxidation stains are present. Layer 13 is structureless clayey silt with a slightly undulating upper boundary. Brown oxidation stains are present. Layer 14 is structureless silty fine sand with black and reddish-brown oxidation stains and a slightly undulating upper boundary. Some small nodules are present. Charcoal sampled from layer 14 (depth 2.00 m) was radiocarbon-dated at 680–560 cal BP (Poz-8047, 725 ±35 BP). Layer 15 is clayey silt with a very diffuse upper boundary. A diffuse, undulating band of sand is present in the middle of layer 15. Layer 16 is clayey silt with a diffuse upper boundary and red-brown oxidation stains. Layer 17 is sandy silt with oxidation stains and an undulating upper boundary. Layer 18 is well-rounded gravel (ca. 60%, up to 5 cm) in a matrix of coarse sand. Red-brown oxidation stains are present. The upper boundary is horizontal. Layer 19 is well-bedded fine to coarse sand with a horizontal upper boundary and ca. 50% red-brown oxidation stains. Layer 20 is well-rounded coarse gravel (up to 20 cm).

Profile LM-2 records both allochthonous and autochthonous sedimentation at the mouth of Quebrada La Muña. The bottom of the profile (depth 3.50–3.05 m) is characterised by low- to medium-energy fluvial sedimentation (sandy silt to gravel) overlying coarse gravel indicative of a high-energy fluvial environment (riverbed). Between depths of 3.05 and 1.43 m, several layers with high clay content (layers 16, 13 and 10) are interpreted as results of low-energy (still- water) deposition. Root molds present in most layers between depths of 1.75 and 0.65 m together with the presence of organic-rich bands indicate plant growth at the site. While autochthonous sedimentation has to be assumed for the top part of the profile due to the surface morphology of the site, the lack of clearly identifiable autochthonous material (angular clasts of local provenance; only present at the top of the profile) inhibits a definition of the transition from allochthonous

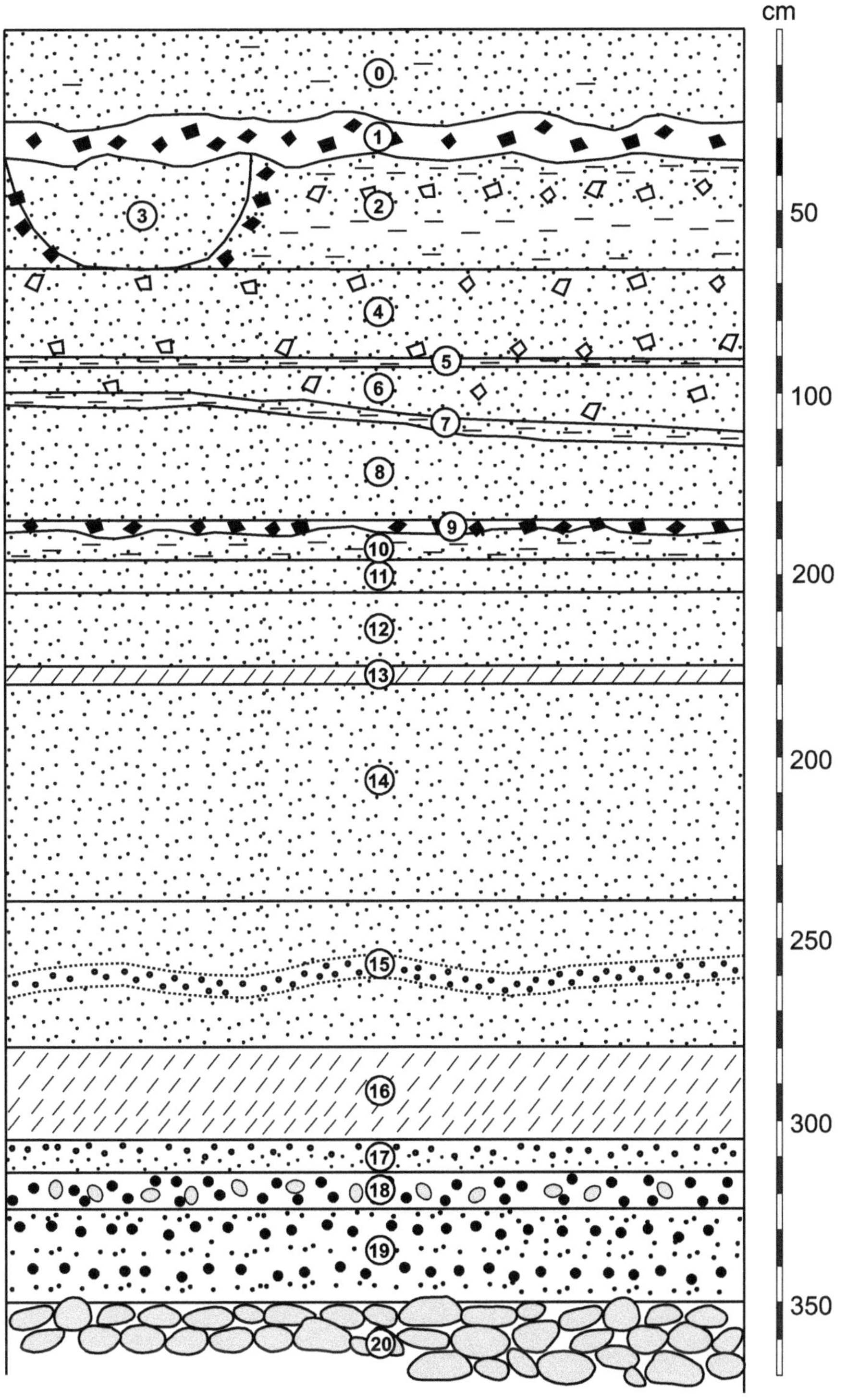

Fig. A-5: Sketch of profile LM-2 (drawing: R. Mendler; vertical scale 1: 20).

to autochthonous dynamics. A series of four stratigraphically consistent radiocarbon dates exists for profile LM-2. Accumulation of the sediments began before 680–560 cal BP (depth 2.00 m) and has since taken place at an average sedimentation rate of 2.72-3.25 mm/a.

Profile LM-3

Profile LM-3 (Figures A-7 and A-8; see Figure 23) was a pit in the field between the mouth of Quebrada La Muña and the Rio Grande. It is situated 15 m north-west of profile LM-4 and 8 m north-east of profile LM-6 The irrigation channel running at the western edge of the field lies at a distance of ca. 20 m.

Layer 0 (the recent A_p / irrigation horizon) is silty fine to medium sand with cohesive clods of silt (ca. 10%, 0.5–2.0 cm in size) and a crumbly structure. A few stones (up to 3 cm) are present. Layer 1 is a strongly disturbed band of medium to fine sand with a broken-up upper boundary. Faint remnant bedding can be discerned locally. Locally, the layer has a lens-like appearance. Layer 2 is silty

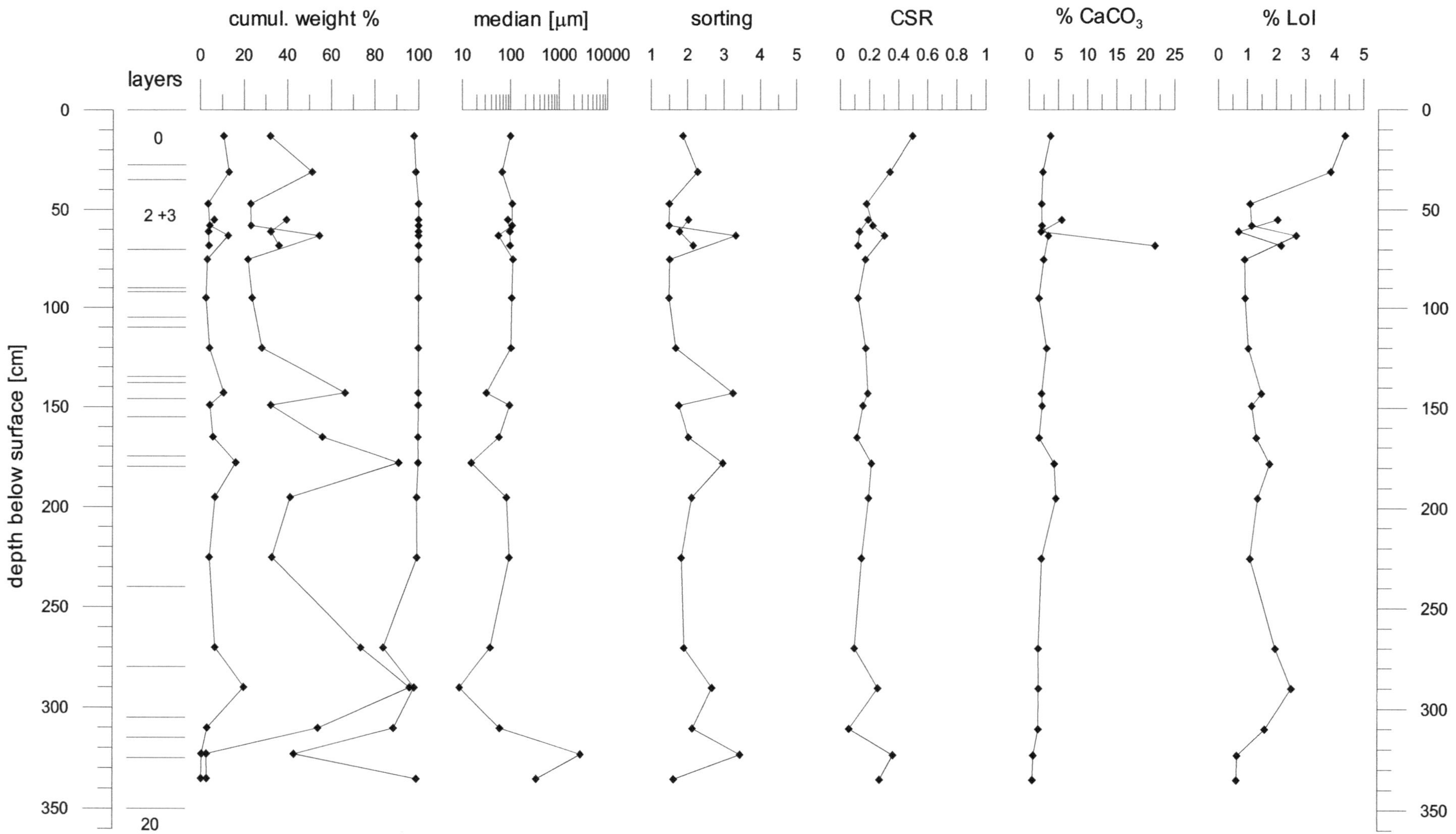

Fig. A-6: Laboratory results of samples from profile LM-2 (vertical scale 1: 30). Cumulative weight percent graph shows clay, silt and sand from left to right. Note that the lines connecting data points do not imply values between data point but serve solely to aid visualisation.

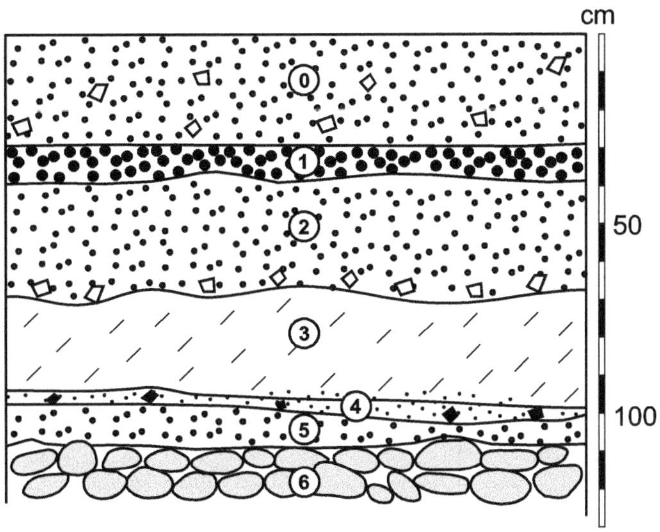

Fig. A-7:Sketch of profile LM-3 (drawing: R. Mendler, vertical scale 1: 20).

fine sand with cohesive clods of silt (1–4 cm in size, ca. 10%, concentration increasing towards the lower boundary). Layer 3 is loamy silt with a slightly undulating upper boundary. Near its base, some blotches of sand are present. Layer 4 is bedded fine sand with a diffuse, undulating upper boundary. Locally, blotches of loam and clay are present. The layer contains abundant charcoal. A charcoal sample from layer 4 (depth 0.98 m) was radiocarbon-dated at ≤260 cal BP (Poz-5266, 105 ±30 BP). Layer 5 is sand with a diffuse, undulating upper boundary. Layer 6 is well-rounded coarse gravel (>90%, up to 30 cm) in a sandy matrix.

Profile LM-3 records rapid sedimentation mainly by fluvial processes on the field between the mouth of Quebrada La Muña and the Rio Grande in historic times. High-energy fluvial channel deposits below a depth of 1.10–1.00 m are overlain by two thin fine-grained layers the upper of which is characterised by abundant charcoal (likely representing nearby human activity). This layer was radiocarbon-dated at ≤260 cal BP. The following layer (depth 0.95/1.00–0.70 m) likely represents agricultural disturbance of fluvial sediments and addition of irrigation sediment; it is characterised by increased content of organic matter (LoI=2.2%). The top 0.70 m of the profile likely represent young, agriculturally disturbed fluvial overbank sediment as indicated by low organic matter content (LoI: 0.5–1.4%), remnant bedding and broken-up layer boundaries as well as cohesive clods of broken-up sediment within layers. Average sedimentation rate for the top 0.98 m of the profile (since ≤260 cal BP) is ≥3.11 mm/a.

Profile LM-4

Profile LM-4 (Figure A-9; see Figure 23) was recorded on a natural exposure on the right bank of the Rio Grande. The top of the exposure is covered by trees and undergrowth. Profile LM-4 is situated 15 m south-east of profile LM-3 and 19 m east of profile LM-6.

Layer 0 is structureless sand. Layer 1 consists of laminated material alternating between fine and coarse sand. Root molds penetrate the layer. Layer 2 is fine sand with faint bedding. Layer 3 is structureless silty fine sand. An up to 6 cm thick lens of laminated sand is present at a depth of ca. 0.60 m. Layer 4 is bedded sand with an upwards-fining tendency from coarse to fine sand. Layer 5 is structureless silty sand. Layer 6 is bedded loam and sand (alternating). Some brown oxidation stains are present. Layer 7 is structureless clay. A 5 cm thick lens of sand without internal structure is present at a depth of 2.00 m. Low water level of the Rio Grande (water depth in the channel ca. 0.3 m) impinges on the exposure at 2.25 m below the top of the profile. Charcoal from layer 7 (depth 2.00 m) was radiocarbon-dated at ≤260 cal BP (Poz-5264, 85 ±30 BP). Coarse, well-rounded gravel is present at a depth of 2.35 m.

Profile LM-4 records rapid near-channel fine-grained aggradation of 2.00 m since ≤260 cal BP. The exclusively low-energy fluvial deposits include mainly bedded to laminated fine to coarse sand as well as structureless silty sand. Indicators of human impact and agricultural disturbance such as broken-up layer boundaries or mixed-in cohesive clods of silt, stones or abundant charcoal are absent. The average sedimentation rate for the top 2.00 m of the profile (since ≤260 cal BP) is ≥6.35 mm/a.

Profile LM-5

Profile LM-5 (Figures A-10, A-11 and A-12; see Figure 23) was a pit near the western edge of the field in front of the mouth of Quebrada La Muña, 10 m south-east of profile LM-8 and 20 m north-west of profile LM-6 with which

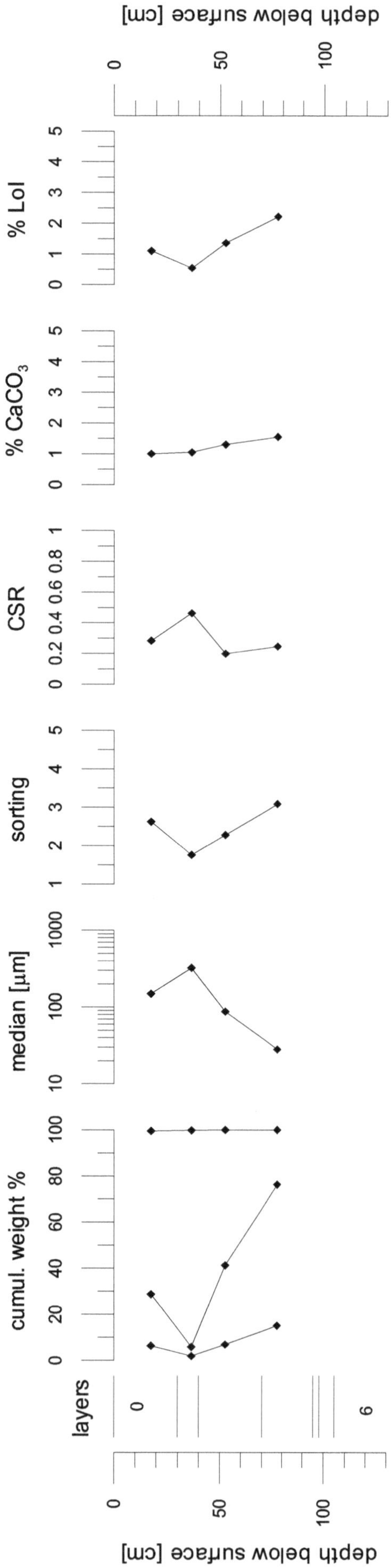

Fig. A-8: Laboratory results of samples from profile LM-3 (vertical scale 1: 30). Cumulative weight percent graph shows clay, silt and sand from left to right. Note that the lines connecting data points do not imply values between data point but serve solely to aid visualisation.

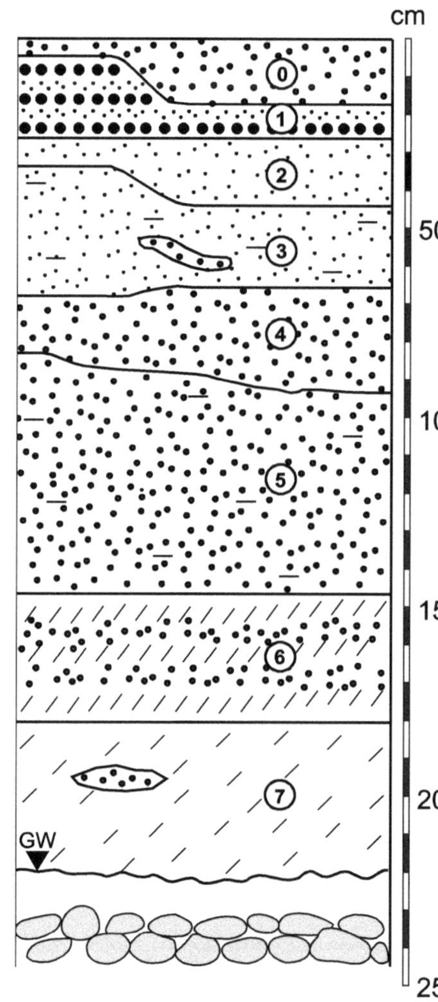

Fig. A-9: Sketch of profile LM-4 (drawing: R. Mendler, vertical scale 1: 20).

it was connected by a trench. Recent irrigation furrows were present at the surface of the field.

Layer 0 (the recent A_p / irrigation horizon) is crumbly loamy sand with some clods of cohesive silt. Abundant roots are present. Layer 1 is loamy sand with some charcoal and clods of silt. Abundant roots are present. Like layer 0, it has a crumbly structure, however, it is more compacted. The upper boundary is diffuse and disturbed / pocketed. Layer 2 is loamy sand with abundant clods of silt. It contains charcoal and has a disturbed, pocketed upper boundary. Charcoal from the base of layer 2 (depth 0.72 m) was radiocarbon-dated at ≤300 cal BP (Poz-8015, 205 ±25 BP). Layer 3 is structureless, well-sorted silty fine sand with a horizontal upper boundary. Some clods of compacted material as well as root molds are present. Layer 4 is structureless sandy silt with an undulating to pocketed upper boundary. Red and black lenses are present along the upper boundary; some charcoal is present. Layer 5 is loamy silt with some charcoal and oxidation stains. Some horizontal lenses of sand are present. The upper boundary is horizontal. Layer 6 is sandy silt with a horizontal upper boundary. Mica content is high and strong oxidation stains as well as loamy disturbances are present. Layer 7 is clayey silt with carbonate nodules and a stepped, erosional upper boundary. Layer 8 is structureless, well-sorted silty fine sand with black and red oxidation stains. Its sharp upper boundary is stepped and erosional. Layer 9 consists of alternating beds of silty loam and silty sand. Charcoal from the base of layer 9 (depth 1.75-1.80 m) was radiocarbon-dated at 320–150 cal BP (Poz-8014, 255 ±25 BP). Pockets of loam are present in the undulating upper boundary of the underlying layer 10. Layer 10 consists of poorly sorted, well-rounded coarse gravel (up to 25 cm) in a matrix of coarse sand and fine gravel. Granites are strongly weathered.

Profile LM-5 records a sequence of mainly fluvial low-energy deposition (abandoned channel fill and/or overbank flow) following the termination of a river channel situation. Erosional events and subsequent infilling are recognisable at depths between 1.78 and 1.43 m. Increased LoI values (2.1–3.1%) in the lower portion of the stratigraphy indicate the accumulation of organic matter, probably in a context of riverine vegetation. Flood deposits are recognisable upwards to a depth of 0.69 m, followed by strongly turbated sediment with poorer sorting and increasing LoI (up to 2.4%) and indicators of human, in particular agricultural, activities such as the presence of cohesive clods of silt and charcoal. However, given the proximity of the Rio Grande and the high sedimentation rate of ≥2.03 mm/a since ≤300 cal BP (depth 0.72 m), fluvial overbank sedimentation may have contributed to the build-up of the stratigraphy. Average sedimentation rate for the top 1.78 m of the profile is 4.73–8.66 mm/a.

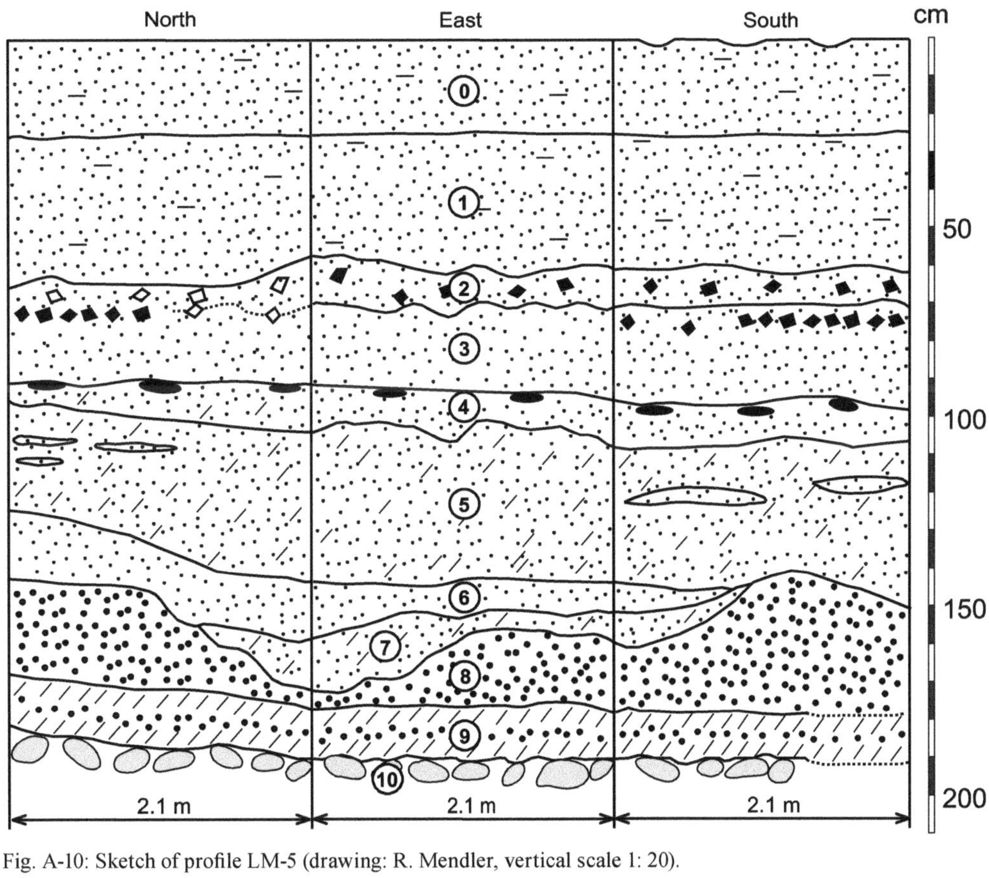

Fig. A-10: Sketch of profile LM-5 (drawing: R. Mendler, vertical scale 1: 20).

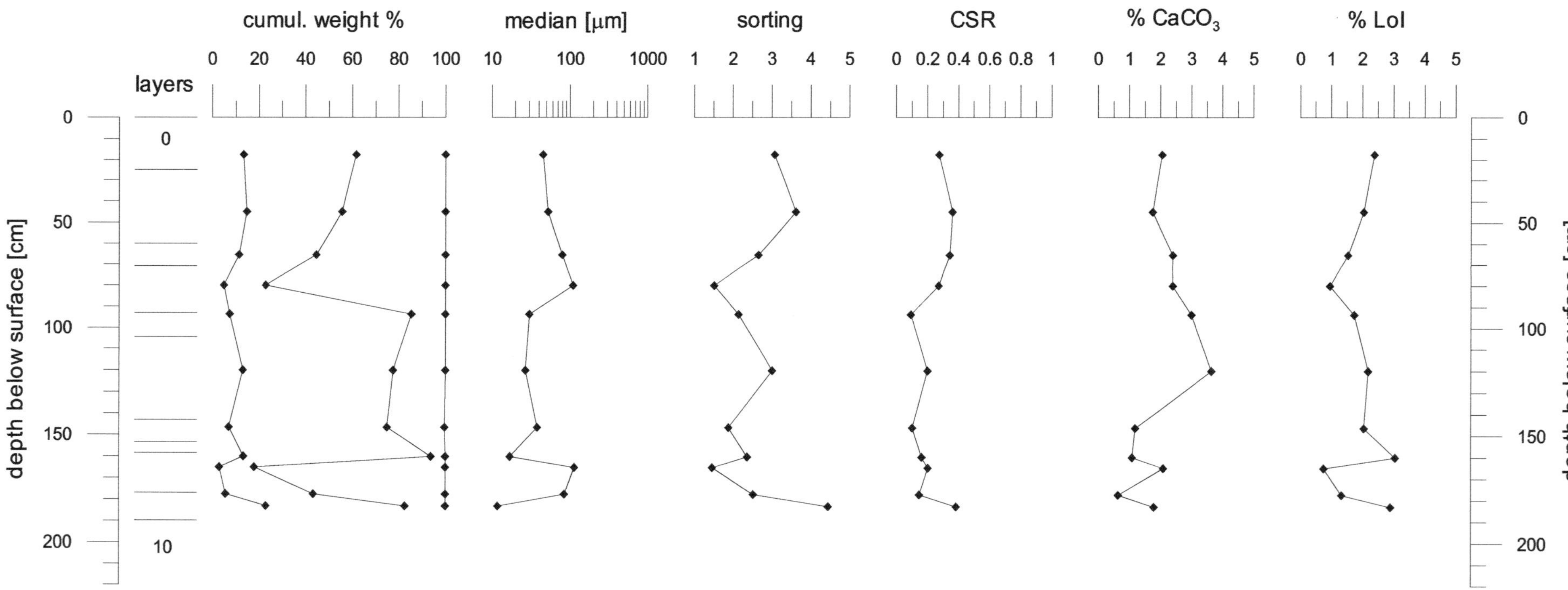

Fig. A-11: Laboratory results of samples from profile LM-5 (vertical scale 1: 30). Cumulative weight percent graph shows clay, silt and sand from left to right. Note that the lines connecting data points do not imply values between data point but serve solely to aid visualisation.

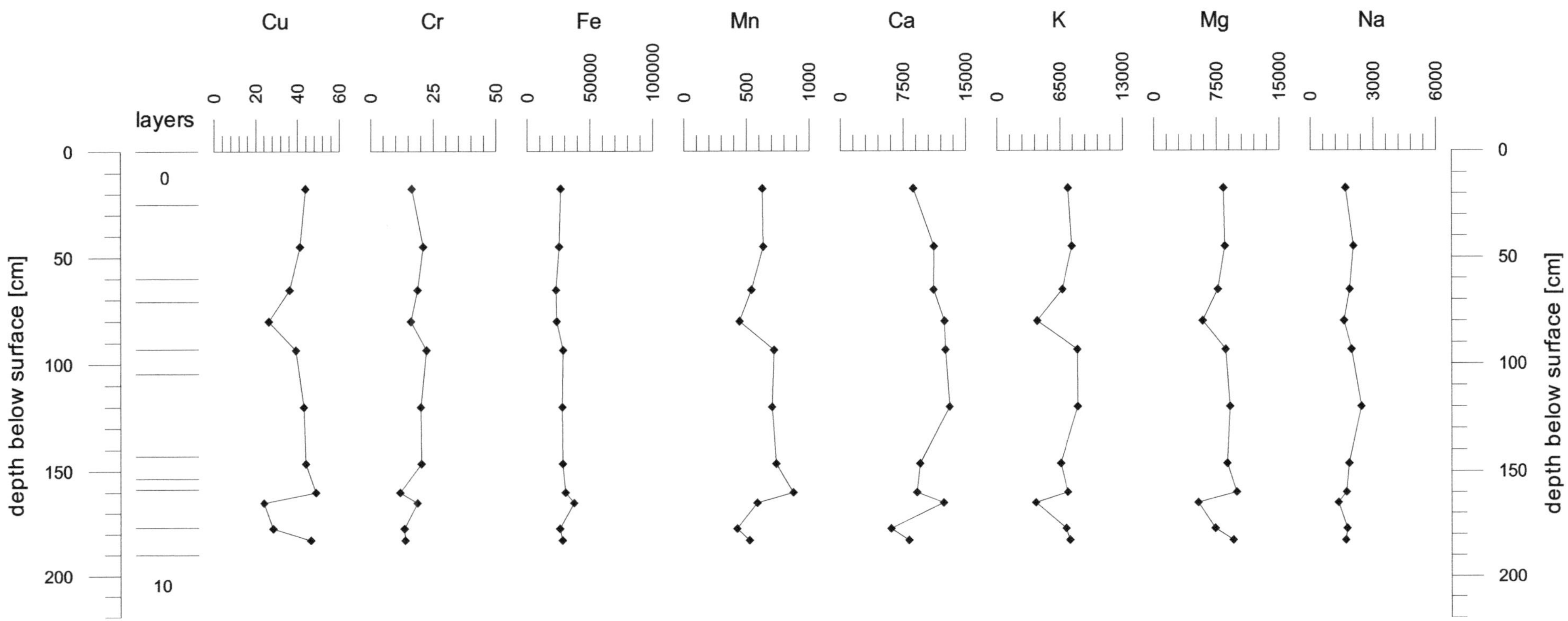

Fig. A-12: AAS measurement results of samples from profile LM-5 (vertical scale 1: 30). Note that the lines connecting data points do not imply values between data points but serve solely to aid visualisation.

Profile LM-6

Profile LM-6 (Figures A-13, A-14, A-15 and A-16; see Figure 23) was a pit on a field between the mouth of Quebrada La Muña and the Rio Grande. It is situated 20 m south-east of profile LM-5 with which it was connected by a trench and 8 m south-west of profile LM-3. Profile LM-4 lies at a distance of 19 m to the east.

Layer 0 is loose silty fine sand with organic debris. Layer 1 is compact silty fine to medium sand with abundant clods of cohesive silt (up to 2 cm) and a sharp, slightly undulating upper boundary. Layer 2 is silty fine sand with sand lenses and a sharp to diffuse upper boundary. Layer 3 is well-sorted silty fine sand with abundant roots and silt clods. Its upper boundary is horizontal. Layer 4 is loamy silt with disperse charcoal and a few silt clods. The upper boundary is diffuse and undulating and locally pocketed. Layer 5 is well-sorted silty fine sand. Layer 6 is well-rounded, poorly sorted coarse gravel. On the south-eastern side of the pit, a hearth is present on top of the gravel. Charcoal from this hearth (depth 1.12 m) was radiocarbon-dated at ≤260 cal BP (Poz-8049, 110 ±30 BP).

Profile LM-6 records rapid aggradation from low-energy fluvial deposition on top of riverbed gravel present at a depth below 1.13/1.03 m. Fine-grained sedimentation began after ≤260 cal BP as indicated by a radiocarbon-dated hearth on top of the basal gravel. Plant growth on very-low-energy (still water, likely abandoned channel fill) sediment (loamy silt, depth 1.00–0.74 m) is indicated by increased LoI (4.0%). The presence of charcoal and clods of silt indicates incipient agricultural use of the site which was followed by two further flood deposits between 0.74–0.36 m depth, the lower of which also shows traces of tillage. The top 0.36 m of the profile represent irrigation sediment likely augmented by flood deposits and strongly turbated by tillage. Average sedimentation rate for the top 1.12 m of the profile (since ≤260 cal BP) is ≥3.56 mm/a. A further radiocarbon date from the base of the trench connecting profiles LM-5 and LM-6 (depth 1.05 m; 8.1 m from LM-5, 11.9 m from LM-6) is ≤300 cal BP (Poz-8016, 210 ±25 BP), resulting in an average sedimentation rate of ≥2.96 mm/a.

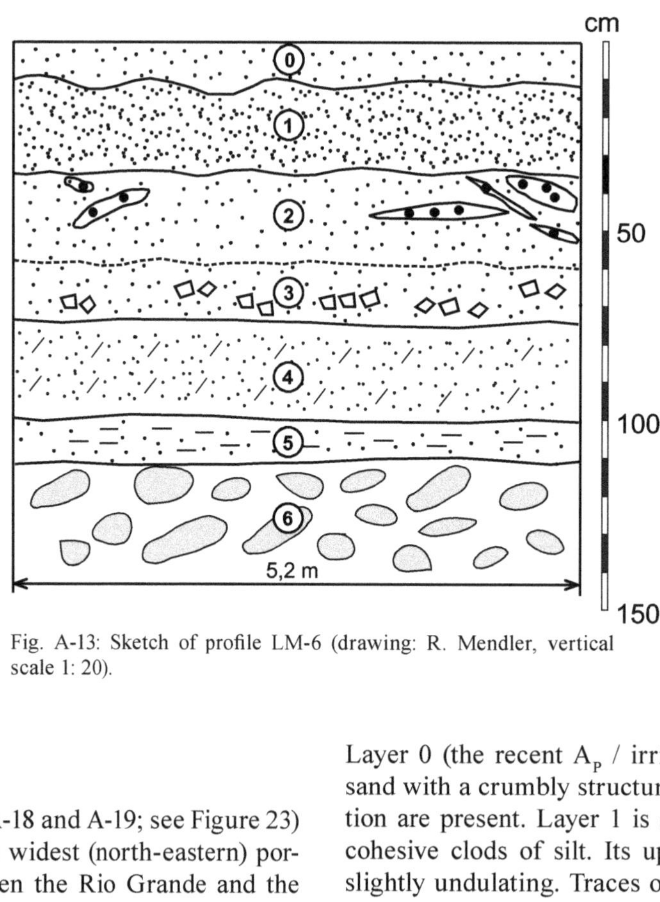

Fig. A-13: Sketch of profile LM-6 (drawing: R. Mendler, vertical scale 1: 20).

Profile LM-7

Profile LM-7 (Figures A-17, A-18 and A-19; see Figure 23) was a pit in the centre of the widest (north-eastern) portion of the field lying between the Rio Grande and the mouth of Quebrada La Muña. It is situated 204 m south of profile RG-Fu (on the opposite side of the river) and ca. 250 m north to north-east of profiles LM-1 to LM-6 as well as LM-8. The Rio Grande runs at a distance of ca. 70 m north to south-east of profile LM-7. Recent irrigation furrows were present on the surface.

Layer 0 (the recent A_P / irrigation horizon) is silty fine sand with a crumbly structure. Roots and surface vegetation are present. Layer 1 is structureless silty sand with cohesive clods of silt. Its upper boundary is sharp and slightly undulating. Traces of a filled-in minor irrigation channel are recognisable. Layer 2 is bedded silty sand with an undulating upper boundary. It appears to cover a former field surface with traces of irrigation furrows on the northern and southern side of the pit. The grain-size sequence of layer 2 (from top to bottom) is medium to coarse sand, fine sand, silt, fine to coarse sand, silt. Layer

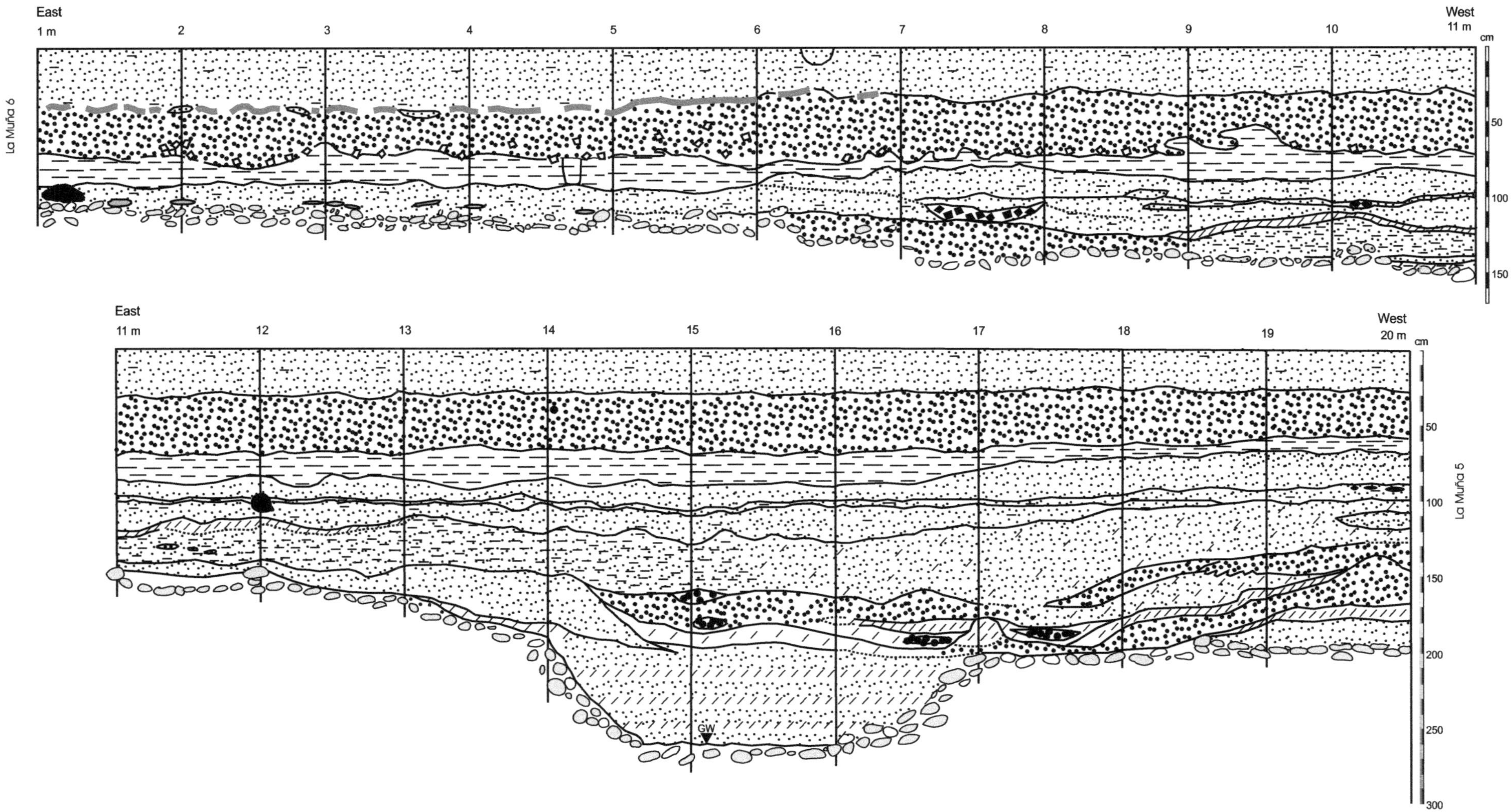

Fig. A-14: Sketch of the cut from profile LM-5 to profile LM-6 (drawing: R. Mendler, scale 1: 40).

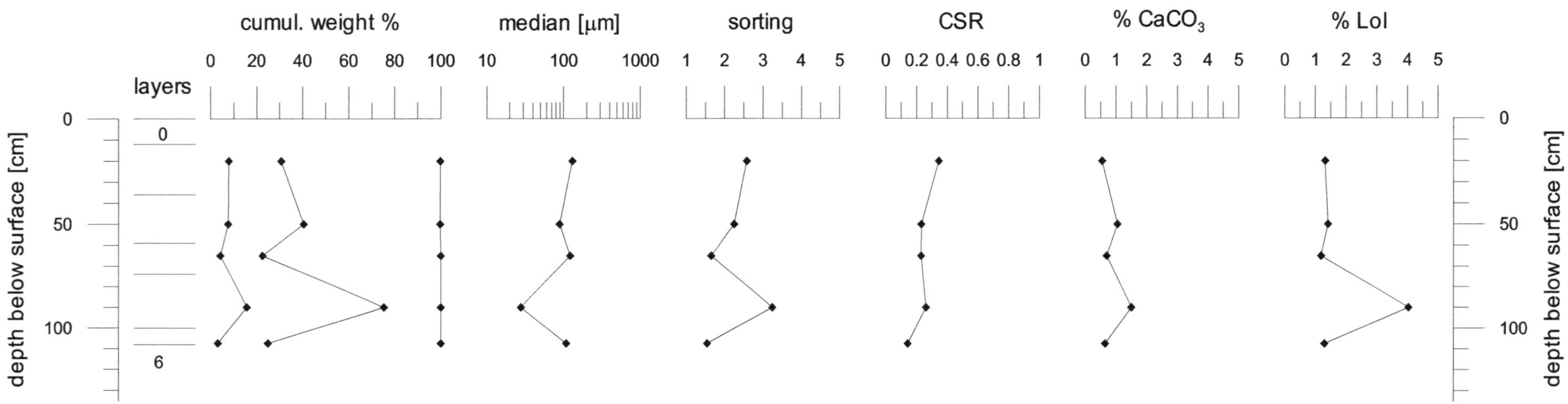

Fig. A-15: Laboratory results of samples from profile LM-6 (vertical scale 1: 30). Cumulative weight percent graph shows clay, silt and sand from left to right. Note that the lines connecting data points do not imply values between data point but serve solely to aid visualisation.

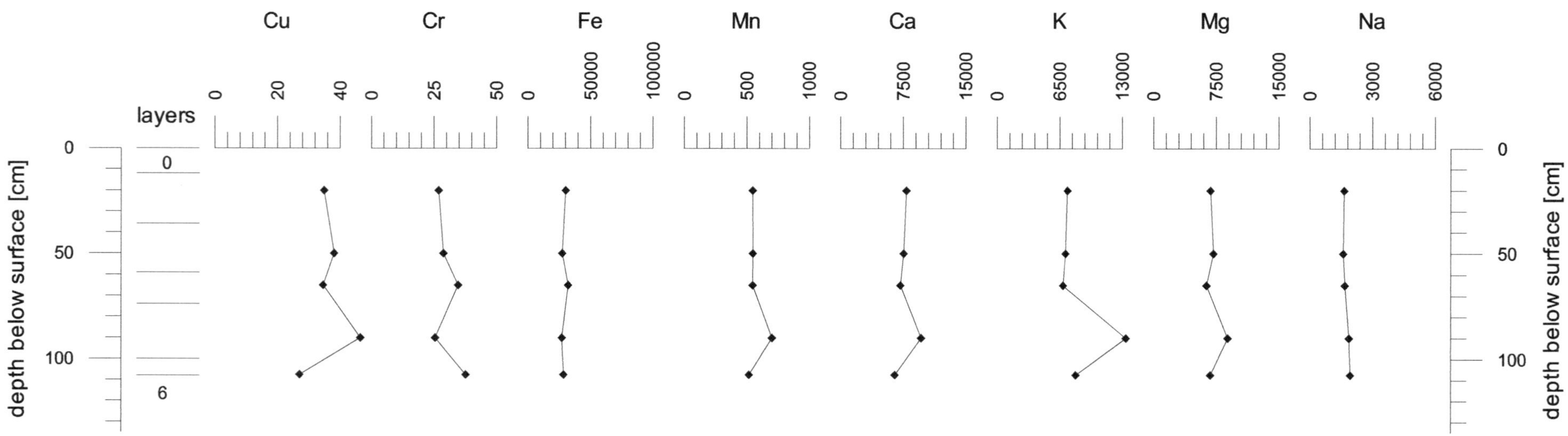

Fig. A-16: AAS measurement results of samples from profile LM-6 (vertical scale 1: 30). Note that the lines connecting data points do not imply values between data points but serve solely to aid visualisation.

3 is crumbly to structureless sandy to loamy silt with a horizontal upper boundary. It contains cohesive clods of silt and some charcoal. Layer 4 is a 2–4 cm thick, undulating band of crumbly loam. Layer 5 is bedded to laminated silty fine sand with undulating bedding planes. Layer 6 is a 1–2 cm thick band of crumbly loam. Layer 7 is bedded loamy sand with an upwards-coarsening tendency of the sand fraction from coarse to fine sand. Its upper boundary is undulating. Layer 8 is silty fine to medium sand with clods to bands of cohesive silt (up to 9 cm long) distributed within the layer. Some charcoal and oxidation stains are present. Its upper boundary is undulating. Layer 9 is bedded, well-sorted fine and medium sand. Cross-bedding and a north-south trending channel structure as well as oxidation stains are present. Layer 10 is well-rounded coarse gravel locally capped by thin loamy deposits.

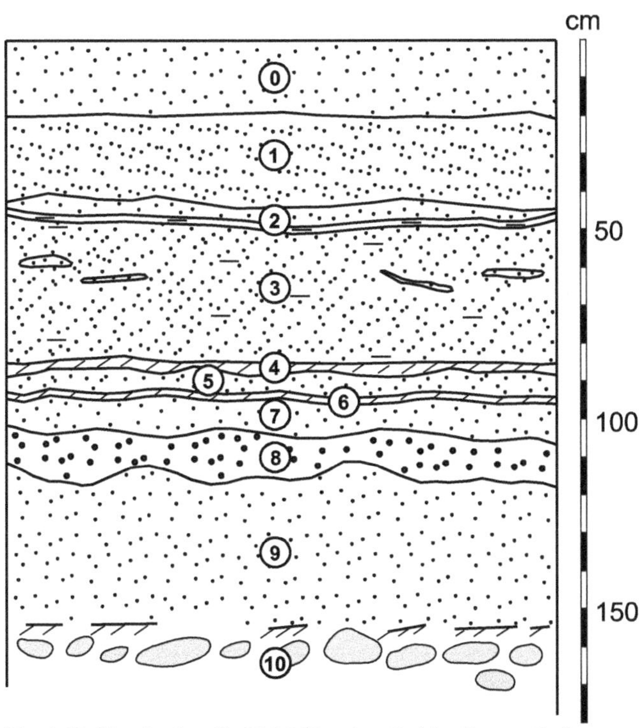

Fig. A-17: Sketch of profile LM-7 (drawing: R. Mendler, vertical scale 1: 20).

Profile LM-7 records at least four fluvial overbank deposition events interspersed and capped by sediments indicative of agricultural activity. Basal gravel is present at a depth of 1.50 m. Like the top 0.40 m of the profile, two layers at depth of 0.83–0.40 and 1.18–0.98 overlie bedded to laminated fluvial sediments and are characterised by strong turbation (with cohesive clods of silt distributed throughout the layers). LoI values are low (0.7–1.5%) throughout the profile and indicate lacking accumulation of organic matter and hence a rapid build-up of the stratigraphy.

Profile LM-8

Profile LM-8 (Figures 20, 21 and 22; see Figure 23) was a pit at the edge of the irrigation channel at the western edge of the field between the mouth of Quebrada La Muña and the Rio Grande. It is situated 8 m south-east of profile LM-2 and 10 m north-west of profile LM-5.

Layer 0 is loose medium sand with abundant organic debris. Traces of recent fire are visible on the surface. Layer 1 is structureless silty fine sand with a sharp upper boundary. It appears to have been erosionally truncated on the eastern side of the pit. Layer 2 is a 6–9 cm thick band of strongly turbated silty fine sand with a sharp, slightly undulating upper boundary. At its base, charcoal-rich fine sand is present. Like layer 1, layer 2 is truncated on the eastern side of the pit. Layer 3 is bedded silty fine sand with a few silt-rich beds. Its upper boundary is disturbed (pocketed). Layer 4 is loamy silt with faint bedding and an undulating upper boundary. Top and base of layer 4 are marked by bands of silt to loam. Layer 5 is loam with disturbed bedding and a disturbed, horizontal upper boundary. It contains abundant charcoal.

Layer 6 is structureless clayey silt with a disturbed upper boundary. Oxidation stains and small carbonate nodules are present. The thickness of layer 6 fluctuates between 8 and 30 cm. Layer 7 is bedded to laminated loamy silt with an undulating, disturbed upper boundary. Oxidation stains and sand lenses (up to 1 m wide and 10 cm thick) are present. Layer 8 is bedded loamy silt with an undulating upper boundary. Oxidation stains and small nodules are present. Layer 9 is silty fine sand with oxidation stains and an undulating upper boundary. Layer 10 consists of lenses of well-sorted fine sand with oxidation stains. Layer 11 is silty fine sand with an undulating upper boundary. Oxidation stains, charcoal and carbonate nodules are present. Layer 12 is silty sand with a sharp, slightly undulating upper boundary. It contains cohesive clods of silt, oxidation stains, blotches of coarse sand and traces of tillage. Layer 13 is well-sorted medium sand with an undulating upper boundary. Oxidation stains are present. Bedding is well-developed on the northern and southern side of the pit but is not discernible on the western side. On the top of layer 13, a wedge-like band of fine sand extends from the south-east and peters out towards the north-west. Layer 14 is bedded clayey silt with a sharp, undulating upper boundary. Carbonate nodules as well as

Fig. A-18 (following page, top): Laboratory results of samples from profile LM-7 (vertical scale 1: 30). Cumulative weight percent graph shows clay, silt and sand from left to right. Note that the lines connecting data points do not imply values between data points but serve solely to aid visualisation.

Fig. A-19 (following page, bottom): AAS measurement results of samples from profile LM-7 (vertical scale 1: 30). Note that the lines connecting data points do not imply values between data points but serve solely to aid visualisation.

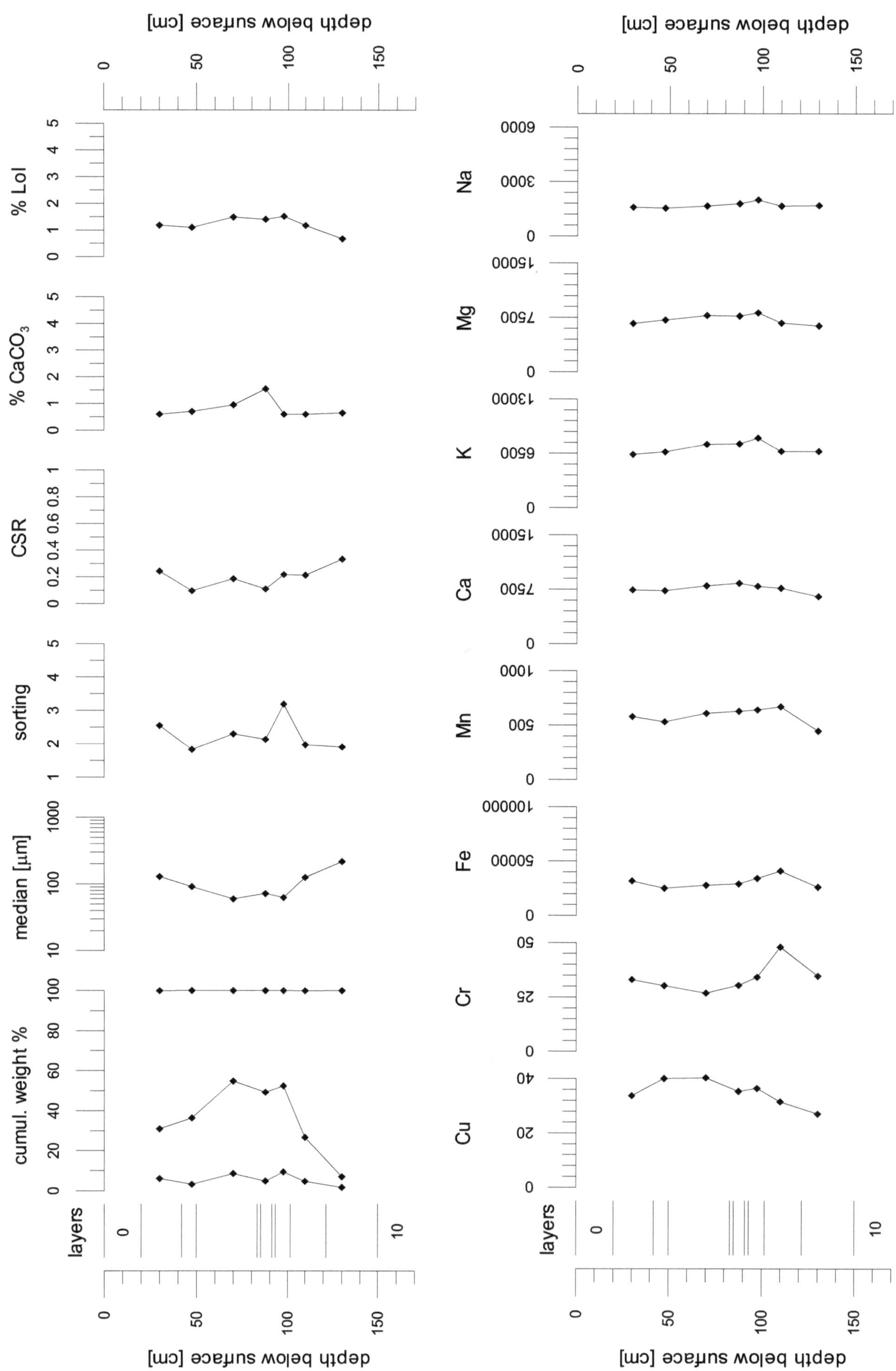

depth below surface [cm]
layers
cumul. weight %
median [µm]
sorting
CSR
% $CaCO_3$
% LoI
Cu
Cr
Fe
Mn
Ca
K
Mg
Na

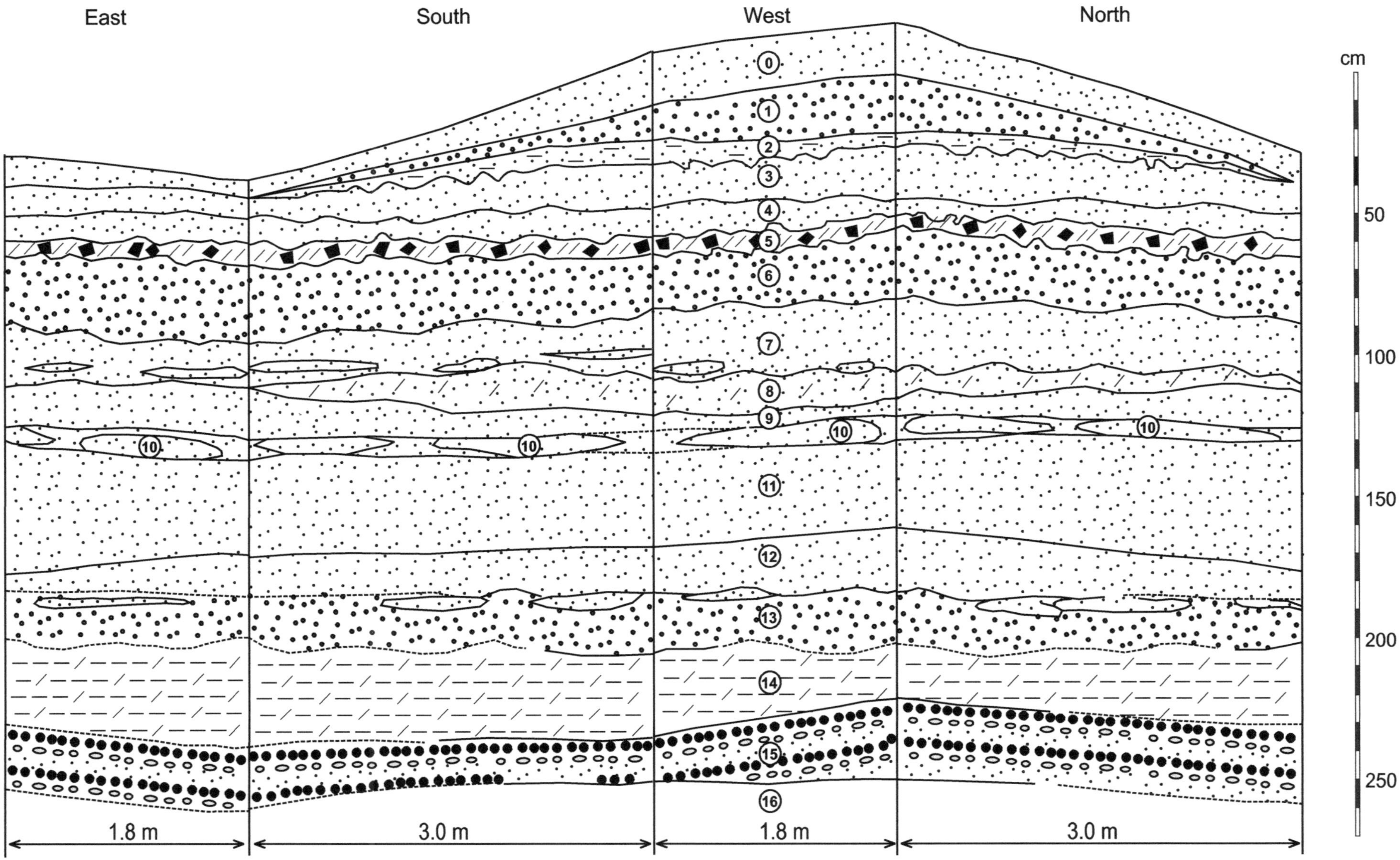

Fig. A-20: Sketch of profile LM-8 (drawing: R. Mendler, vertical scale 1: 20).

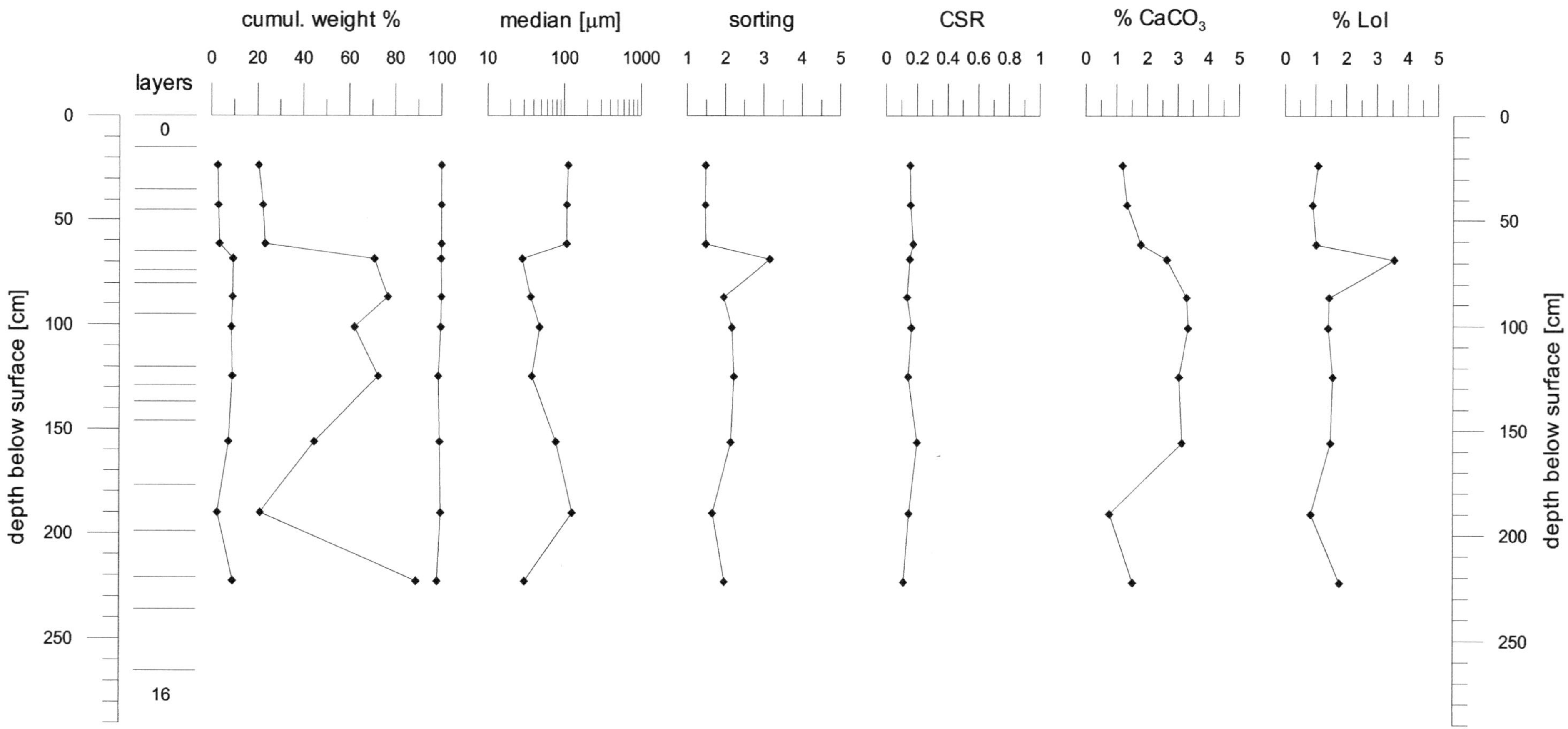

Fig. A-21: Laboratory results of samples from profile LM-8 (vertical scale 1: 30). Cumulative weight percent graph shows clay, silt and sand from left to right. Note that the lines connecting data points do not imply values between data points but serve solely to aid visualisation.

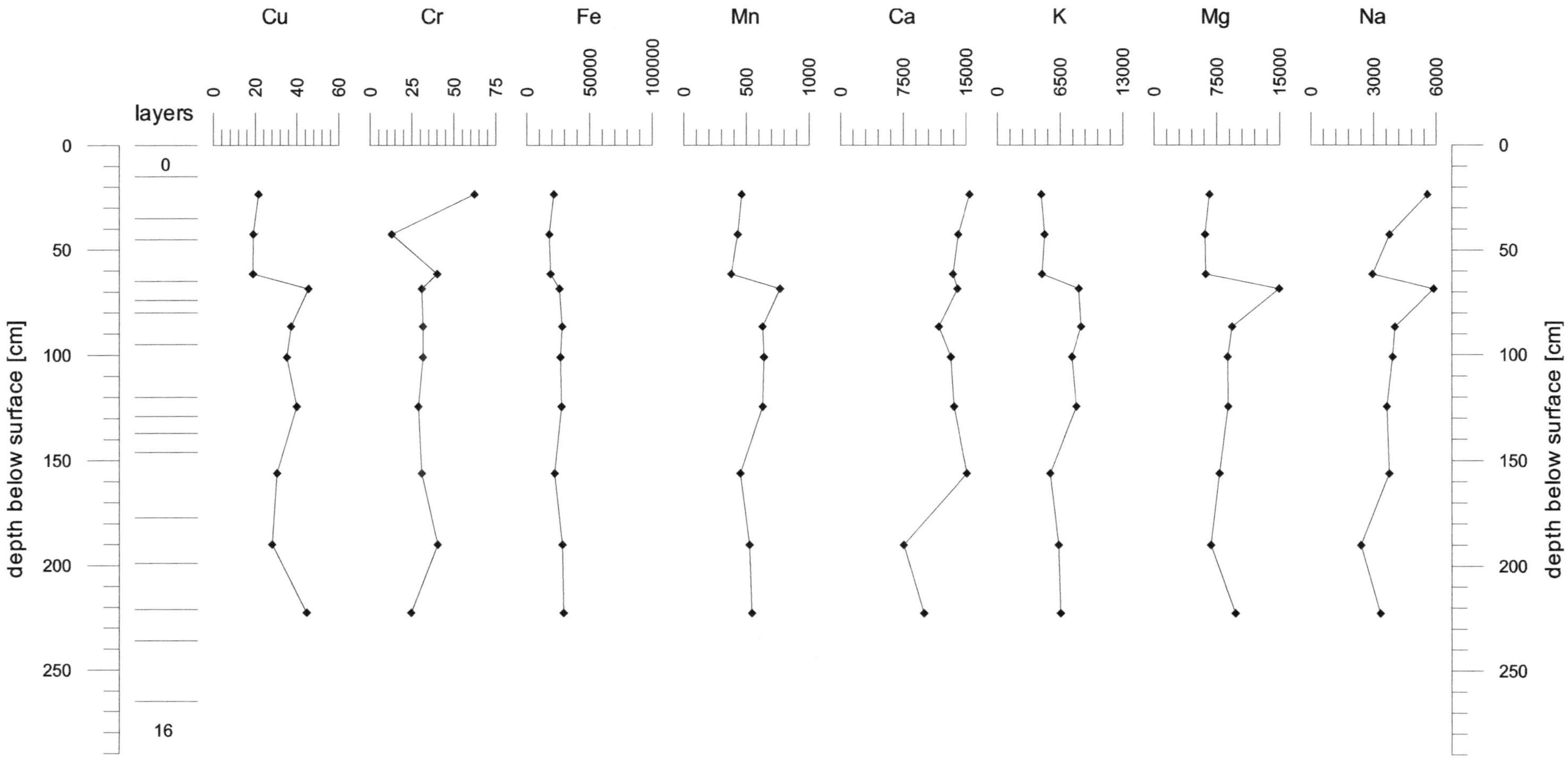

Fig. A-22: AAS measurement results of samples from profile LM-8 (vertical scale 1: 30). Note that the lines connecting data points do not imply values between data points but serve solely to aid visualisation.

strong oxidation stains are present. Layer 15 is bedded, well-sorted gravely coarse sand with bands of medium and fine sand and oxidation stains. Its upper boundary slopes to the east. Layer 16 consists of well-rounded coarse gravel.

Profile LM-8 records sedimentation at a site which is today the edge of an irrigated field. Coarse fluvial gravel is present at a depth of 2.65 m and is overlain by a sequence of low-energy fluvial sediments up to a depth of 0.69 m. The top 0.69 m of the profile are characterised by traces of human impact (charcoal) and sedimentation related to the adjacent irrigation channel (proximal irrigation sedimentation and locally erosion) as well as the deposition of excavated material which had accumulated in the irrigation channel. Carbonate precipitation (2.7–3.4%) has affected the profile at depths between 0.60 and 1.77 m.

Profiles Rio Grande

Tab. A-5: Positions of profiles: Rio Grande.

Profile	Depth [m]	Northing [m]	Easting [m]	Altitude [m]
RG-FL-1	2.15	8,392,761.41	475,858.49	304.95
RG-Fu	2.20	8,390,996.50	475,897.97	286.67
RGB-1	2.90	8,391,181.70	475,852.18	288.12
RGB-2	2.40	8,391,187.10	475,813.37	288.54
RGB-3	2.05	8,391,194.47	475,780.26	289.23
RG-Mo	3.75	8,392,249.60	476,179.24	304.77
RG-BG	3.46	8,393,976.71	477,186.25	322.07

Profile RG-FL-1

Profile RG-FL-1 was a natural exposure of coarse angular gravel and silty fine sand west of the Rio Grande opposite of the archaeological site of Los Molinos and a few metres above the riverbed. The stratigraphy consists mainly of debris flow deposits. The top 0.60 m of the profile (layers 0 through 2) are characterised by very high carbonate content (categories c3 to c4 in the field test, cf. AG Boden 2005) and carbonate nodules. In the top 0.30 m, oxidation stains are present. Layer 3 (0.60–1.00 m) contains somewhat less carbonate in the form of small nodules. Layer 4 (1.00–1.25 m) consists of structureless silty fine sand. Layer 5 (1.25–1.45/1.60 m) is a reddish horizon characterised by strong heat (fire) impact. Layer 6 (1.45/1.60-1.95 m) is well-sorted fine sand with up to 10% angular gravel. Charcoal from a hearth at a depth of 1.68 m was radiocarbon-dated at 1,290–1,140 cal BP (Poz-8024, 1,330 ±25 BP), i.e. the Middle Horizon (1,300–950 cal BP, cf. Unkel 2006: 111). The base of the profile below 1.95 m (layer 7) consists of angular slope debris and debris flow deposits and contains carbonate.

Profile RG-FL-1 records the occurrence of debris flows interspersed with human activity during the Middle Horizon indicated by the repeated presence of hearths in the stratigraphy. The precipitation of high amounts of carbonate since 1,290-1,140 cal BP indicates a possible late-Holocene genesis of caliche layers in the study area.

Profile RG-Fu

Profile RG-Fu (Figure A-23) was a pit on the left side of the Rio Grande, ca. 70 m from the bank of the river. It was situated between 192 and 231 m south to south-east of profiles RGB-1, RGB-2 and RGB-3.

Layer 0 (the recent A_p horizon) is fine sand with a crumbly structure. Layer 1 is well-sorted fine to medium sand with clods of cohesive silt dispersed throughout the layer. Its upper boundary is sharp and undulating. Layer 2 is structureless silty fine sand with a few clasts larger than 2 mm. Layer 3 is well-sorted, bedded medium sand with faint ripple marks. Its upper boundary is sharp and undulating. Layer thickness diminishes towards the east (away from the Rio Grande). Layer 4 is well-laminated silty fine sand with a sharp, undulating upper boundary. Like the overlying layer 3, layer 4 is characterised by diminishing thickness towards the east. Layer 5 is well-bedded fine to medium sand with some coarse sand. Ripple marks are present on the western and eastern side of the pit, while the northern wall shows foreset structures and a grading from finer to coarser and back to finer grain sizes. Oxidation stains and carbonate precipitation are present. The upper boundary of layer 5 is sharp and slightly undulating. Layer 6 is a band of silty fine sand with oxidation stains and carbonate precipitation; its upper boundary is sharp and undulating. Layer 7 is well-bedded sand with some fine gravel. Crossbedding is present. The upper boundary is diffuse and slightly undulating. Layer 8 is medium to coarse sand with faint bedding and distinct oxidation

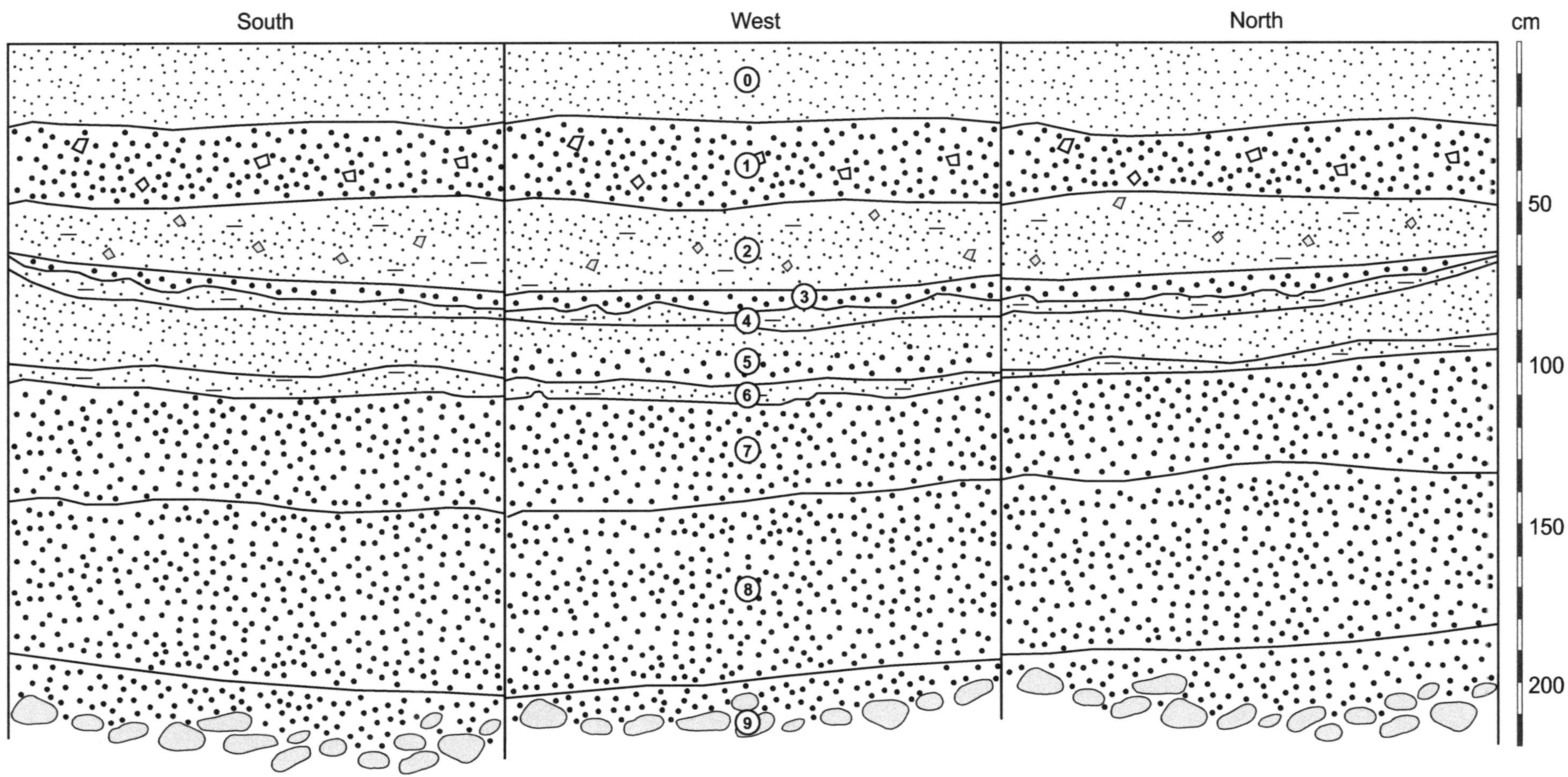

Fig. A-23: Sketch of profile RG-Fu (drawing: R. Mendler, vertical scale 1: 20).

stains. It has a sharp, horizontal upper boundary. Layer 9 is medium sand with well-rounded coarse gravel to boulders (up to 30 cm).

Profile RG-Fu records a sequence of fine-grained fluvial sedimentation overlying coarse fluvial gravel. Rapid deposition is indicated by thick beds of sands with foreset structures and ripple marks. Although the profile is situated in an irrigated field, the entire stratigraphy between depths of 2.20 to ca. 0.70 m is of undisturbed fluvial origin. The upper 0.70 m of profile RG-Fu shows traces of agricultural activity; however, the sediment likely represents agriculturally reworked fluvial deposits with only a minor contribution of irrigation sediment.

Profile RGB-1

Profile RGB-1 (Figure A-24) was a pit on a field on the right side of the Rio Grande, 25 m from the river. It is situated 39 m east of profile RGB-2.

Layer 0 (the recent A_p/irrigation horizon) is silty fine sand with a crumbly structure. Layer 1 is structureless silty fine sand with an undulating upper boundary. It contains clods of cohesive silt as well as charcoal. Layer 2 is structureless silty fine sand with a somewhat diffuse, horizontal upper boundary. A few lenses of coarse and medium sand (likely related to tillage) are present. Layer 3 is well-bedded to laminated sand. Layer 4 is well-sorted, laminated silt to medium sand. Ripple marks are present. Its lower boundary is sloping to the east (towards the Rio Grande). Layer 5 is well-bedded fine to medium sand with a horizontal upper boundary. It is only present on the northern side of the pit. Layer 6 is a band of well-bedded silt with oxidation stains. Layer 7 is a well-bedded sequence of medium to fine to medium sand. Its upper boundary is undulating. Layer 8 is well-bedded fine, medium and coarse sand (16 beds of varying thickness). Its upper boundary is sharp. Layer 9 is medium sand with an undulating upper boundary. A radiocarbon date from charcoal from the top of this layer (depth 1.45 m) in the south-eastern corner of the pit is ≤270 cal BP (Poz-8021, 145 ±25 BP). Layer 10 is well-bedded fine and medium sand (horizontal bedding, bed thickness ca. 5 mm) with an undulating upper boundary. Oxidation stains are present. Layer 11 is diffusely-bedded loam and sand with oxidation stains and a sharp upper boundary sloping to the east. Layer 12 is bedded loam and sand with strong oxidation stains and a sharp horizontal upper boundary. Layer 13 is well-bedded loam to coarse sand. Oxidation stains are very strong in the coarser-grained beds.

Profile RGB-1 records a river-proximal sequence of fluvial overbank sedimentation characterised by well-sorted and well-bedded to laminated loam to coarse sand. This sequence is capped by ca. 0.70 m of structureless fine-grained material interpreted to represent either irragric anthrosol or irrigation sediment turbated with flood deposits by agricultural activity. The upper 1.45 m of the stratigraphy were deposited since ≤270 cal BP. The high average sedimentation rate of ≥4.46 mm/a calculated from this age (≥2.15 mm/a for the agriculturally disturbed top portion if fluvial aggradation between depths of 1.45 and 0.70 m is assumed to have been instantaneous) strongly suggests that there has been a contribution of fluvial overbank sedimentation to the agriculturally used layers.

Profile RGB-2

Profile RGB-2 (Figures A-25, A-26 and A-27) was a pit on a field on the right side of the Rio Grande, 60 m from the river. It is situated 39 m west of profile RGB-1 and 34 m to the east of profile RGB-3.

Layer 0 (the recent A_p / irrigation horizon) is silty fine sand with a crumbly structure. Layer 1 is structureless silty sine sand with silt clods and charcoal and a sharp, undulating upper boundary. Layer 2 is fine sandy silt containing abundant charcoal. It has a broken-up upper boundary and is not present on the eastern side of the pit. Layer 3 is well-sorted, bedded fine sand with some charcoal and faint oxidation stains. Its upper boundary is diffuse and strongly disturbed. Layer 4 is bedded silt to fine sand with an undulating upper boundary. Layer 5 is silty to loamy fine sand with horizontal oxidation stains perhaps indicating faint bedding. Layer 6 is structureless fine sandy silt with a horizontal upper boundary. Oxidation stains and charcoal are present. Layer 7 is loamy fine sand with a horizontal upper boundary which is marked by an up to 1 cm thick, charcoal-rich band. Layer 8 is loamy silt with a diffuse horizontal upper boundary. Its top is marked by a 2 cm thick, charcoal-rich band underlain by reddish, heat-induced oxidation stains. Layer 9 is structureless fine sandy silt with some oxidation stains and carbonate nodules and a diffuse upper boundary. Layer 10 is a wedge of loam with charcoal, heat-induced oxidation and carbonate nodules. Layer 10 is present only on the northern side of the pit. Layer 11 is a wedge of fine sand with oxidation stains and carbonate nodules. It is present only on the northern side of the pit. Layer 12 is loam with oxidation stains and carbonate nodules. Its upper boundary is slightly and its lower boundary strongly sloping to the east (towards the Rio Grande). Layer 13 is a bedded sequence of well-sorted fine gravel, medium sand, loamy sand, medium sand, loamy sand, coarse sand and fine gravel (from top to bottom). The bands of loamy sand contain carbonate nodules. The base of layer 13 is well-rounded coarse gravel.

Fig. A-24 (following page): Sketch of profile RGB-1 (drawing: R. Mendler, vertical scale 1: 20).

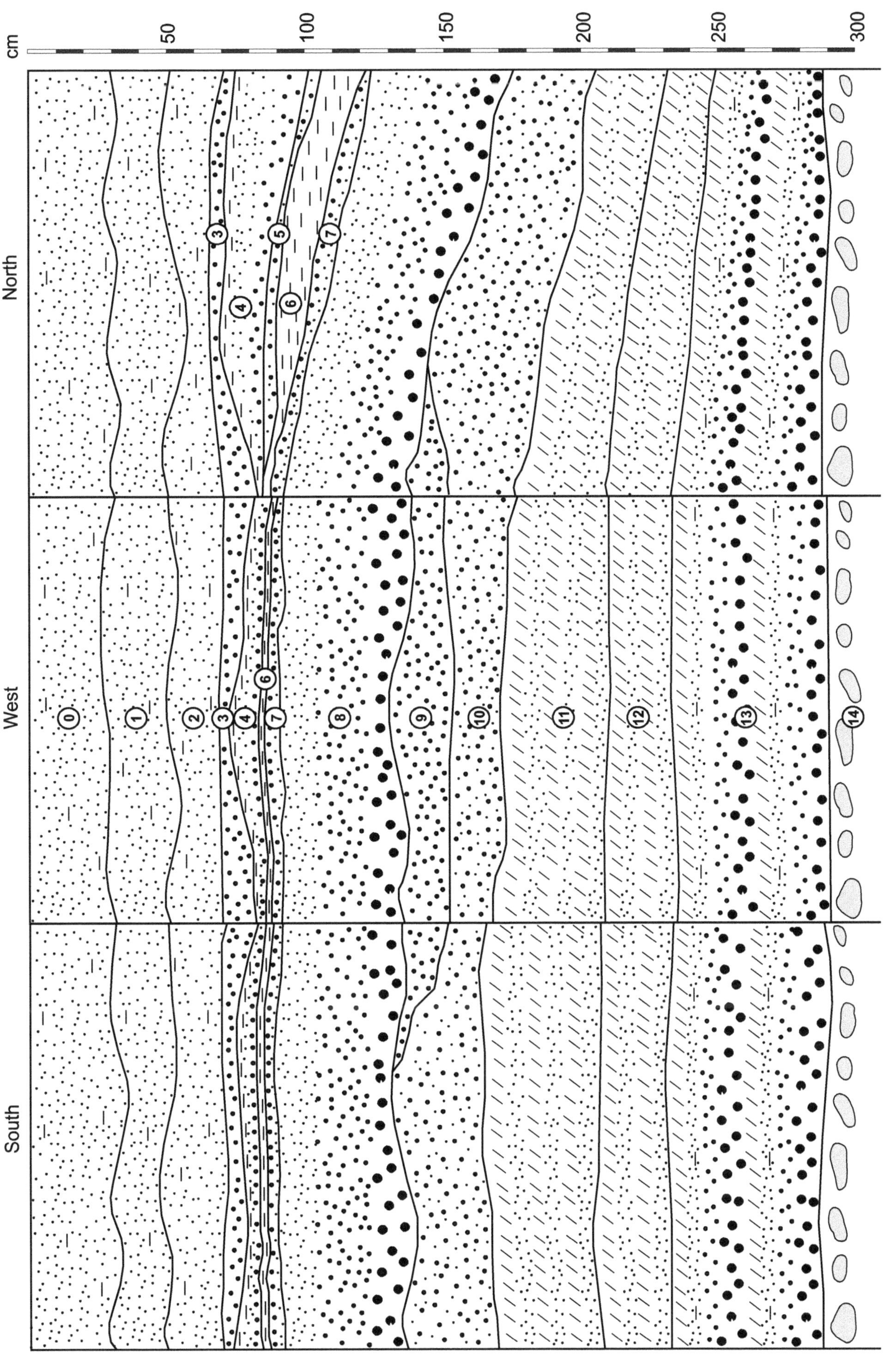
cm
50
100
150
200
250
300
North
West
South
0
1
2
3
4
5
6
7
8
9
10
11
12
13
14

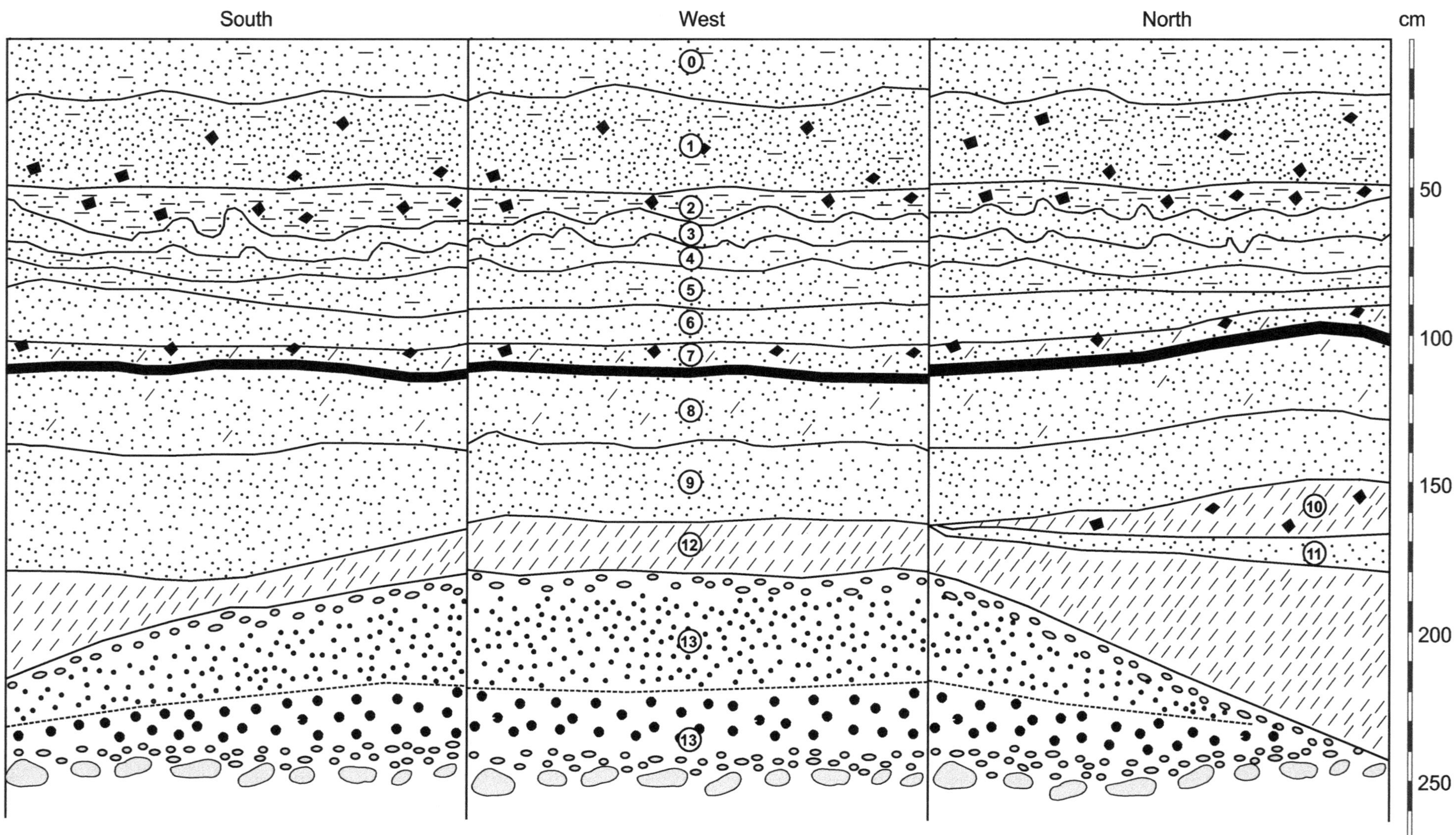

Fig. A-25: Sketch of profile RGB-2 (drawing: R. Mendler, vertical scale 1: 20).

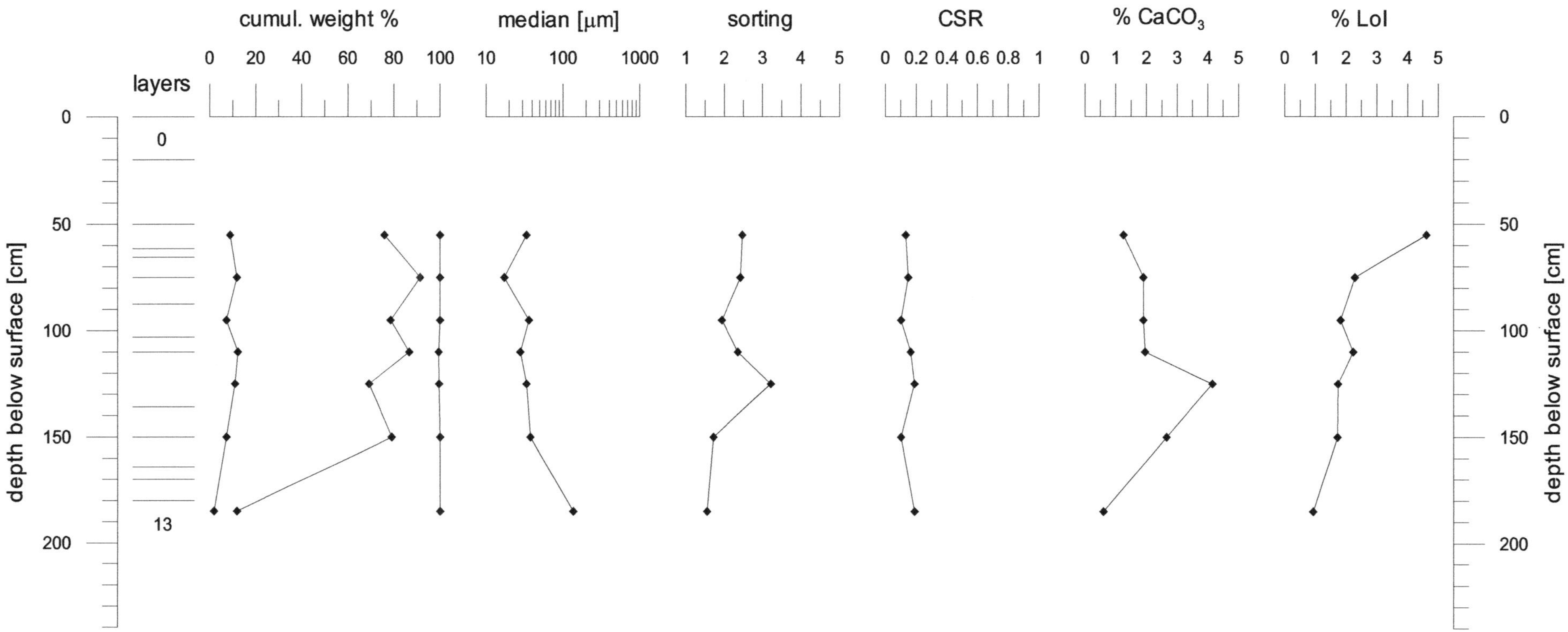

Fig. A-26: Laboratory results of samples from profile RGB-2 (vertical scale 1: 30). Cumulative weight percent graph shows clay, silt and sand from left to right. Note that the lines connecting data points do not imply values between data point but serve solely to aid visualisation.

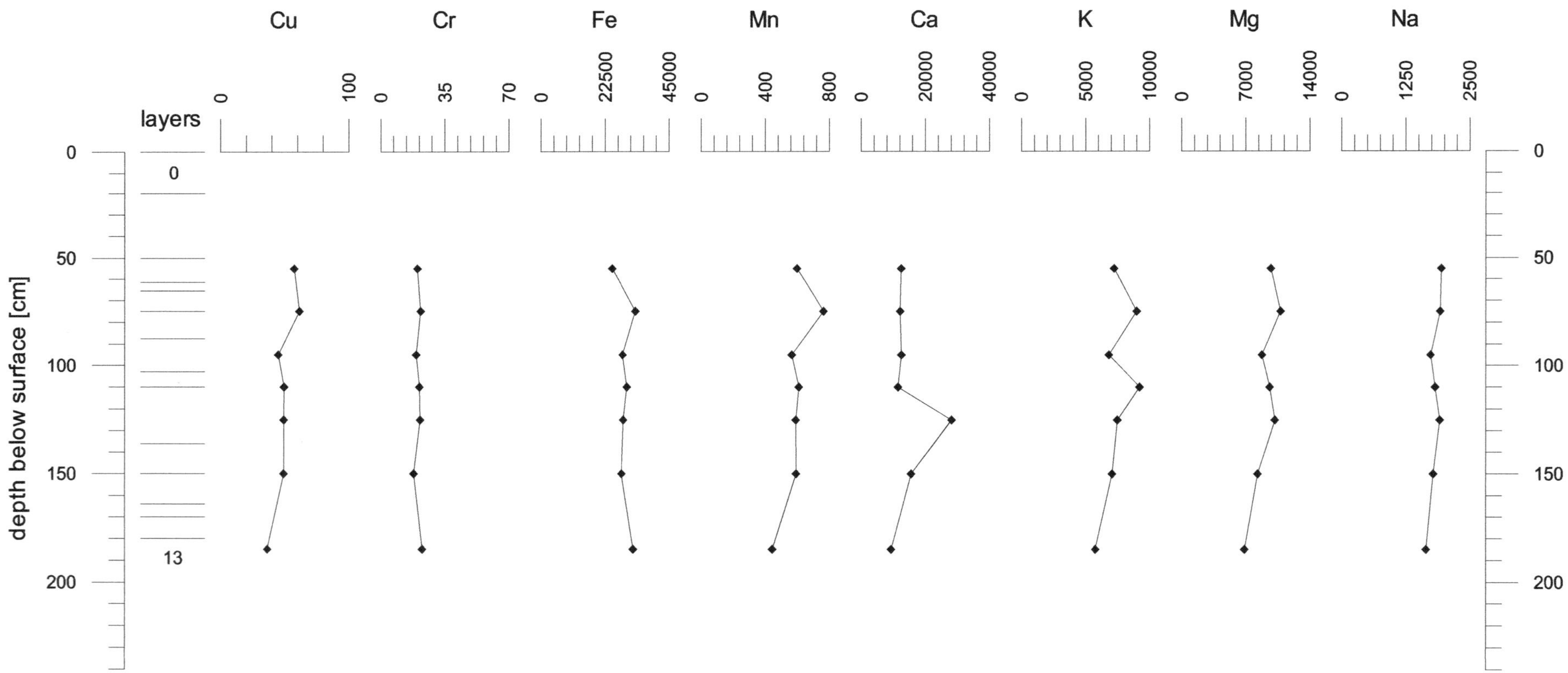

Fig. A-27: AAS measurement results of samples from profile RGB-2 (vertical scale 1: 30). Note that the lines connecting data points do not imply values between data points but serve solely to aid visualisation.

Profile RGB-2 shows a sequence of sediments dominated by fluvial sedimentation in high-energy (basal gravel indicating a former riverbed situation) and low-energy (well-sorted and bedded fine-grained overbank sediments) environments. This sequence is interspersed with indicators of human impact (abundant charcoal). While some of the fluvial sediments both in the lower and upper part of the stratigraphy may have been influenced by agricultural disturbance and irrigation, the relative influence of irrigation sedimentation and fluvial sedimentation can not be discerned.

Profile RGB-3

Profile RGB-3 (Figures A-28, A-29 and A-30) was a pit on a field on the right side of the Rio Grande, 100 m from the river. It is situated 34 m to the west of profile RGB-2.

Layer 0 (the recent A_p / irrigation horizon) is loamy sand with a crumbly structure. It contains roots and is covered by organic debris. Layer 1 is structureless loamy sand with cohesive clods of silt (increasing towards the base of the layer). The upper boundary is horizontal. Charcoal is present. Layer 2 is loamy sand with a diffuse to sharp and somewhat undulating upper boundary. Charcoal is present. Layer 3 is structureless, well-sorted silty sand with a horizontal upper boundary. Silt clods and other traces of disturbance are present. Layer 4 is a band of bedded fine sand and clayey silt with a slightly undulating upper boundary. Layer 5 is structureless, well-sorted sandy silt with a diffuse horizontal upper boundary. Abundant clods of silt are present; the base of the layer is marked by a broken-up layer of silt. Layer 6 is turbated silt and fine sand with some lenses of fine sand and a disturbed, undulating upper boundary.

Layer 7 is bedded, well-sorted silty fine to medium sand (upwards-coarsening) with a sharp horizontal upper boundary. At the top of the layer, charcoal is present. A radiocarbon assay of charcoal from the western side of the pit (depth 0.78 m) gave an age of 480–310 cal BP (Poz-8022, 370 ±25 BP) which falls into the Late Horizon (550–418 cal BP, cf. Unkel 2006: 111) to early historic period. Layer 8 is bedded silty fine sand with a slightly undulating upper boundary. A loamy band is present at the base of layer 8. Layer 9 is well-sorted silty fine sand. Charcoal as well as some loamy bands are present. Layer 10 is sandy silt with a slightly undulating upper boundary. Oxidation stains are present. The top of the layer is characterised by clods of loamy material. Layer 11 is fine sand with an undulating upper boundary which is marked by a 1–3 cm thick band of sandy loam. Charcoal is present. Layer 12 is bedded, well-sorted silty fine sand (top) to medium sand (bottom). Its upper boundary is horizontal. Charcoal from the base of layer 12 (depth 1.55 m) was radiocarbon-dated at 910–680 cal BP (Poz-8023, 905 ±30 BP) which corresponds to the Late Intermediate Period (950–550 cal BP, cf. Unkel 2006: 111). Layer 13 is well-sorted silty fine sand with oxidation stains. Its undulating upper boundary is marked by lenses of loamy material. Layer 14 is bedded sandy silt with horizontal oxidation stains and a sharp, slightly undulating upper boundary. Layer 15 is structureless silty fine sand with a sharp, slightly undulating upper boundary. Horizontal lines of oxidation stains hint at the presence of bedding planes. Layer 16 is structureless, well-sorted silty fine sand with oxidation stains. Layer 17 is structureless, well-sorted fine sand with a horizontal upper boundary. Layer 18 is well-rounded coarse gravel (up to 20 cm) in a sandy matrix.

Profile RGB-3 represents both fluvial and anthropogenic influences. At the base of the profile, well-rounded fluvial gravel is present. Most of the profile is characterised by well-sorted, fine-grained sediment with low CSR, low LoI and the presence of bedding. These layers (depth 1.95–0.58/0.67 m) are attributable to fluvial overbank deposition. Human impact is indicated by the presence of charcoal down to a depth of 1.55 m (radiocarbon-dated at 910–680 cal BP). Strongly turbated layers with poor sorting, increased CSR (≥0.27) and increased LoI (≥2.0%) as well as commonly occurring charcoal are interpreted as irragric anthrosol. Inception of irrigation agriculture is indicated by the turbation of a fluvial layer and the commencing sedimentation of irrigation sediments at a depth of 0.67–0.58 m after 480–310 cal BP (depth 0.78 m). Six flood events are recorded between 910–680 cal BP and 480–310 cal BP, i.e. floods reached this site at a rate of 1–3 events per century. Average sedimentation rates in profile RGB-3 are 1.28–3.85 mm/a for the lower (exclusively fluvial) and 1.46–2.14 for the upper (fluvial and irrigation) part of the profile.

Profile RG-Mo

Profile RG-Mo (Figure A-31; see Figure 25) was a pit near the lower edge of the highest field terrace left of the Rio Grande near the present settlement of Molinos, ca. 100 m south-west of the uppermost irrigation channel and ca. 300 m east of the river. The archaeological site Los Molinos begins ca. 330 m north of the profile. The mouths of several small quebradas descending from the Cresta Sacramento lie at distances between 120 and 200 m.

Layer 18 (the recent A_p / irrigation horizon) consists of crumbly silty fine sand. Layer 17 is structureless silty sand with a diffuse horizontal upper boundary. Layer 16 is poorly sorted, structureless silty sand with angular gravel and granules. Its upper boundary is diffuse and horizontal. Layer 15b is poorly sorted sand with a sharp horizontal upper boundary. On the eastern side of the pit, a large (ca. 40 cm long) rounded, flat stone is present. Charcoal from the top of layer 15b (depth 0.90 m) was radiocarbon-dated at 745–655 cal BP (Poz-15234, 815 ±35 BP), i.e. the Late Intermediate Period (950–550 cal BP, cf. Unkel 2006: 111). Layer 15a is poorly sorted silty sand with angular gravel (ca. 50%, up to 5 cm) and a diffuse, hor-

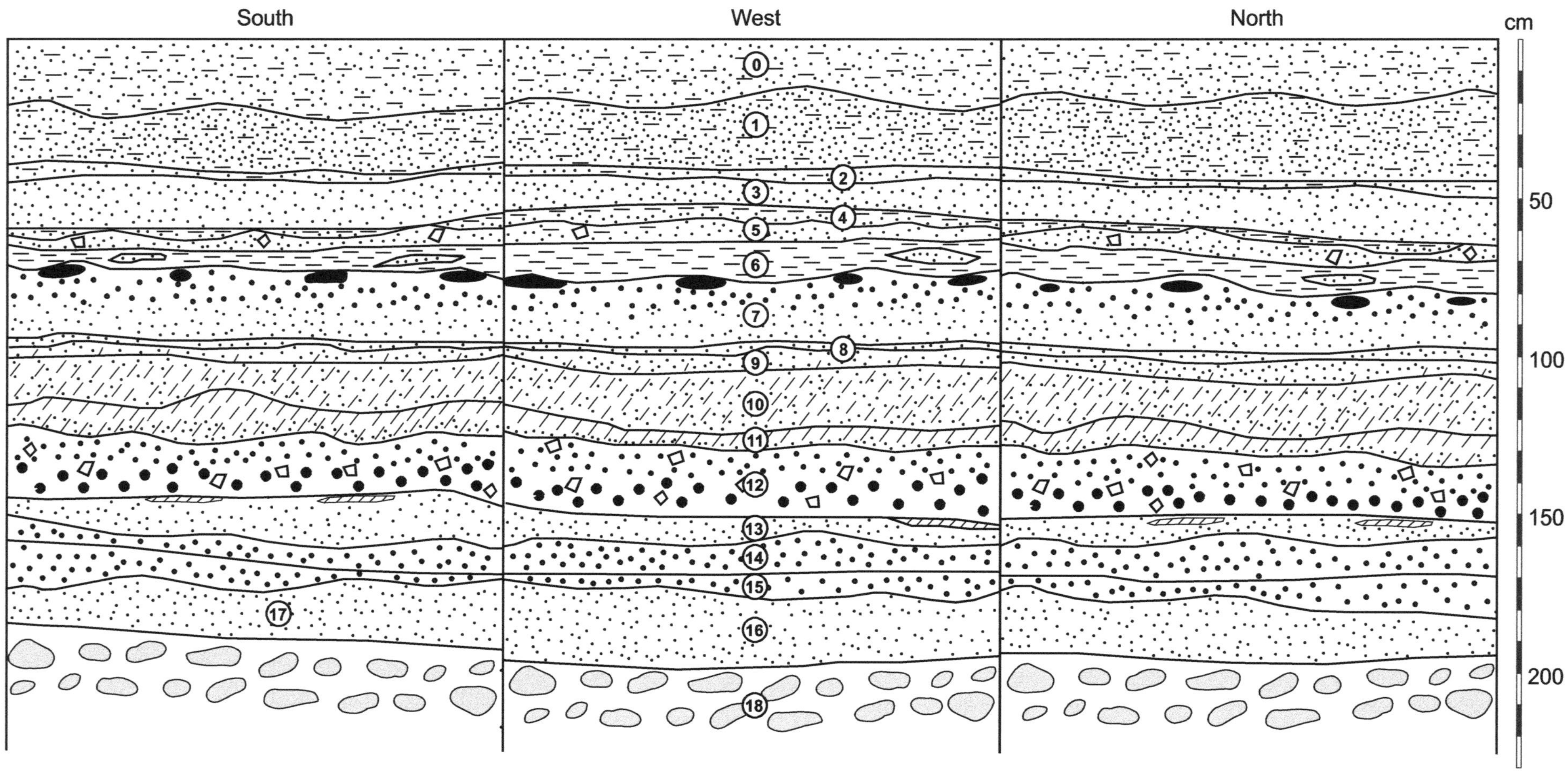

Fig. A-28: Sketch of profile RGB-3 (drawing: R. Mendler, vertical scale 1: 20).

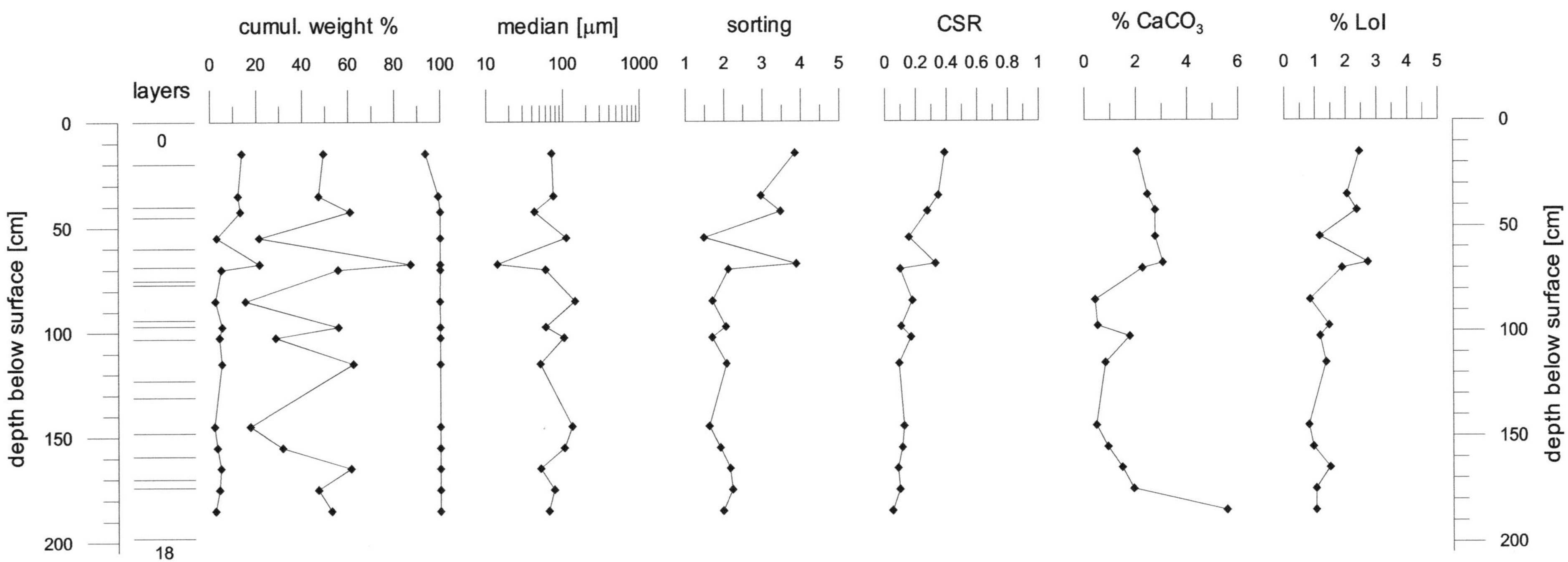

Fig. A-29: Laboratory results of samples from profile RGB-3 (vertical scale 1: 30). Cumulative weight percent graph shows clay, silt and sand from left to right. Note that the lines connecting data points do not imply values between data points but serve solely to aid visualisation.

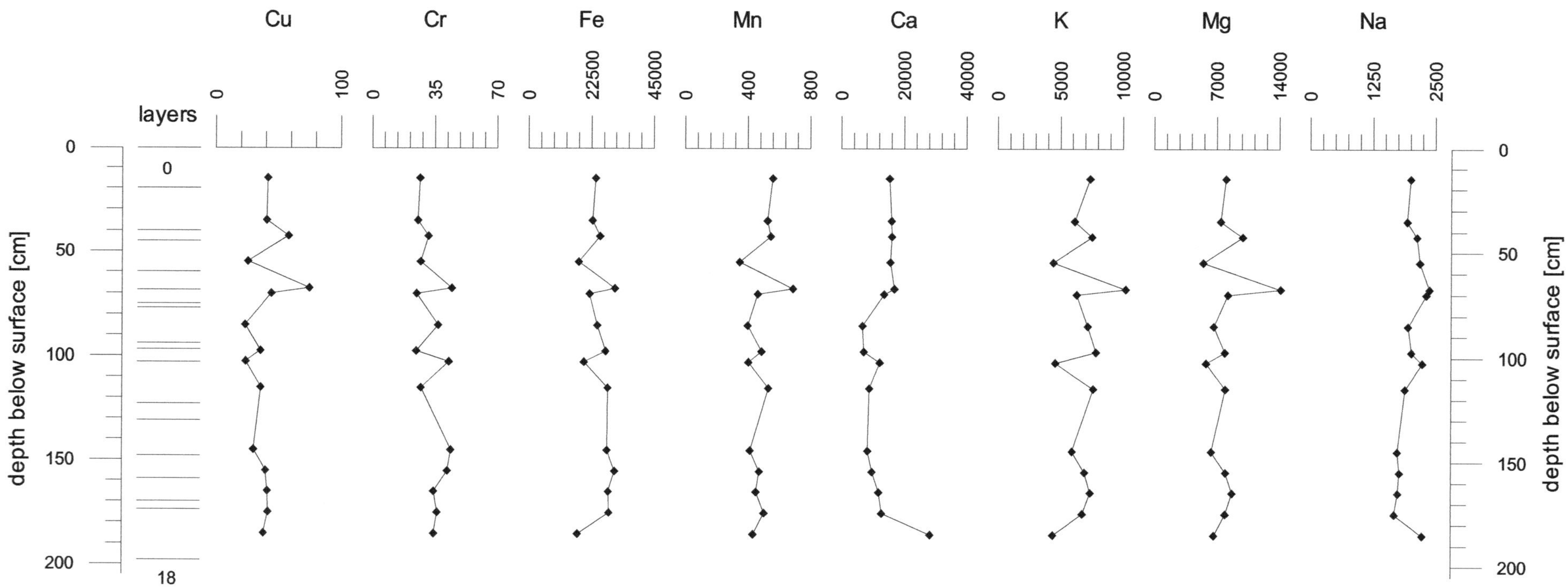

Fig. A-30: AAS measurement results of samples from profile RGB-3 (vertical scale 1: 30). Note that the lines connecting data points do not imply values between data points but serve solely to aid visualisation.

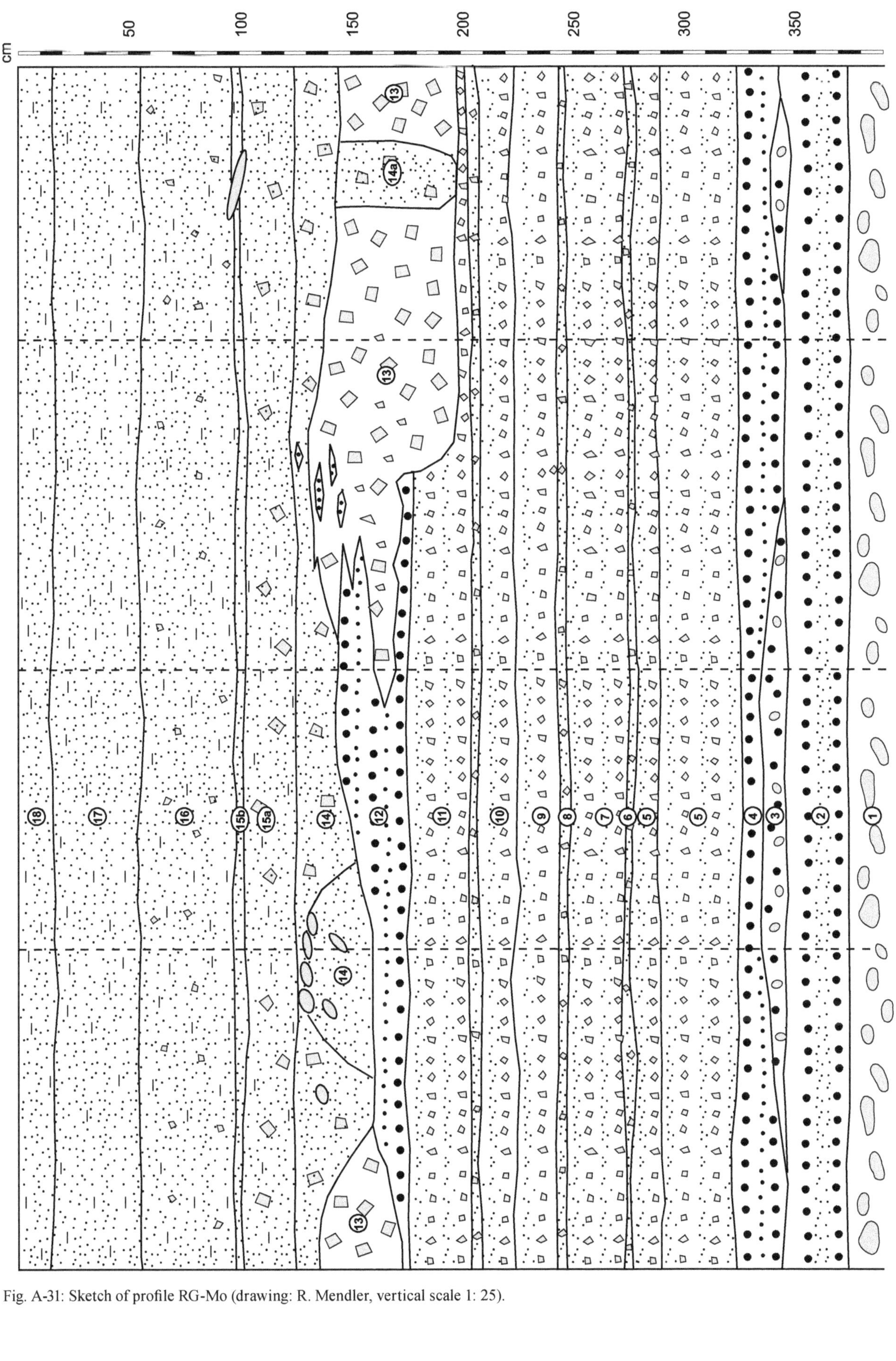

Fig. A-31: Sketch of profile RG-Mo (drawing: R. Mendler, vertical scale 1: 25).

izontal upper boundary. Layer 14 is poorly sorted sand with angular, platy stones (up to 4 cm). In the south-western corner of the pit, a mass of ceramic-rich (Nasca period, shards up to 40 cm in size), poorly sorted sand with increased concentration of stones (up to 15 cm) is present . The top of this mass is covered by large rounded and angular stones (up to 20 cm). Layer 14 and in particular the ceramic-rich mass contain abundant charcoal. The upper boundary of layer 14 is diffuse and undulating. Layer 14a is a ca. 20 cm wide block of material similar to layer 14 (poorly sorted sand with angular gravel). It is a filled-in pit with nearly vertical sides extending 0.55 m through the underlying layer 13. It contains Nasca period ceramic shards and charcoal. Layer 14a may be a looted pit grave, the excavation spoil of which may have been heaped up to form the ceramic-rich mass in layer 14.

Layer 13 consists of bedded, imbricated angular stones (up to 5 cm) in a sandy matrix. Its sharp upper boundary has been broken by layer 14a. The beds of layer 13 interfinger from east to west with layers 14 and 12; on the northern side, some small sand lenses are present. Layer 12 is well-bedded medium to coarse sand with a sharp, undulating (erosional) upper boundary. On the northern side of the pit, the beds have been erosionally truncated and interfinger with layer 13. Layer 11 consists of bedded and imbricated platy stones (up to 5 cm, a few rounded clasts are present) with a sandy matrix. Its upper boundary is generally sharp and horizontal; on the eastern and northern side of the pit the upper portion of layer 11 has been eroded. Layer 10 is well-sorted medium sand with mixed-in platy stones (up to 10 cm, a few rounded clasts are present). Its upper boundary is diffuse and horizontal. Layer 9 consists of bedded and imbricated platy stones (up to 5 cm, a few rounded clasts are present) with a sandy matrix. Its upper boundary is diffuse and horizontal. Layer 8 is well-sorted medium sand with mixed-in platy stones (up to 10 cm, a few rounded clasts are present). Its upper boundary is diffuse and undulating. Layer 7 consists of bedded and imbricated platy stones (up to 5 cm, a few rounded clasts are present) with a sandy matrix. Its upper boundary is diffuse and undulating. Layer 6 well-sorted medium sand with mixed-in platy stones (up to 10 cm, a few rounded clasts are present). Its upper boundary is diffuse and undulating. Layer 5 consists of bedded and imbricated platy stones (up to 5 cm, a few rounded clasts are present) with a sandy matrix. Its upper boundary is diffuse and horizontal. Layer 4 is well-sorted, bedded medium to coarse sand with a diffuse horizontal upper boundary. Some oxidation stains are present. Layer 3 consists of well-rounded, imbricated gravel (up to 5 cm) in a matrix of coarse sand. Oxidation stains are lacking. Its upper boundary is sharp and undulating. Layer 2 is bedded sand (from top to bottom: coarse sand, loamy sand, coarse sand) with strong oxidation stains. Its upper boundary is sharp, undulating and clearly erosional. Layer 1 is well-rounded gravel (up to 20 cm) with an undulating upper boundary.

Profile RG-Mo represents a transition from fluvial dynamics controlled by the Rio Grande through debris flow dynamics from a nearby quebrada to anthropogenically dominated (irrigation) sedimentation. High-energy riverbed sediments (coarse, well-rounded gravel) are present at a depth of 3.75 m (i.e. 4.60 m above the present riverbed), followed by medium- to low-energy fluvial sediments between 3.75 and 3.25 m. Between 3.25 and 1.25 m, at least seven major debris flow deposits are recorded at the site, most of them represented by ca. 20 cm of imbricated angular, platy stones overlain by a thin bed of sand containing similar clasts. The platy stones are of local origin (Cresta Sacramento). At depths between 1.75 and 1.35 m, fluvial and debris flow facies interfinger, followed by strong human impact present as a filled-in pit (possibly a looted grave) and abundant Nasca period ceramic shards as well as charcoal. A further major debris flow set (depth 1.35–0.90 m) was deposited before the accumulation of ca. 0.90 m of irragric anthrosol with a basal date of 745–655 cal BP. Average sedimentation rate for the irragric anthrosol is 1.12–1.27 mm/a.

Profile RG-BG

Profile RG-BG (Figure A-32) was an exposure on the side of a deeply incised irrigation channel ca. 350 m south-east of the Rio Grande. The tree-lined edge of the channel is ca. 30–40 cm higher than the adjacent field, apparently due to the deposition of organic refuse, charcoal and ash. The top of the profile (layer 4) consists of bedded silt with high amounts of organic material and charcoal. Layer 5 is a layer of organic material with charcoal with a sharp, horizontal upper boundary. Layer 6 is sandy silt with a sharp horizontal upper boundary. The layer contains ash; faint bedding is present in its top portion. Layer 7 is sandy silt with a crumbly structure and a sharp horizontal upper boundary. Layers 4 through 7 represent the accumulation of material along the edge of the irrigation channel. Layer 8 is sandy silt with a polyedric to columnar structure. Some rounded stones are present throughout the layer, their concentration increases from top to bottom. The upper boundary is sharp and horizontal. A radiocarbon assay on charcoal from near the base of layer 8 (depth 1.25 m, i.e. 0.87 m below the surface of the adjacent field) is 1,880–1,710 cal BP (Poz-15230, 1905 ±30 BP). This age pertains to the Early Nasca period (1,870–1,690 cal BP, cf. Unkel 2006: 111). Layer 9 is silty sand with disperse charcoal and some small rounded stones (up to 1 cm). Its upper is slightly dome-shaped and diffuse. Layer 10 is upwards-fining coarse to fine sand with some angular stones (up to 5 cm).

At a depth of 1.95 m, a concentration of stone is present. The layer contains small (up to 0.5 cm) soft carbonate nodules. Its upper boundary is slightly diffuse. Layer 11 is a 4–20 cm thick bed of imbricated gravel (up to 4 cm) with a matrix of medium to coarse sand and a sharp, horizontal upper boundary. Layer 12 is structureless silty sand

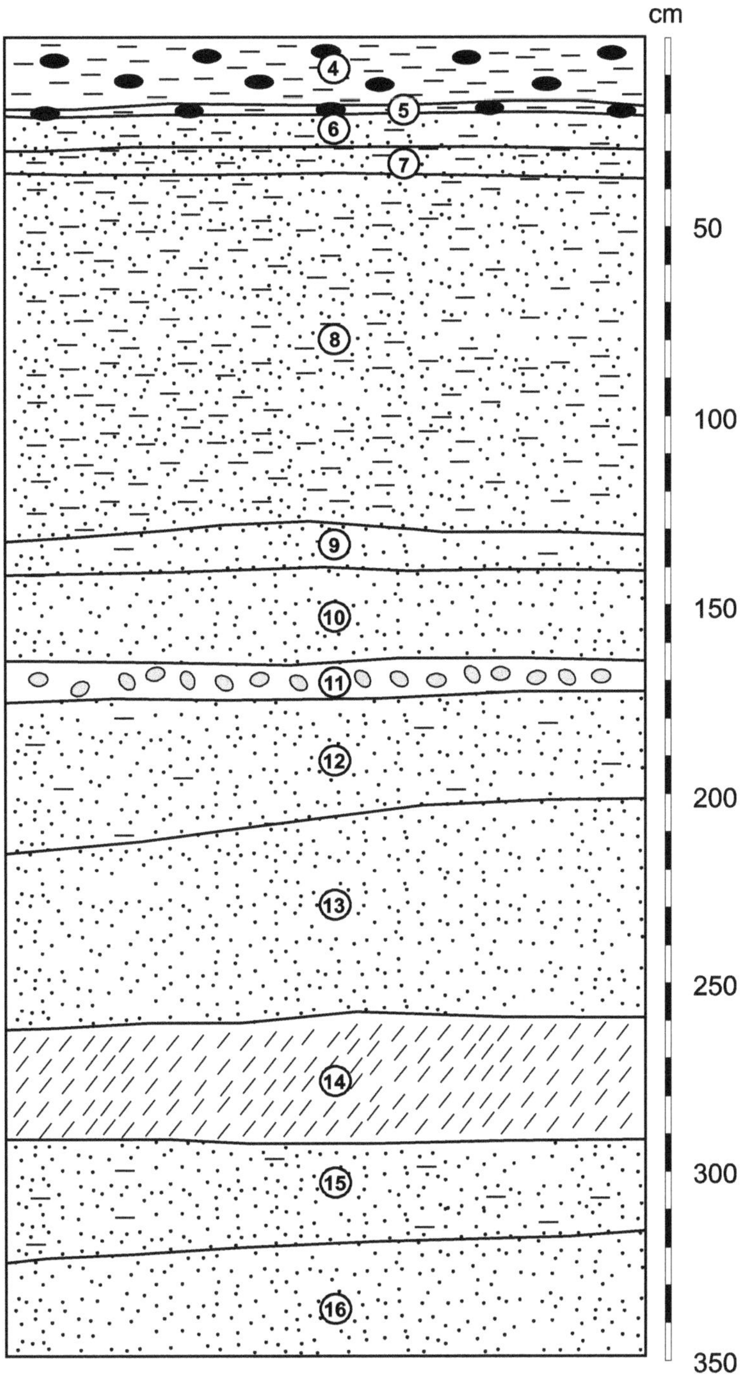

Fig. A-32: Sketch of profile RG-BG (drawing: R. Mendler, vertical scale 1: 20).

with soft carbonate nodules (up to 2 cm) and a sharp, horizontal upper boundary. Layer 13 is structureless fine sand with larger nodules (up to 5 cm) and a sharp, north-east-sloping upper boundary. Layer 14 is structureless loam with increasing concentration of nodules and a diffuse upper boundary. Oxidation stains are present. Layer 15 is structureless silty fine sand to loam with strong oxidation stains, carbonate nodules (up to 5 cm) and a horizontal, broken-up to diffuse upper boundary. Layer 16 is well-sorted sand and probably represents recent infilling in the undercut bank of the irrigation channel.

Profile RG-BG represents both natural and human-induced sedimentation. Coarse fluvial gravels corresponding to a riverbed situation were not reached in this profile, and the basal sand likely represents modern infilling of an undercut bank. Sediments at depths between ca. 3.20 and 1.41 m are interpreted as five separate low-energy fluvial overbank deposits and one (depth 1.72–1.64 m) overbank gravel splay deposit. The upper portion of the stratigraphy is characterised by 1.03 m of irragric anthrosol and 0.38 m of field-edge deposition. An age of 1,880–1,710 cal BP near the base of the irragric anthrosol indicates incipient irrigation agriculture at the site by the Early Nasca period

(1,870–1,690 cal BP, cf. UNKEL 2006: 111). The average sedimentation rate since 1,880–1,710 cal BP is 0.45–0.49 mm/a (referring to the recent field surface elevation) or 0.65–0.71 mm/a (including the field-edge deposit).

Profiles north-west of the Rio Palpa

Tab. A-6: Positions of profiles north-west of the Rio Palpa.

Profile	Depth [m]	Northing [m]	Easting [m]	Altitude [m]
RPE	2.80	8,391,748.75 – 8,391,699.46	476,804.88 – 476,788.31	297.25 – 297.53
RPF-1	1.65	8,391,760.28	476,778.17	299.60
RPF-2	2.00	8,391,844.12	476,687.92	301.78
RPF-3	1.90	8,391,964.07	476,540.63	306.20
RPQ	2.25	8,391,710.62 – 8,391,695.00	476,837.96 – 476,912.00	295.60 – 296.10
RPW	2.50	8,391,599.64	476,740.60	298.28
YN-1	2.85	8,392,373.65	477,450.03	308.05
YN-2	2.20	8,392,447.77	477,407.96	311.33
Ma-T	3.40	8,394,281.90	480,203.24	357.44

Profile RPE

Profile RPE (Figure A-33) was recorded on the ca. 2.80 m (up to 3.60 m) high erosional right bank of the Rio Palpa adjacent to profile RPQ and ca. 30 m south-east of profile RPF-1. Along a 30-m-segment of the bank exposure, stratigraphy was recorded at varying intervals, with positions measured down-river relative to the begin of the well-defined exposure. The exposure has been fluvially cut into a lower terrace which had been 55 to 80 m wide in 1944 as evidenced by aerial photographs and of which only a narrow strip remains. At a distance of ca. 15 to the north-west, the main agricultural field surface rises in a terrace-like step of ca. 1.5 m.

At position 1.0 m, 12 layers were recorded. The profile here has a height of 2.70 m. Layer 0 (0.00–0.25 m) is crumbly silty sand with some gravel. Layer 1 (0.25–0.44 m) is structureless silty sand with ca. 10% gravel (up to 6 cm, well-rounded to angular). Layer 2 (0.44–0.45 m) is a strongly undulating band of charcoal feathering out beyond a position of 3.0 m. Layer 3 (0.45–0.91 m) is well-bedded silt and sand with an undulating upper boundary. Beds are partially undulating. Charcoal is present. At a depth of 0.45–0.58 m, an 80 cm wide channel cut into the bedded material and filled-in with unsorted silty sand to gravel (10%, up to 6 cm) is present. Layer 4 (0.91–0.95 m) is laminated, upwards-coarsening fine to medium sand with ripple marks. Layer 5 (0.95–1.45 m) is a depression or channel cut into the underlying layers 6 and 7 and filled-in with unsorted silty sand and gravel (up to 8 cm). Charcoal is present. Gravel content increases from base to top. The upper boundary is distinct and undulating. Layer 6 (1.10/1.15–1.35/1.40 m) is sandy silt with carbonate nodules and an undulating upper boundary. Layer 7 (1.42–1.47 m) is well-cemented caliche with a diffuse, undulating upper boundary. Layer 8 (1.47–1.65/1.73 m) is sand with a high mica content and abundant carbonate nodules (ca. 25%). At the base of layer 8, a thin, well-cemented but broken-up band of caliche is present. Layer 9 (1.65/1.73–2.08/2.33 m) is broken-up caliche (large blocks, 40–50%) in a matrix of sand and silt. Layer 10 (2.08/2.33–2.65 m) is well-sorted fine sand with some carbonate nodules. Its upper boundary is undulating. It appears to represent the fill of an undercut bank. Layer 11 (below 2.65 m) is well-developed caliche with an undulating upper boundary.

Between 1.0 and 8.6 m, 14 layers were recorded. Layer 0 (0.00–0.10/0.15 m) is silty sand with a crumbly structure, some stones (ca. 2%, up to 10 cm) and abundant roots. Layer 1 (0.10/0.15–0.40 m) is structureless silty sand with 20–25% gravel. Its upper boundary is undulating. Layer 2 (0.40–0.45 m) is a layer of bedded silt to coarse sand which is present between positions 2.9 and 8.6 m. Layer 3 (0.45–0.55 m) is silty sand with 20–25% gravel. It is present between positions 0.0 and 4.0 m. At the base of layer 3, a band of charcoal is present. Layer 4 (0.55–0.75 m) is structureless silty fine sand with cohesive clods of silt. It has a broken-up appearance which is likely due to agricultural turbation. Layer 5 (0.75–1.00 m) consists of bedded silt to sand and is present between positions 0.0 and 3.7 m. Layer 6 (1.00–1.05/1.15 m) is silty fine sand with a crumbly structure, some stones, cohesive silt clods and abundant charcoal. Faint bedding is discernibly. Layer 6 is present between positions 3.2 and 6.5 m.

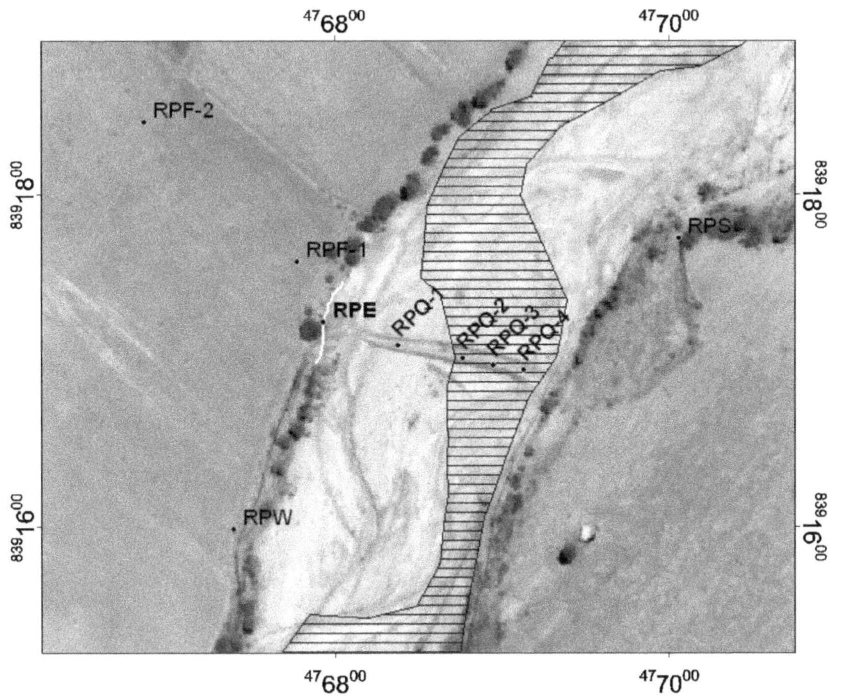

Fig. A-33: Satellite image of March 25, 2005 (Quickbird) of the location of profile RPE (white line) as well as RPF-1, RPF-2, RPQ, RPS and RPW. Hatched area shows riverbed location deduced from aerial photographs of 1944.

Layer 7 (1.05/1.15–1.25 m) is gravel to boulders with a crumbly matrix. It is present between positions 4.3 and 8.2 m. Layer 8 (1.25–1.35 m) is sandy silt with ca. 10% gravel, some silt clods and abundant charcoal. Its upper boundary is undulating; its lower boundary is marked by a charcoal-rich band. Layer 8 is present between positions 5.8 and 8.4 m. Layer 9 (1.25–1.55/1.60 m) is sandy silt with ca. 10% gravel and has an undulating upper boundary. Charcoal is present at low concentrations. Close to the base of layer 9, there are abundant cohesive clods of silt which are interpreted as resulting from agricultural use of an incipient irragric anthrosol deposited upon the silt-capped underlying gravel. Layer 10 (below 1.55/1.60 m) is well-rounded coarse gravel (up to 15 cm) in a matrix of coarse sand. Layer 10 is present between positions 4.6 and 6.1 m; where it is absent, the base of the exposure is formed by layer 13. Layer 13 is well-cemented caliche. Layer 11 is a lens of sandy silt with ca. 40% gravel which is present between positions 6.4 and 8.4 m at depths between 0.45 and 0.62 m. Layer 12 is a depression filled with bedded silt to coarse sand which is present between positions 6.3 and 8.3 m at depths between 0.65 and 1.10 m.

Between 8.6 and 15.0 m, 9 layers were recorded. The profile here has a height of 2.75 m. Layer 0 (0.00–0.20 m) is silty sand with a crumbly structure, abundant roots and ca. 15% rounded stones. Layer 1 (0.20–0.45/0.50 m) is structureless silty sand with 15–20% angular and rounded stones (up to 8 cm). At the base of the layer, clast concentration increases. The upper boundary of layer 1 is undulating. Layer 2 (0.45/0.50–0.65 m) is structureless silty sand with an undulating upper boundary and ca. 10% angular to rounded stones (up to 10 cm). Cohesive silt clods are present. Layer 3 (0.65–1.37 m) is bedded silt to fine sand with ripple marks. The silt band marking the upper boundary of layer 3 is partially broken-up. Layer 4 (1.37–1.42 m) is a charcoal-rich band of silty sand with ca. 10% gravel. Its thickness varies between 1 and 5 cm. Layer 5 (1.42–1.55 m) is structureless, upwards-coarsening silt to fine sand with oxidation stains. Layer 6 (1.55–2.05 m) is silty sand with gravel (up to 12 cm) grading into the underlying gravel (layer 7). Layer 7 (2.05–2.45 m) consists of imbricated, well-rounded coarse gravel to boulders (up to 30 cm) overlying strongly weathered, cemented coarse gravel to boulders. Layer 8 (below 2.45 m) is silty fine sand present between near position 9.6 m and interpreted as secondary fill of the undercut bank.

At position 15.5 m, 7 layers were recorded. Layer 0 (0.00–0.20/0.25 m) is crumbly silty sand with 5-10% rounded and angular gravel. Layer 1 (0.20/0.25–0.40/0.45 m) is silty sand to sandy silt with 10–20% gravel (up to 15 cm) and an undulating upper boundary. Layer 2 (0.40/0.45–0.60 m) is sandy silt to silty sand with 10–20% gravel; its undulating upper boundary is locally marked by increased concentrations of gravel. Layer 3 (0.60–1.00/1.15 m) consists of bedded silt to sand with an undulating upper boundary. The bedding is broken up. Layer 4 (1.00/1.15–1.20 m) is a strongly broken-up band of silt with oxidation stains. Layer 5 (1.20–1.70 m) is coarse angular and rounded gravel (80%, up to 12 cm) in a sandy matrix. It contains ceramic shards as well as charcoal. Layer 6 (below 1.70 m) is well-weathered and cemented coarse gravel to boulders with an undulating upper boundary.

Between 16.0 and 18.0 m, 7 layers were recorded. Layer 0 (0.00–0.30 m) is silty sand with some gravel and a crumbly structure. Layer 1 (0.30–0.55 m) is silty sand with ca. 5% gravel and some silt clods. At a depth of 0.45–0.50 m, faint bedding is present. Layer 2 (0.55–0.70 m) is silty sand with some gravel, 5–10% silt clods and a concentration of gravel at the base of the layer. The upper boundary is diffuse and undulating. Layer 3 (0.70–1.10 m) consists of strongly disturbed silt to sand with 10% silt clods, some charcoal and locally present remnant bedding. Layer 4 (1.10–1.15/1.20 m) is a broken-up lens of silty sand which is present between positions 16.0 and 18.0 m. Layer 5 (1.15/1.20–1.60/1.70 m) is matrix-supported coarse gravel (50%, up to 15 cm) in a matrix of silty sand to fine gravel. Charcoal is present. Layer 6 (below 1.60/1.70 m) is rounded, well-weathered, cemented coarse gravel to boulders.

Between 19.9 and 24.0 m, 7 layers were recorded. Layer 0 (0.00–0.25 m) is crumbly silty sand with ca. 5% gravel. Layer 1 (0.25–0.55/0.60 m) is structureless silty sand with 20% gravel. The layer contains ceramic shards. Its upper boundary is undulating, its lower boundary is marked by a concentration sand, gravel and silt clods. Layer 2 (0.55/0.60–0.76/0.80 m) is silty sand with 5% gravel (up to 6 cm) and some silt clods. At the lower boundary, there is a broken-up, ca. 5 mm thick band of sand. Layer 3 (0.75/0.80–1.05 m) is bedded silt and sand which is locally disturbed. Layer 4 (1.05–1.15 m) is a band of silt which is present between positions 20.0 and 22.3 m. Layer 5 (1.15–1.55 m) is charcoal-rich silty sand with 10–50% gravel (partially weathered). Layer 6 (below 1.55 m) is strongly weathered, cemented coarse gravel to boulders.
At position 24.0 m, a gully cuts to the north-west into the exposure by up to 2.50 m. 12 layers were recorded at the right (north-east) side of the gully (ca. 1.00 m high). Layer 0 (0.00–0.10 m) is crumbly silty sand with less than 5% gravel. Locally, laminated sandy silt is present. Layer 1 (0.10–0.20 m) is silty sand with 10% gravel. Layer 2 (0.20–0.40/0.60 m) is structureless silty sand with 25% gravel (angular to rounded). Layer 3 (0.40–0.60 m) is a wedge of poorly sorted, structureless sand to gravel containing charcoal. Layer 3 extends wedge-like from the front of the exposure ca. 1.0 m into the gully. Layer 4 (0.60–0.85 m) consists of strongly disturbed, bedded silt to sand with some clods of silt. The layer extends wedge-like into the gully. Layer 6 (0.25/0.85–0.50/1.00 m) is a wedge of silty sand with <5% gravel which becomes thinner towards the north-west (into the gully). It contains abundant ceramic shards and disperse charcoal. Layer 7 (0.50–1.10 m) is a block of strongly weathered, cemented, rounded coarse gravel to boulders which is present in the gully at 1.0–2.0 m from the front of the exposure. Layer 8 (1.00–1.50 m) is silty sand with 30% gravel which lies wedge-like against layer 7. Layer 9 (1.10–1.50 m) is very loose, structureless coarse sand with ca. 1% larger clasts (up to 4 cm). The sand grains are angular and appear to be derived from weathered granite (i.e. the strongly weathered coarse gravel present at the base of the exposure) without long-distance transport. Layer 10 (below 1.50 m) is coarse gravel (up to 13 cm) in a matrix of coarse sand. Some clasts are weathered. At the upper boundary, brown to black oxidation stains are present. Faint bedding is discernible. Layer 11 (0.33–0.95 m) is a structureless lens of silty sand with 5% gravel and a diffuse upper boundary. Charcoal is present.

At position 26.0 m (the left, north-western side of the gully; ca. 1.30 m high), 8 layers were recorded. Layer 0 (0.00–0.10/0.15 m) is crumbly silty sand with 5% gravel. Layer 1 (0.10/0.15–0.35/0.70 m) is structureless silty sand with 15% angular to rounded gravel and extends wedge-like into the gully. Layer 2 (0.35/0.70–0.28/1.00) is structureless, charcoal-rich silty sand with 5% rounded gravel which extends wedge-like into the gully. Both its upper and lower boundary are undulating. Layer 3 (0.50–0.90 m) is a lens of silty sand which is present at 1.5–2.3 m from the front of the exposure. Layer 4 (0.30–0.90 m) is a block of partially strongly weathered, rounded coarse gravel to boulders (up to 20 cm) in a matrix of cemented coarse sand. It is present between 2.3 and 2.9 m from the front of the exposure. Layer 5 (0.55–1.30 m) is a second block of partially strongly weathered, rounded coarse gravel to boulders (up to 20 cm) in a matrix of cemented coarse sand. It is present between 0.5 and 1.4 m from the front of the exposure. Layer 6 (0.90–1.25 m) is bedded silty fine sand to coarse sand with horizontal upper and lower boundaries. Layer 6 lies locally under the block of weathered gravel (layer 5). Layer 7 (below 1.10/1.25) is coarse sand with well-preserved crystals and a few rounded gravel-sized clasts (up to 5 cm) as well as some strongly weathered clasts. The upper boundary is undulating.

Between 26.0 and 29.0 m, 6 layers were recorded. Layer 0 (0.00–0.15 m) is crumbly silty sand. Layer 1 (0.15–0.40 m) is structureless silty sand with 15–20% rounded to angular gravel (up to 20 cm). Layer 2 (0.40–0.80 m) is structureless silty sand with 5% gravel (up to 10 cm) and a diffuse upper boundary. Ceramic shards are present in this layer. Layer 3 (0.80–1.00/1.55 m) is structureless, charcoal-rich silty sand with 10% gravel (up to 17 cm). Generally, the lower boundary of layer 2 is between depths of 1.00 to 1.30 m; however, at position 27.6 m and between positions 27.5 and 27.7 m, it fills channels or depressions down to a depth of 1.55 m. Between positions 26.6 and 27.5 m, a poorly sorted gravel lens is present. Layer 4 (1.00/1.15–1.25 m) consists of pockets of coarse sand with up to 50% rounded and partially weathered gravel (up to 12 cm) which extend down into layer 5. Layer 5 (below 1.00/1.55 m) is poorly sorted, strongly weathered coarse gravel to boulders (90%) in a cemented matrix.

Profile RPE reflects a complex history of erosion and deposition of a terrace near the Rio Palpa. Well-weathered and cemented coarse gravel interpreted to be of pre-Holocene age is exposed locally at the base of the profile, suggesting that erosion and accumulation have remained within the limits of a few metres during the Holocene. High carbonate concentrations and well-developed caliche are interpreted as being due to preferential carbonate precipitation from ground or irrigation water. Human impact at the site is indicated by charcoal (both distributed throughout the stratigraphy and concentrated in charcoal-rich bands). A high age of the upper part of the stratigraphy is suggested by the Nasca 7 ceramic shards at positions 4.2 m (depth 0.55 m), 4.5 m (depth 0.25 m), 17.3 m (depth 0.40 m) and 17.6 m (depth 0.90 m). However, this high age is contradicted by two radiocarbon assays of charcoal from position 4.90 m (depth 1.50 m) and position 5.10 m (depth 0.45 m) which both gave ages of ≤260 cal BP (Poz-5270, 115 ±30 BP and Poz-5271, 100 ±30 BP, respectively). The presence of even earlier Nasca ceramic shards at comparable depths (position 8.9 m, depth 0.82 m: Nasca 3 or 5; position 15.9 m, depth 0.60 m: Middle Nasca; position 22.6 m, depth 0.75 m: Nasca 4) in layers stratigraphically younger than historic period artefacts (position 13.8 m, depth 1.30 m: glass shards in gravel layer) further shows that the upper ca. 1.50 m of profile RPE consists of massively disturbed, mixed-up material.

It is probable that this disturbance was caused by either fluvial reworking and short-distance hyperconcentrated flow transport (e.g. slumping of upstream river banks) or mechanised agricultural activity. In any case, despite their secondary deposition during late historic times, the abundance of ceramic shards (108 pieces were recovered from profile RPE) from a range of periods including Early, Middle and Late Nasca as well as the Middle Horizon (Loro) and possibly Initial Nasca strongly suggests long-term and perhaps uninterrupted settlement activity in the vicinity of the site. Two Nasca 3 shards found at position 20.9 m (depth 3.56 m) may reflect primary deposition as judged from their placement very low in the stratigraphy. While traces of fluvial activity (bedded overbank deposits) are present at many positions throughout profile RPE and indicate that flooding of the lower terrace must have occurred since ≤260 cal BP, the strongly disturbed character and chronologically inconsistent findings do not allow for a detailed interpretation of the sedimentary history at this site.

Profile RPF-1

Profile RPF-1 (Figures A-33, A-34 and A-35) was a pit on a field on the right (north-western) side of the Rio Palpa, ca. 30 m north-west of the erosional bank exposure recorded as profile RPE. Profile RPF-2 lies at a distance of 123 m to the north-west. Profile RPF-1 is situated near (ca. 10 m) the lower edge of the main agricultural field terrace which is here ca. 300 m wide and rises from ca. 5 to ca. 8 m above the current riverbed. The stratigraphy of all four sides of the pit was recorded.

On the north-eastern side of the pit, 13 layers were recorded. Layer 0 is silty sand with a crumbly structure. Layer 1 is structureless sandy silt with ca. 5% gravel; its upper boundary is undulating. Layer 2 is structureless silty sand with some charcoal and up to 5% gravel. Its upper and lower boundary are undulating. Layer 3 is a lens of poorly sorted, angular and rounded gravel in a matrix of silt to coarse sand. Its upper boundary is undulating. Layer 4 is silt with faint remnant lamination and some oxidation stains. Charcoal is present; some rounded stones mark its upper and lower boundary. Layer 5 is a structureless wedge of silty sand with ca. 5% rounded and angular gravel (up to 10 cm) which extends from the northern corner to ca. the middle of the north-eastern side of the pit. It contains abundant charcoal. Layer 6 is structureless silty sand with 10% mainly rounded gravel (up to 10 cm) and abundant charcoal. Layer 7 is a lens of matrix-supported, rounded to subrounded gravel (50%, up to 10 cm) in a sandy matrix. It is present on the north-eastern and south-eastern side of the pit and lies stratigraphically within layer 6. Layer 8 is a north-sloping band of imbricated, rounded to angular gravel (40%, up to 10 cm) in a matrix of silty sand. Imbricated ceramic shards are present. Layer 9 is a band of silty sand with abundant disperse charcoal and some gravel (<5%, up to 5 cm). Layer 10 is a wedge of structureless, compacted silty sand with <5% gravel (up to 4 cm). It contains charcoal and ceramic shards. Layer 11 is a poorly sorted mass of rounded and angular gravel (70%, up to 15 cm) in a sandy matrix. It contains abundant ceramic shards pertaining to Nasca 7 (identified by J. Isla). Layer 12 is cemented coarse gravel (90%, up to 17 cm) in a matrix of coarse sand. Some clasts are strongly weathered.

On the south-eastern side of the pit, 9 layers were recorded. Layers 0 through 2 correspond to layers 0 through 2 on the north-eastern side. At a depth of 0.30 m, a rusted piece of iron wire was found in layer 2. Layer 3 is a broken-up band of silt corresponding to layer 4 on the north-eastern side. Layer 4 is structureless silty sand containing abundant charcoal. Its upper boundary is marked by a concentration of gravel (up to 8 cm). Layer 5 is silty sand and corresponds to layer 6 on the north-eastern side. It is interrupted by layer 6, a gravel layer which corresponds to layer 7 on the north-eastern side. Layer 7 is silty sand corresponding to layer 10 on the north-eastern side. Layer 8 is rounded gravel corresponding to layer 12 on the north-eastern side.

On the north-western side of the pit, 12 layers were recorded. Layer 0 through 2 correspond to layers 0 through 2 on the north-eastern side. Layer 3 is silty sand and corresponds to layer 5 on the north-eastern side. Layer 4 is a band of gravel corresponding to layer 8 on the north-eastern side. Layer 5 is a lens of silty sand with 10% rounded

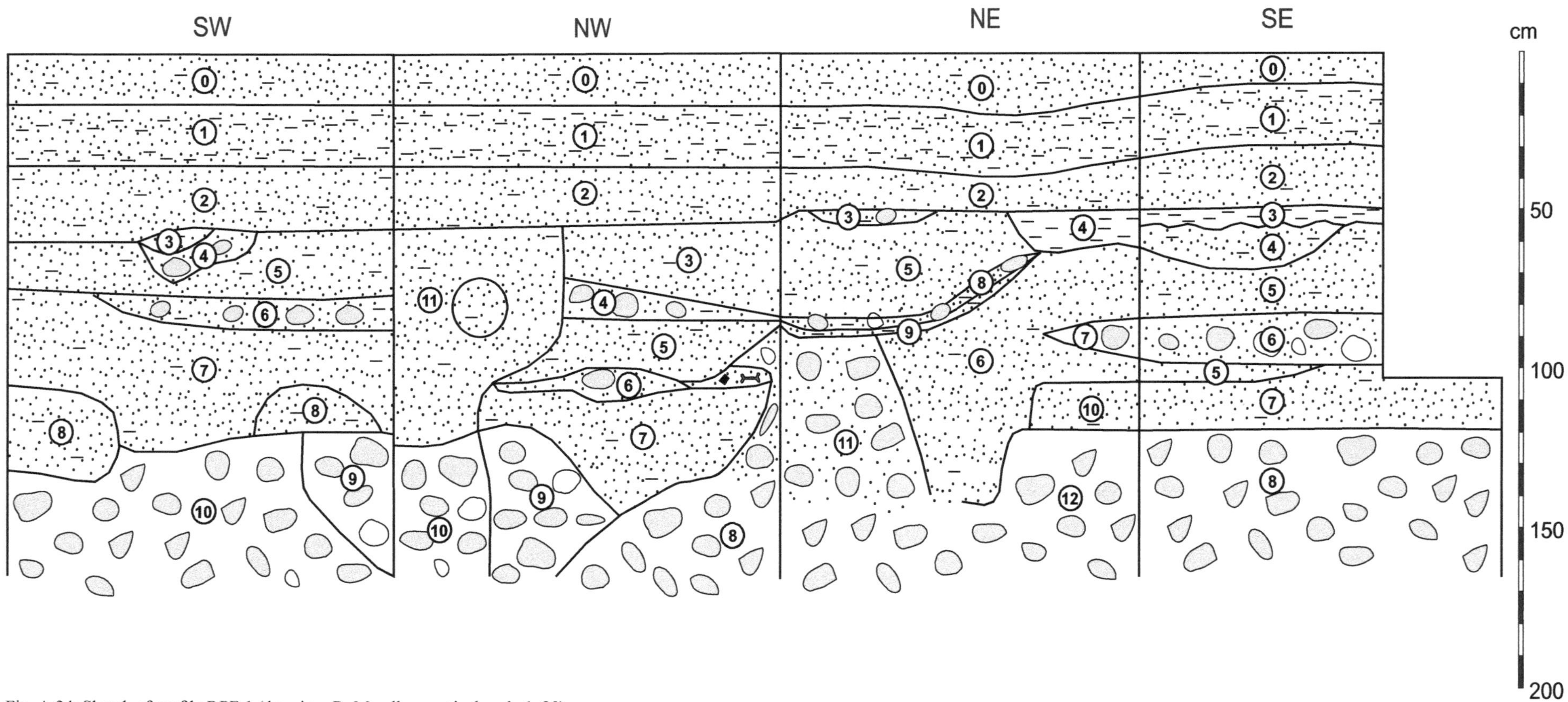

Fig. A-34: Sketch of profile RPF-1 (drawing: R. Mendler, vertical scale 1: 20).

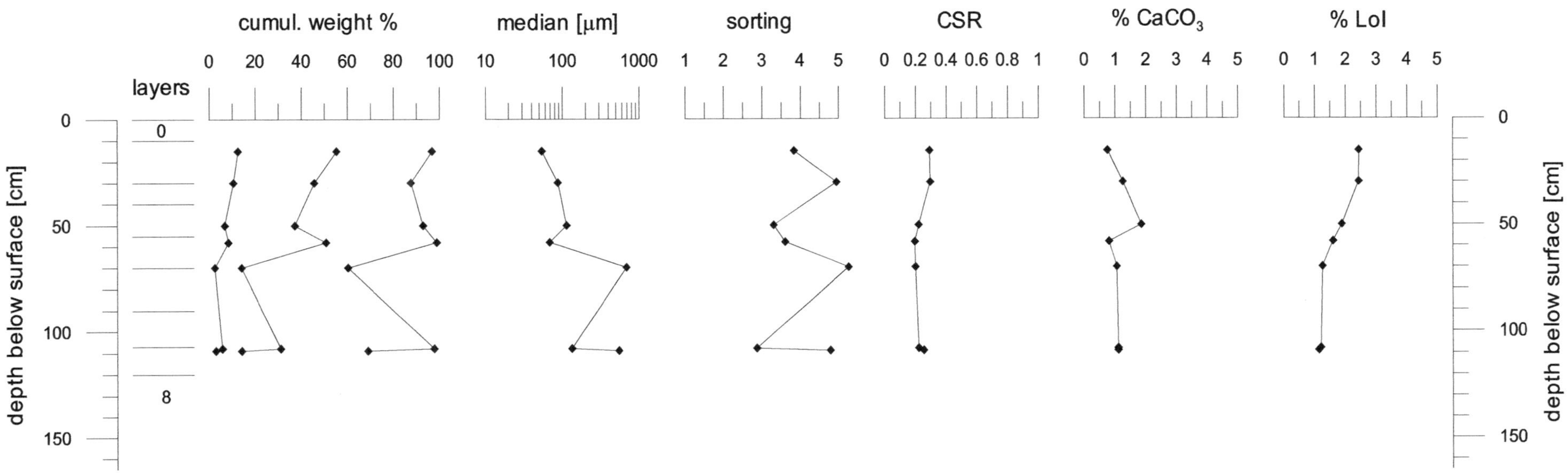

Fig. A-35: Laboratory results of samples from profile RPF-1 (vertical scale 1: 30). Cumulative weight percent graph shows clay, silt and sand from left to right. Note that the lines connecting data points do not imply values between data points but serve solely to aid visualisation.

and angular gravel (up to 8 cm) with ceramic shards and charcoal. Layer 6 is a poorly sorted band of silty sand with 40% gravel (up to 8 cm). Near the northern corner of the pit, it contains a large bone fragment embedded in charcoal-rich silty sand. Layer 7 is a lens of structureless silty sand with 1% gravel (up to 4 cm). Layer 8 is structureless coarse rounded and angular gravel (60%, up to 11 cm). Layer 9 is a wedge of partially weathered coarse, rounded gravel with a cemented upper boundary. Layer 10 is structureless coarse rounded and angular gravel (60%, up to 11 cm) similar to layer 8. Layer 11 is a structureless mass of silty sand with 10% gravel (up to 10 cm) apparently filling a depression penetrating layers 3 through 7. It contains abundant ceramic shards, charcoal and charred stones. A circular block of compact loamy sand (possibly a rounded adobe) is present at a depth between 0.72 and 0.90 m. Ceramic shards from depth between 0.80 and 1.55 pertain to the periods Nasca 5 and Nasca 7 (identified by J. Isla) and are not stratigraphically consistent (Nasca 5 shards occurring at 1.10 m and Nasca 7 shards at 1.55 m depth).
On the south-western side of the pit, 11 layers were recorded. Layers 0 through 2 correspond to layers 0 through 2 on the north-eastern side. Layer 3 is a small lens of silty sand in the middle of the profile side. Layer 4 is a small lens of laminated silty fine sand beside coarse sand with ca. 50% rounded gravel (up to 5 cm). Layer 5 is structureless silty sand with 1% rounded to angular gravel (up to 4 cm) and some charcoal. Layer 6 is a wedge of silty sand with 30–40% rounded to angular gravel (up to 10 cm). At the base of the layer, clasts appear imbricated. Layer 7 is structureless silty sand with 20% rounded to angular gravel (up to 8 cm). Layer 8 is structureless silty sand with 1% rounded gravel (up to 3 cm) and some clods of silt (up to 3 cm). It is present as 2 separate lenses at the base of layer 7. Layer 9 is gravel corresponding to layer 10 on the north-western side. Layer 10 is coarse rounded gravel with a cemented upper boundary. It corresponds to layer 9 on the north-western side. Where layer 9 is intrusive into the upper boundary, the cemented top of layer 10 is lacking.

Profile RPF-1 reflects significant anthropogenic disturbance throughout the stratigraphy. The surface of older (perhaps pre-Holocene) fluvial gravel has been affected by erosion and locally been covered by younger fluvial gravel. Between depths of 0.5 and 1.5 m, the stratigraphy reflects repeated horizontal deposition of poorly sorted material as well as the excavation and infilling of steep-sided pits, with most of the deposited material containing charcoal and/or ceramic shards pertaining to the Middle to Late Nasca period (1,620–1,300 cal BP, cf. Unkel 2006: 111). The chronologically inconsistent position of Nasca 5 and Nasca 7 shards in the stratigraphy indicates that much of the disturbance took place during or after Nasca 7. The abundance of ceramic shards (in total, 161 pieces and fineware and ca. 6 kg of plainware were recovered from the 2 by 3 m pit) strongly suggests the presence of a nearby settlement.

Profile RPF-2

Profile RPF-2 (Figures A-33, A-36 and A-37) was a pit on a field on the right (north-western) side of the Rio Palpa, 146 m from the river and approximately in the middle of the main agricultural field terrace which is here ca. 300 m wide and rises from ca. 5 to ca. 8 m above the current riverbed. It is situated between profiles RPF-1 (123 m) and RPF-3 (190 m). The stratigraphy of the north-eastern side and the left part of the south-western side of the pit was recorded.

On the north-eastern side, 11 layers were recorded. Layer 0 (the recent A_p / irrigation horizon) is crumbly loamy silt with abundant roots. Layer 1 is structureless loamy silt grading downwards to silty fine sand; it contains ca. 1% disperse gravel (up to 6 cm). Layer 2 is silty fine sand with some gravel and some ceramic shards; locally, faint bedding is recognisable from the occurrence of small silt lenses. Layer 3 is sand with ca. 20% gravel (rounded and angular) containing ceramic shards and marked by a blackish, charcoal-rich band at its base. The upper boundary is slightly undulating. Layer 4 is a band of silt with a horizontal upper boundary. It is only present on the north-eastern and north-western side of the pit. Layer 5 is structureless sand with 5% gravel and a horizontal upper boundary. Layer 6 is silty sand with disperse charcoal. Layer 7 is sand with 15% fine gravel and a few larger clasts (up to 5 cm). Layer 8 is sand with ca. 15% gravel. At its base, a silt band is present locally. A radiocarbon date of charcoal from the boundary between layers 8 and 9 (depth 1.50 m) is 2,070–1,820 cal BP (Poz-13540, 2,050±40 BP). Layer 9 is sand with 10% gravel (up to 5 cm). Layer 10 is subrounded to well-rounded coarse gravel (up to 17 cm) in a matrix of coarse sand and fine gravel.

In the left part of the south-western side, 5 layers were recorded. Layer 0 is crumbly loamy silt corresponding to layer 0 on the north-eastern side. Layer 1 is fine sand with some gravel and corresponds to layers 1 and 2 of the north-eastern side without recognisable internal differentiation. At its base, ceramic shards are present. Layer 2 is structureless, poorly sorted silty sand with angular and rounded gravel. It fills a steep-sided pit extending from a depth of 0.90 to 1.48 m (corresponding to layers 3 through 8 on the north-eastern side). At the base of the filled-in pit, human bones were found under the fragments of a large plainware ceramic vessel. According to J. Isla, this is a Nasca period grave of a ca. 10-year-old child. Layer 3 is silty sand with 10% rounded gravel. At a depth of 1.55 m, a disturbed band of silt is present. The upper boundary of layer 3 was disturbed by the bottom part of the anthropogenic pit (layer 2). Layer 4 is rounded so subrounded coarse gravel.

Profile RPF-2 reflects mainly anthropogenic influences on the middle terrace on the right side of the Rio Palpa. Riv-

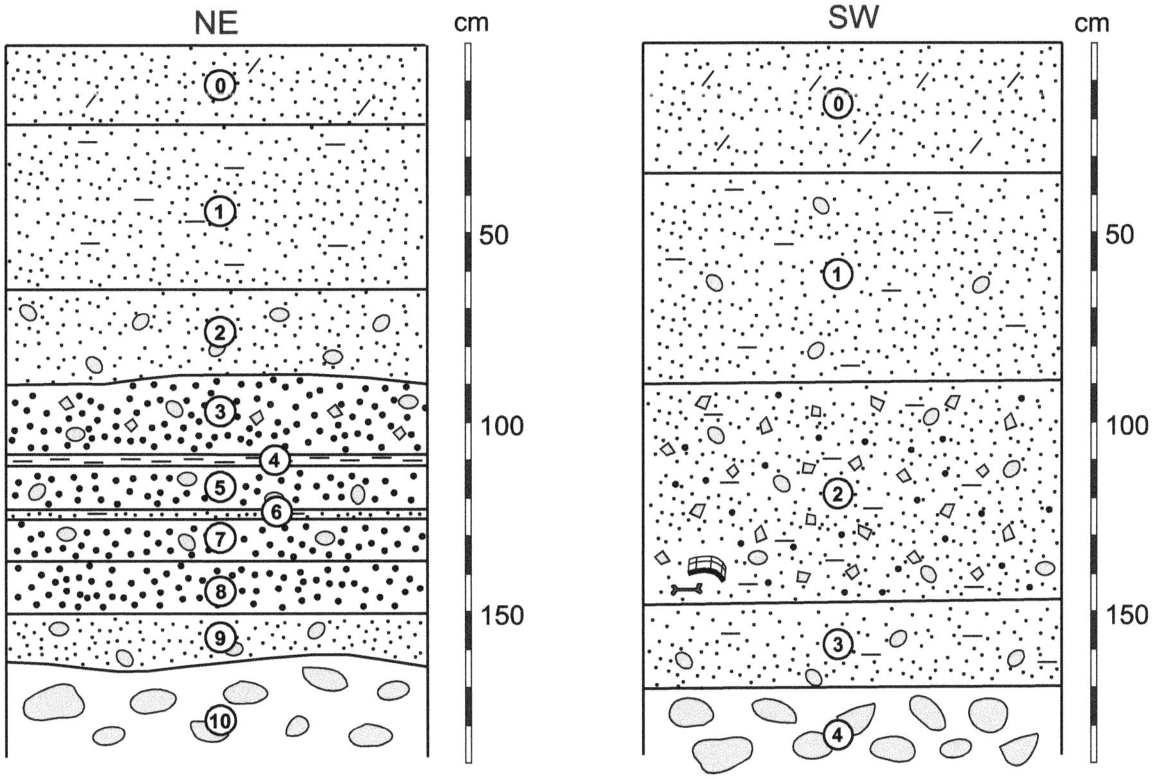

Fig. A-36: Sketch of profile RPF-2 (drawing: R. Mendler, vertical scale 1: 20).

erbed sediments (coarse, well-rounded gravel) are present at a depth of 1.60–1.70 m below the surface. This fluvial gravel is overlain by structureless silty to sandy sediments interpreted as irrigation sediments and generally containing disperse gravel-sized clasts. Anthropogenic influence is evident from abundant ceramic shards which were recovered from the excavated material (from depth ≥1.45 m: 9 pieces Middle Nasca, 1 piece Late Paracas; from depths ≤1.45 m [mainly from 0.63–1.07 m]: 40 pieces Middle Nasca, 2 pieces Nasca 7). Below 1.45 m, three mollusc shell fragments were recovered. Charcoal is present mainly at 0.85–1.07 m and at ca. 1.50 m. A Nasca period pit grave extends from a charcoal- and ceramic-bearing paleosurface at 0.90 m to a depth of 1.48 m in the left (southern) part the south-western side. The stratigraphy on the opposite (north-eastern) side of the pit was not affected by the grave. From charcoal recovered from a depth of 1.50 m, an age of 2,070–1,820 cal BP (Poz-13540) was determined. This age pertains to the Initial Nasca or Early Nasca period (2,150–1,690 cal BP, cf. UNKEL 2006: 111). Sedimentation related to irrigation agriculture therefore had commenced by Initial Nasca or Early Nasca or perhaps as early as Late Paracas times. Human use of the site (possibly settlement-related) of a paleosurface at ca. 0.90 m depth and deposition of a grave probably occurred during the Middle Nasca period (1,620–1,520 cal BP, cf. UNKEL 2006: 111). Average sedimentation rates are 1.09–3.00 mm/a between 2,070–1,820 cal BP and the Middle Nasca Period and 0.54–0.57 mm/a since the Middle Nasca period.

Profile RPF-3

Profile RPF-3 (Figures A-38 and A-39) was a pit on a field on the main agricultural terrace on the right side of the Rio Palpa, ca. 340 m from the river and ca. 12 m above the current riverbed. The uppermost irrigation canal on this side of the Rio Palpa (along the edge of the Cresta Sacramento) runs at a distance of 70 m north-west of the pit. The mouth of a small quebrada draining from the Cresta Sacramento (catchment area ca. 5.5 ha) lies ca. 150 m to the north-west; runoff from three weakly-developed gullies (ca. 0.5 ha each, ca. 25 m elevation difference) could also potentially affect the site. Profile RPF-3 is situated 190 m north-west of profile RPF-2. The south-eastern, north-eastern and south-western sides of the pit were recorded.

On the south-eastern side, 12 layers were recorded. Layer 0 (the recent A_p / irrigation horizon) is sandy silt with ca. 20% angular gravel (up to 5 cm) and a crumbly structure. Layer 1 is structureless sandy silt with 10% angular gravel (up to 10 cm) and a slightly undulating upper boundary. Layer 2 is poorly sorted sand with 60% angular gravel with a diffuse upper boundary. Layer 3 is well-bedded sand to fine gravel (up to 4 cm, mostly angular and a few rounded clasts). Its well-defined upper boundary is marked by a ca. 1 cm thick band of silt. Layer 4 is a layer of angular stones (80%, mostly ca. 3 cm and up to 7 cm) in a sandy matrix. Its upper boundary is marked by a thin band of sand. Layer 5 is structureless silty fine sand with 10% angular gravel (up to 7 cm, imbricated). Its upper

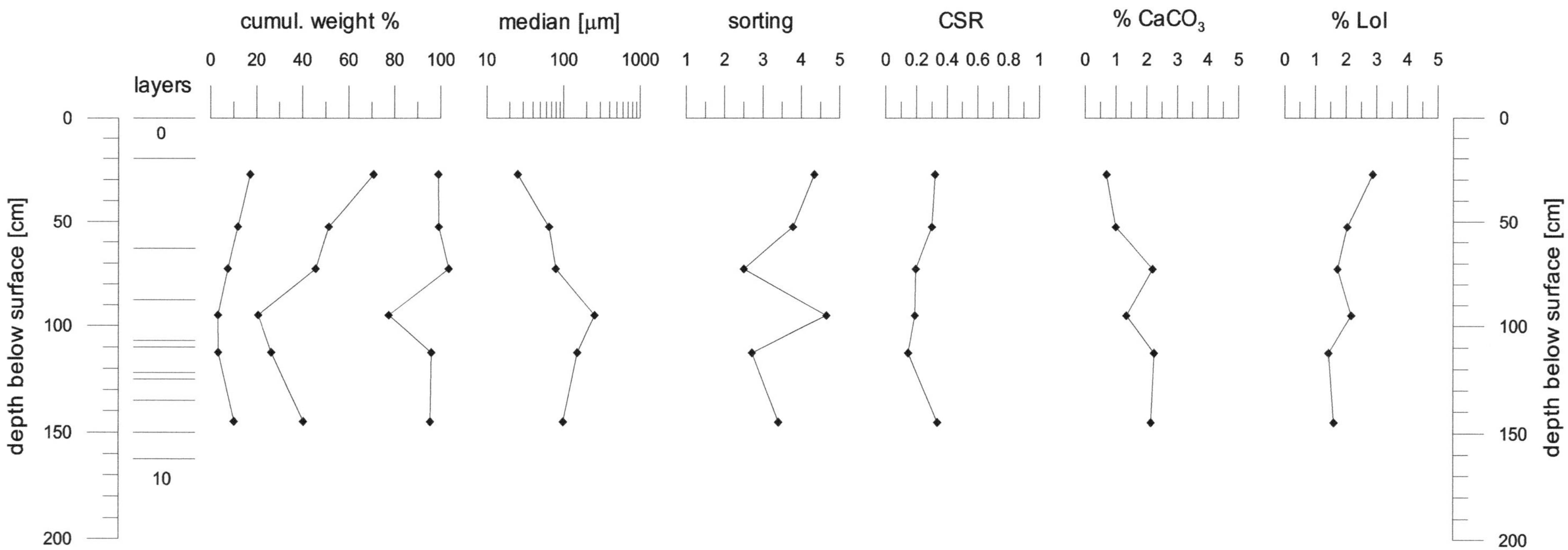

Fig. A-37: Laboratory results of samples from profile RPF-2 (vertical scale 1: 30). Note that the lines connecting data points do not imply values between data points but serve solely to aid visualisation.

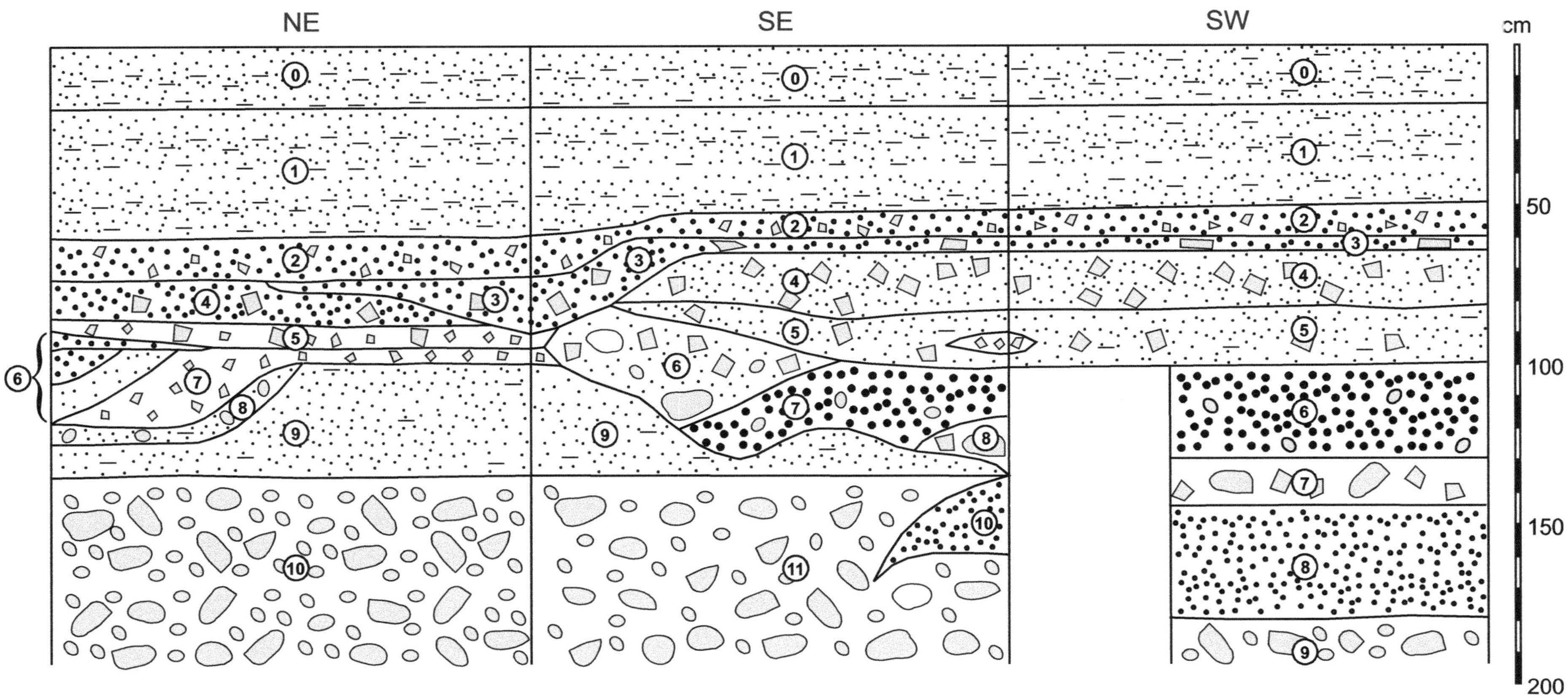

Fig. A-38: Sketch of profile RPF-3 (drawing: R. Mendler, vertical scale 1: 20).

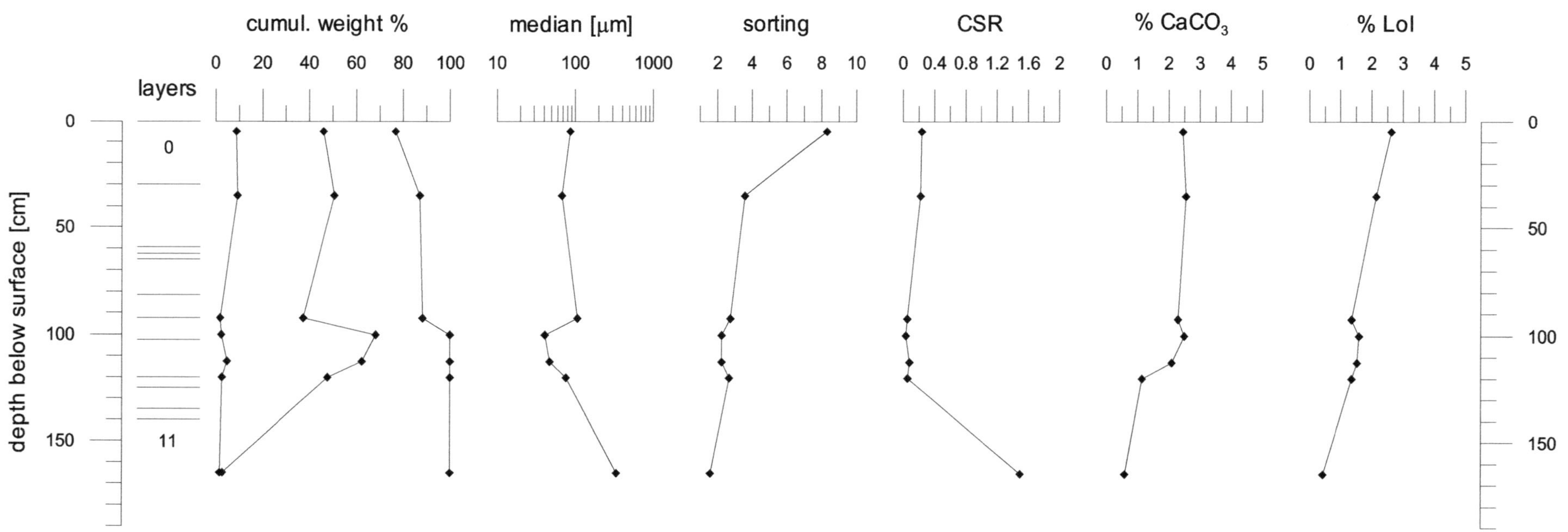

Fig. A-39: Laboratory results of samples from profile RPF-3 (vertical scale 1: 30). Cumulative weight percent graph shows clay, silt and sand from left to right. Note that the lines connecting data points do not imply values between data points but serve solely to aid visualisation.

boundary is horizontal. In the southern corner of the pit, a 7 cm thick lens of angular fine gravel with faint bedding is present. Layer 6 is a pocket of clast-supported, imbricated coarse gravel (up to 16 cm, mostly angular with a few rounded clasts) in a matrix of sand to fine angular gravel. Its upper boundary is undulating. Layer 7 is well-bedded coarse sand to fine gravel practically devoid of material smaller than coarse sand. Layer 8 is a pocket of imbricated rounded (ca. 60%) and angular (ca. 40%) coarse gravel (up to 16 cm). Layer 9 is silty fine sand with high mica content and a downward tendency towards higher sand content. It is vertically structured mainly by colour changes with diffuse boundaries: The top 7–8 cm are mottled reddish/brown/yellowish, the middle ca. 5 cm are dark brown and the bottom ca. 25 cm are yellowish-brown with oxidation stains. At the lower boundary of layer 9, there is an up to 10 cm thick band of bedded, well-sorted sand and fine gravel. Layer 10 is well-sorted, bedded sand. Layer 11 consists of mainly well-rounded and occasional angular coarse gravel (80%, ca. 10 cm).

On the north-eastern side, 11 layers were recorded. Layers 0 and 1 correspond to layers 0 and 1 on the south-eastern side of the pit. Layer 2 is poorly sorted, bedded sandy silt with imbricated angular gravel (up to 5 cm). Its upper boundary is horizontal and is marked by a 2–3 cm thick band of silt. Layer 3 is well-bedded sand to fine gravel (up to 4 cm, mostly angular and a few rounded clasts). Its well-defined upper boundary is marked by a ca. 1 cm thick band of silt. Layer 4 is poorly sorted sand to gravel (up to 7 cm, mainly angular and some subangular clasts). Faint inclined bedding is present. Layer 5 is imbricated coarse angular gravel (80%, up to 13 cm) in a sandy matrix. Layer 6 is a wedge of well-sorted sand with faint bedding. It can be divided into 3 stacked wedges, the top 2 of which are mainly medium sand while the bottom one is mainly fine sand. Layer 7 is well-sorted fine angular gravel with low sand content and faint inclined bedding. Layer 8 is a pocket of matrix-supported subrounded gravel (80%, up to 11 cm) in silty sand. Layer 9 is silty fine sand with high mica content and a downward tendency towards higher sand content. It is vertically structured mainly by colour changes with diffuse boundaries: The top 7–8 cm are mottled reddish/brown/yellowish, the middle ca. 5 cm are dark brown and the bottom ca. 25 cm are yellowish-brown with oxidation stains. At the lower boundary of layer 9, there is an up to 10 cm thick band of bedded, well-sorted sand and fine gravel. Layer 10 is well-sorted and mainly well-rounded coarse gravel (80%, ca. 10 cm) in a matrix of coarse sand. A few angular clasts are present.

On the south-western side, 10 layers were recorded. Their sedimentological properties correspond to those in the right part of the south-eastern side of the pit with some changes in layer thickness in the lower portion of the profile.

Profile RPF-3 reflects geomorphic processes close to the Cresta Sacramento. While high-energy (channel) fluvial deposits characterised by well-rounded coarse gravel are present at the base of the pit at a depth of only 1.35 m, most of the profile is characterised by poorly-sorted hyperconcentrated flow deposits with angular clasts derived from the nearby Cresta Sacramento. At least six probable such events (layers 2–6 and 8 on the south-eastern side of the pit) are recorded at the site. Low- to medium-energy fluvial sediments (bedded, well-sorted sand to fine gravel) occur in the lower third of the profile. Charcoal and other indicators of human activity at the site are missing except for a single fragment of Nasca 5 ceramics recovered from the excavated material. A possible incipient soil development indicated by a slight increase of LoI values (1.4–1.6%) is present below the hyperconcentrated flow deposits at a depth of 0.95–1.35 m. Irragric anthrosol is present only in the top of the profile from a depth of 0.52/0.65 m to the surface. Assuming average sedimentation rates of 0.54–0.57 mm/a as derived at the nearby profile RPF-2 indicates that irrigation agriculture at the site may have been initiated upon the uppermost hyperconcentrated flow deposit between 1,150 and 860 cal BP (i.e. during the Middle Horizon, 1,300–950 cal BP, or Late Intermediate Period, 950–550 cal BP, cf. Unkel 2006: 111).

Profile RPQ

Profile RPQ (see Figure A-33) was recorded in March 2004 in a trench dug perpendicularly across the dry bed of the Rio Palpa adjacent to profile RPE. The trench extended across most of the here 146 m wide riverbed. The stratigraphy of the riverbed sediments was recorded at four stations (RPQ-1, RPQ-2, RPQ-3 and RPQ-4) numbered from north-west to south-east. RPQ-1 is 48.2 m from the nearest portion of profile RPE on the right bank of the Rio Palpa. The distances within RPQ are 39.4 m between RPQ-1 and RPQ-2, 18.2 m between RPQ-2 and RPQ-3 and 18.2 m from RPQ-3 to RPQ-4. Sketches and descriptions pertain to the north-eastern side of the trench. According to aerial photographs, RPQ-1 was not part of the riverbed but instead covered by a later-eroded field in 1944. RPQ-2 and RPQ-4 were situated in flow channel situations in 1944 while RPQ-3 was situated on a large longitudinal bar. In 1977, RPQ-1 was in a shrub-covered, marginal situation of the riverbed while RPQ-4 was still part within a main channel and both RPQ-2 and RPQ-4 were on longitudinal bars.

RPQ-1 (48.2 m from RPE, Figures A-40, A-44 and A-45): Layer 0 is loose gravel (ca. 50%) in a matrix of medium to coarse sand. Layer 1 is well-sorted, horizontally bedded medium to coarse sand with an undulating upper boundary. Layer 2 is poorly sorted gravel (up to 7 cm) in a matrix of medium to coarse sand. The upper boundary is undulating. Layer 3 is imbricated, upwards-fining coarse gravel to boulders (up to 23 cm) in a matrix of coarse sand

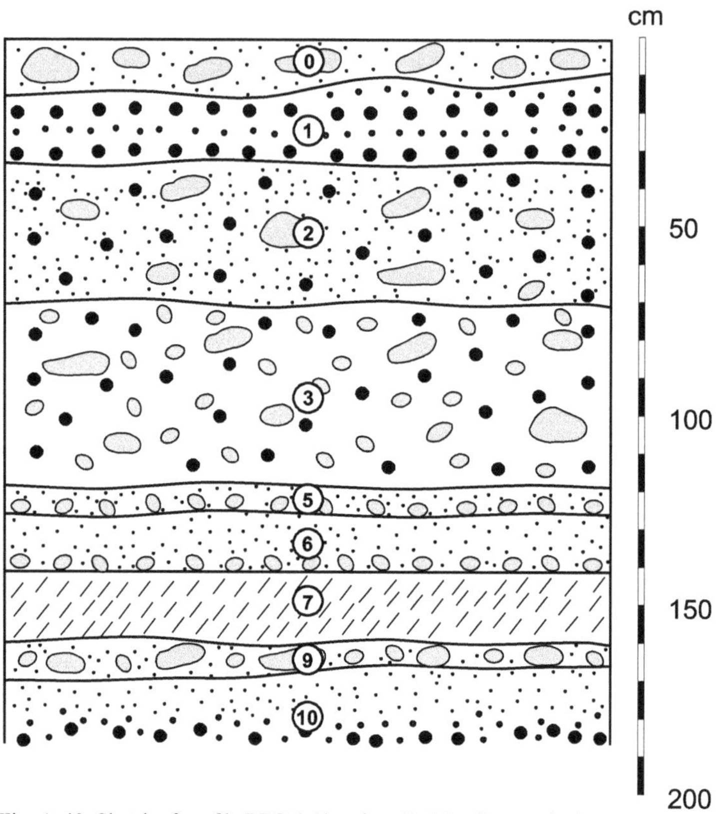

Fig. A-40: Sketch of profile RPQ-1 (drawing: R. Mendler, vertical scale 1: 20).

to fine gravel. The upper boundary is undulating. Layer 4 is well-bedded and well-sorted coarse sand with an undulating upper boundary. Layer 4 commences slightly north-east of station RPQ-1 (where is replaces layers 5 through 7) and is therefore not present in the sketch. Layer 5 is an upwards-fining layer of fine gravel to sand. Layer 6 is an upwards-fining layer of well-sorted fine gravel to sand. Its upper boundary is distinct and undulating. Layer 7 is a 19 cm thick sequence of well-bedded silty loam with beds ca. 1 cm thick and differentiated by colour (yellowish, brown, black). LoI is high (3.3%). The upper boundary of layer 7 is horizontal. Layer 8 is only present to the north-east of station RPQ-1. Layer 9 consists of loosely-packed, sub-rounded coarse gravel to boulders (≥15 cm). A matrix of sandy silt appears to be intrusive between the large clasts from the undulating top of the layer. Layer 10 is well-sorted and upwards-fining sand with oxidation stains and a slightly west-dipping upper boundary.

RPQ-2 (87.6 m from RPE, Figures A-41, A-46 and A-47): Layer 0 is hard-packed, well-bedded coarse sand. Layer 1 is well-rounded coarse gravel to boulders (up to 20 cm) in a matrix of coarse sand. Layer 2 is coarse sand with rounded gravel (up to 10 cm); its upper boundary is undulating. Layer 3 is horizontally bedded and well-sorted medium to coarse sand with a few larger clasts; the upper boundary is undulating. Layer 4 consists of well-rounded coarse gravel to boulders (up to 15 cm) with an undulating upper boundary. Layer 5 is bedded loamy silt with an undulating upper boundary. Bedding is marked by colour changes (red and brown). LoI is high (3.0%). Layer 6 is gravel in a matrix of sandy loam which appears to be intrusive from the overlying layer 5. Towards the base of layer 6, the matrix material becomes increasingly sandy. Layer 7 is a band of loamy sand. Layer 8 is gravel in a sandy matrix.

RPQ-3 (106 m from RPE, Figures A-42, A-48 and A-49): Layer 0 is fine sand with oxidation stains and abundant roots. Layer 1 is fine to coarse sand with crossbedding and a sharp, slightly undulating upper boundary. Layer 2 is clast-supported coarse gravel (up to 17 cm). Layer 3 is well-sorted fine to medium gravel with an undulating upper boundary. At the base of the layer, larger clasts (up to 17 cm) are concentrated. Layer 4 is a pocket of gravel (up to 10 cm) in a matrix of coarse sand and with an undulating upper boundary. Layer 5 is poorly sorted, matrix-supported gravel (up to 30 cm) with south-dipping bedding. Oblong clasts are imbricated. The matrix consists of medium to coarse sand. Towards the base of the layer, matrix content diminishes and the material is clast-supported. At the base there is a thin band of imbricated gravel. Layer 5 is capped by up to 5 cm of silty loam; the upper boundary of the sequence is undulating. Layer 6 is bedded, well-sorted medium and coarse sand to fine gravel with inclined bedding. At the top of the layer, up to 10 cm thick and 30 cm wide cohesive lumps of sandy silt are embedded in sand. Layer 7 is a north-west to south-east stacked sequence of clast-supported gravel in a matrix of sandy fine gravel through clayey silt to mica-

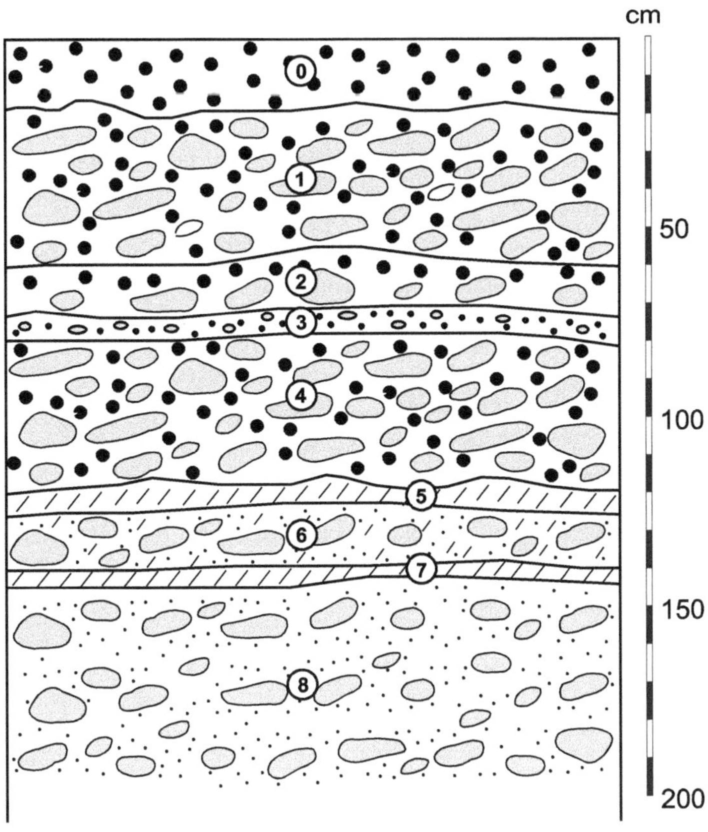

Fig. A-41: Sketch of profile RPQ-2 (drawing: R. Mendler, vertical scale 1: 20).

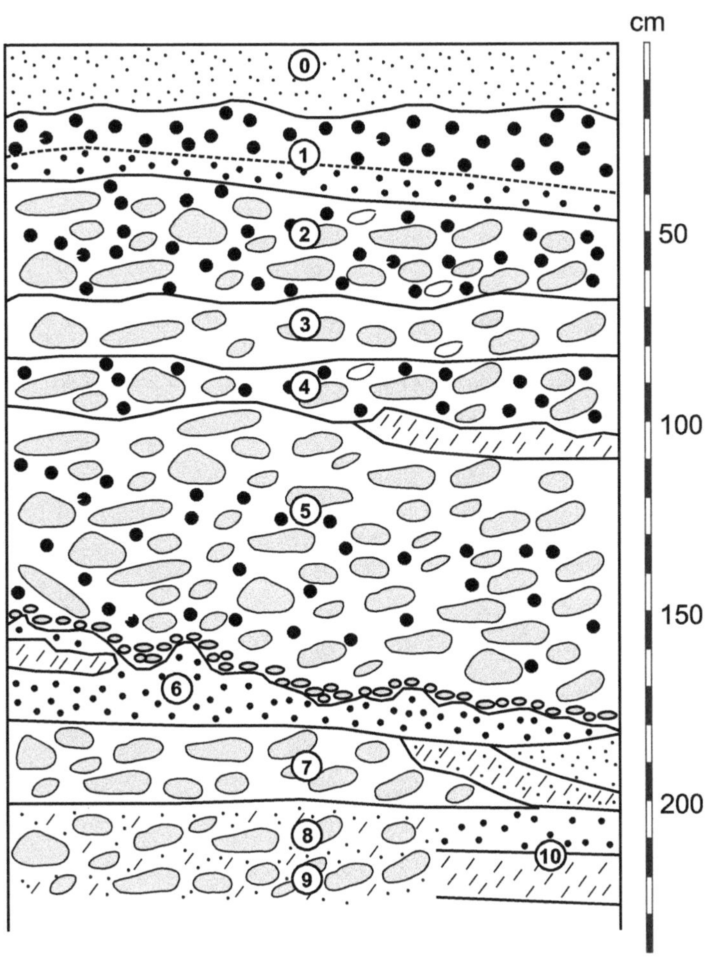

Fig. A-42: Sketch of profile RPQ-3 (drawing: R. Mendler, vertical scale 1: 20).

rich, well-sorted fine sand. The upper boundary is undulating. Layer 8 is clast-supported gravel in a matrix of sandy fine gravel with an undulating upper boundary. Layer 9 is poorly sorted, not imbricated gravel in a matrix of sandy loam. Towards the south-east, layer 9 grades into layer 10, a fine to medium sand with inclined bedding underlain by a 5 cm thick band of loamy silt. The common upper boundary of layers 9 and 10 is horizontal.

RPQ-4 (124 m from RPE, Figures A-43, A-50 and A-51): Layer 0 is bedded fine to medium sand with abundant roots. Layer 1 is horizontally bedded, clast-supported, imbricated gravel (up to 20 cm) in a matrix of coarse sand. The upper boundary is sharp and undulating. Layer 2 is poorly sorted, not imbricated coarse gravel to boulders with south-dipping bedding. The matrix is coarse sand; the upper boundary is sharp. Layer 3 is well-bedded fine to medium sand with a downward-declining content of coarse sand. Its upper boundary is sharp. Layer 4 is a band of matrix-supported gravel in medium to coarse sand. Its upper boundary is sharp. Layer 5 is bedded, upwards-fining medium to coarse sand with some gravel and an undulating upper boundary. A plastic bottle and a piece of a plastic bag were found in layer 5 at depths of 1.43 and 1.53 m below the surface of the dry riverbed. Layer 6 is medium to coarse sand with a sharp upper boundary. At the base of layer 6, two thin (ca. 1 cm) bands of loam are present. Layer 7 is loamy sand. Layer 8 is fine gravel in a matrix of fine to medium sand.

Profile RPQ records more than two vertical meters of fluvial sediments and serves as a type example for the material deposited in the beds of Rio Palpa and Rio Vizcas. The observed grain size parameters and sedimentary structures (bedding, imbrication, cut-and-fill features) are characteristic for braided, gravel-bed rivers. Coarse, imbricated gravel sheets with horizontal or inclined bedding indicate sedimentation in high-energy flow environments such as channels and bar structures. Finer-grained (silt to fine gravel) and commonly well-bedded and/or upwards-fining sediments indicate waning-flow or bar accumulation (a) in primary channels close to the end of the discharge season, (b) in secondary and higher-order channels or (c) on interchannel flow surfaces. The presence of loam with high LoI values at stations RPQ-1 and RPQ-2 which is interpreted as eroded soil hints at lateral erosion affecting fine-grained riverbanks upstream. Repeated erosion and accumulation of sediment during discharge periods and possibly the movement of sediment pulses through the reach within late historic times (second half of the 20th century) are clearly shown by the presence of a plastic bottle and a piece of plastic bag at a depth of 1.43–1.53 m below the surface of the riverbed.

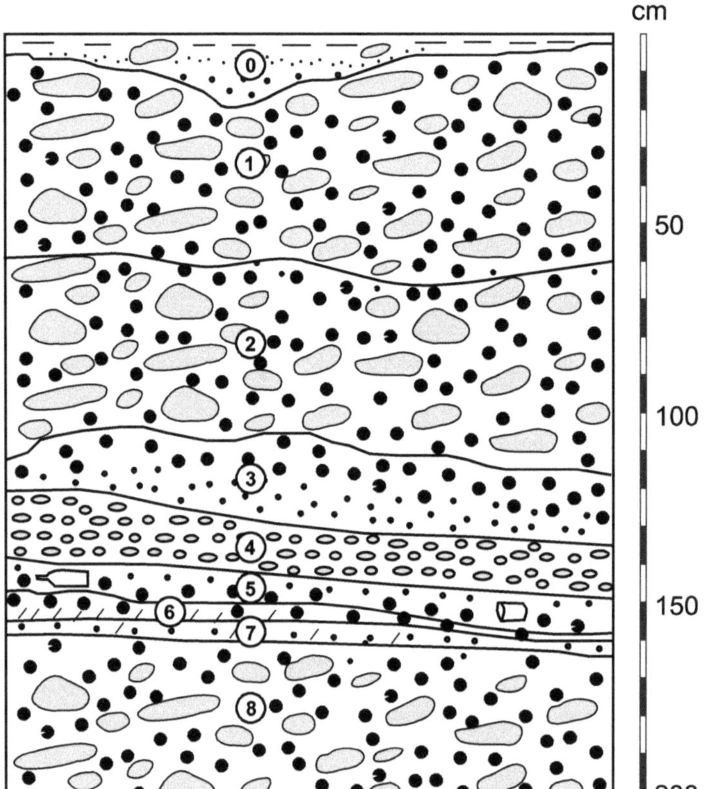

Fig. A-43: Sketch of profile RPQ-4 (drawing: R. Mendler, vertical scale 1: 20).

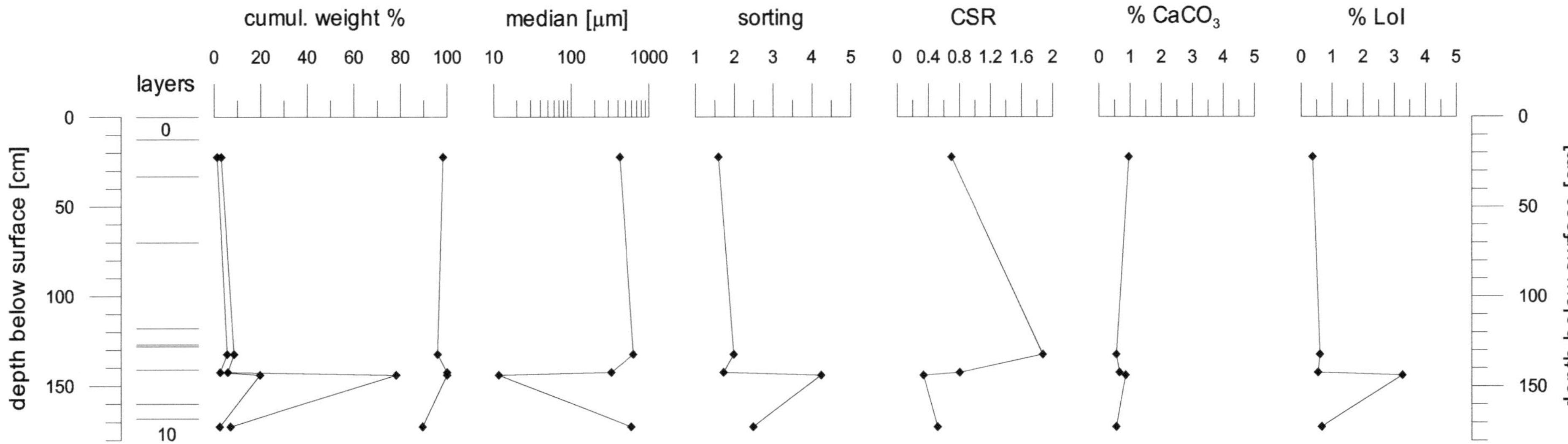

Fig. A-44: Laboratory results of samples from profile RPQ-1 (vertical scale 1: 30). Cumulative weight percent graph shows clay, silt and sand from left to right. Note that the lines connecting data points do not imply values between data points but serve solely to aid visualisation.

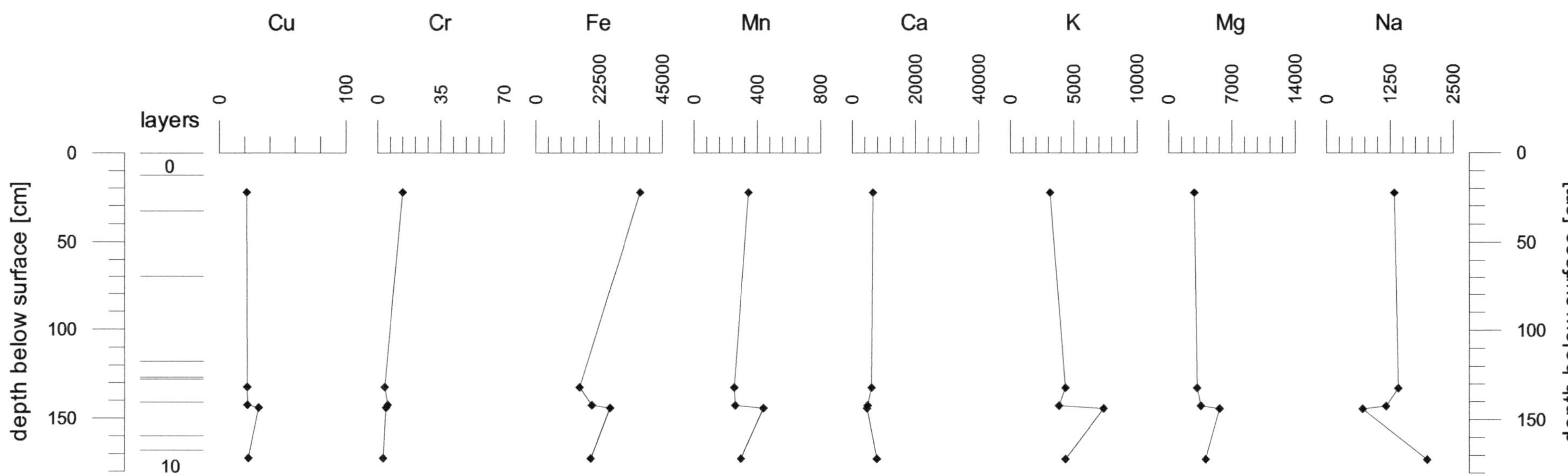

Fig. A-45: AAS measurement results of samples from profile RPQ-1 (vertical scale 1: 30). Note that the lines connecting data points do not imply values between data points but serve solely to aid visualisation.

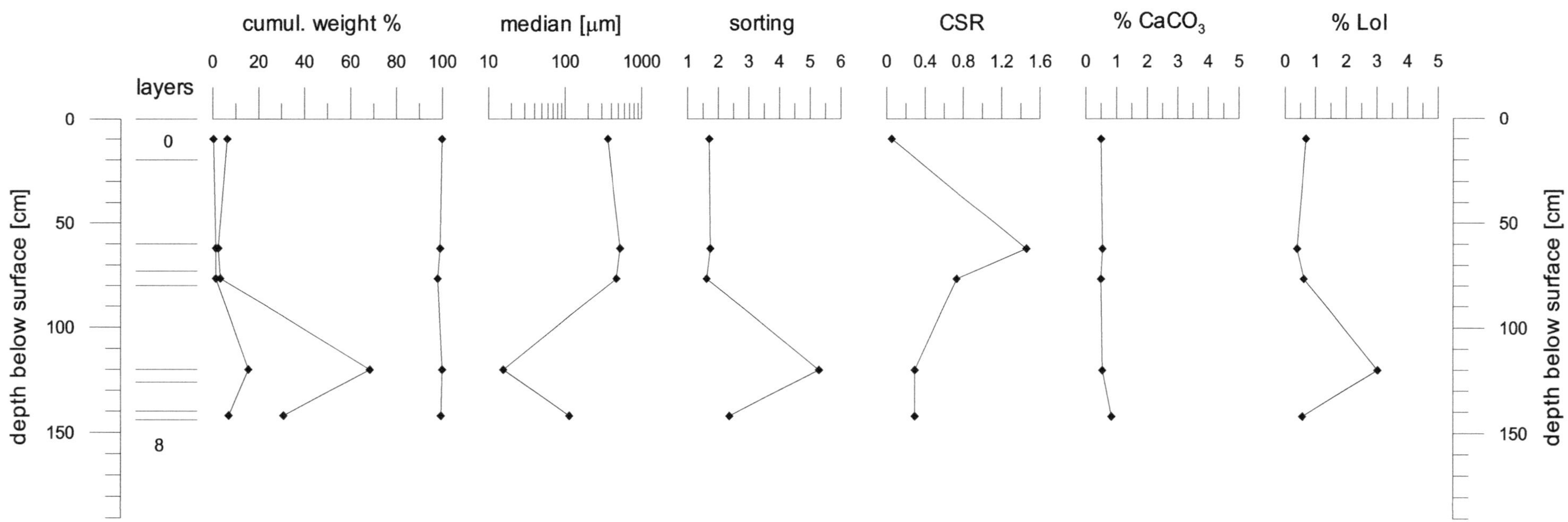

Fig. A-46: Laboratory results of samples from profile RPQ-2 (vertical scale 1: 30). Cumulative weight percent graph shows clay, silt and sand from left to right. Note that the lines connecting data points do not imply values between data points but serve solely to aid visualisation.

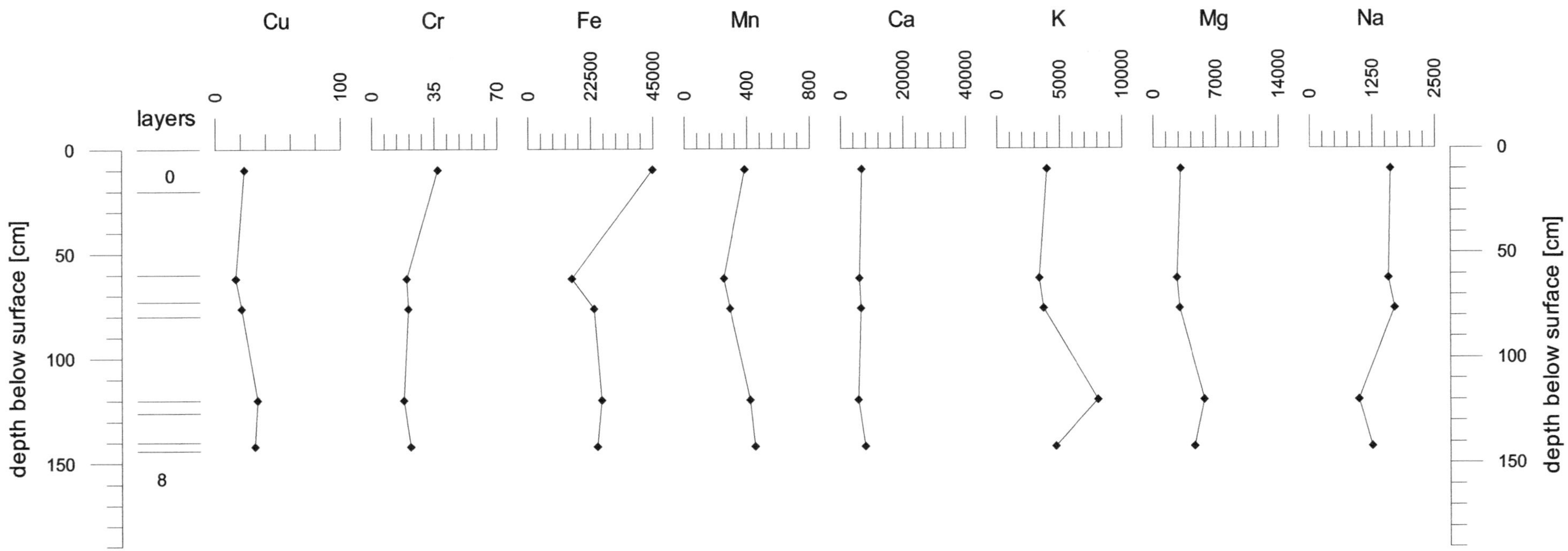

Fig. A-47: AAS measurement results of samples from profile RPQ-2 (vertical scale 1: 30). Note that the lines connecting data points do not imply values between data points but serve solely to aid visualisation.

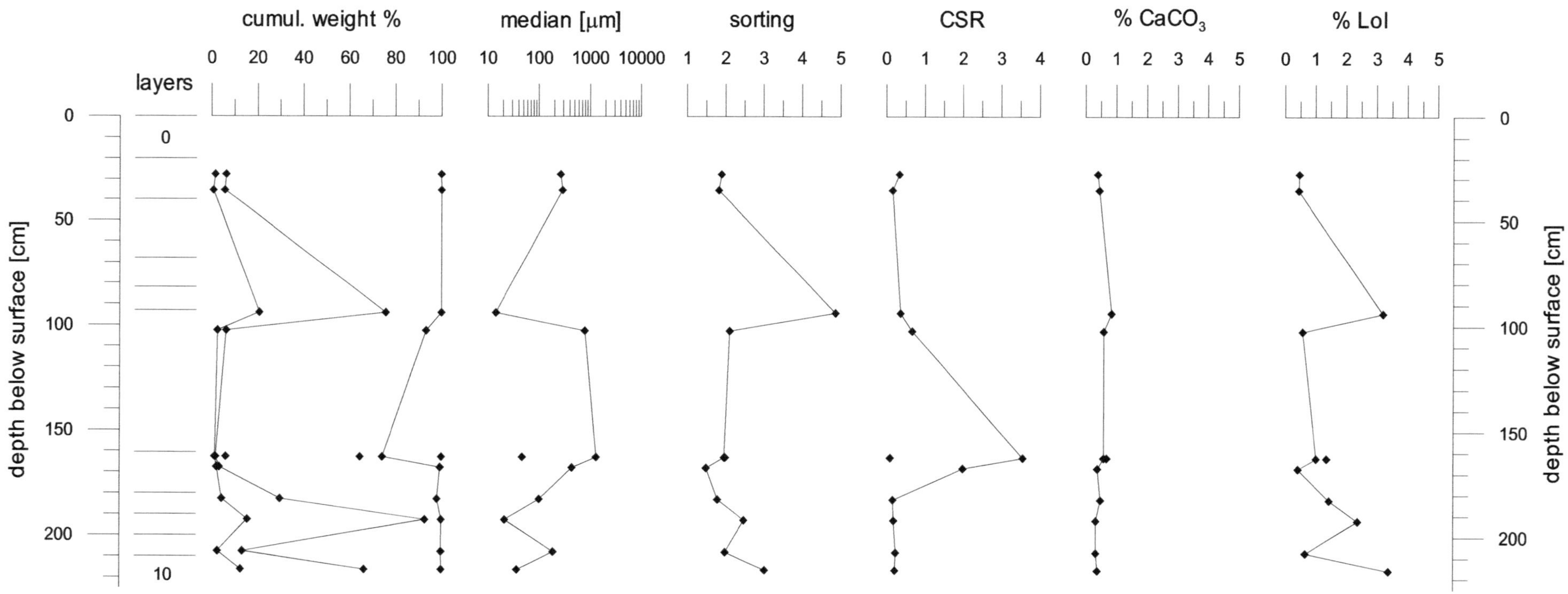

Fig. A-48: Laboratory results of samples from profile RPQ-3 (vertical scale 1: 30). Cumulative weight percent graph shows clay, silt and sand from left to right. Note that the lines connecting data points do not imply values between data points but serve solely to aid visualisation.

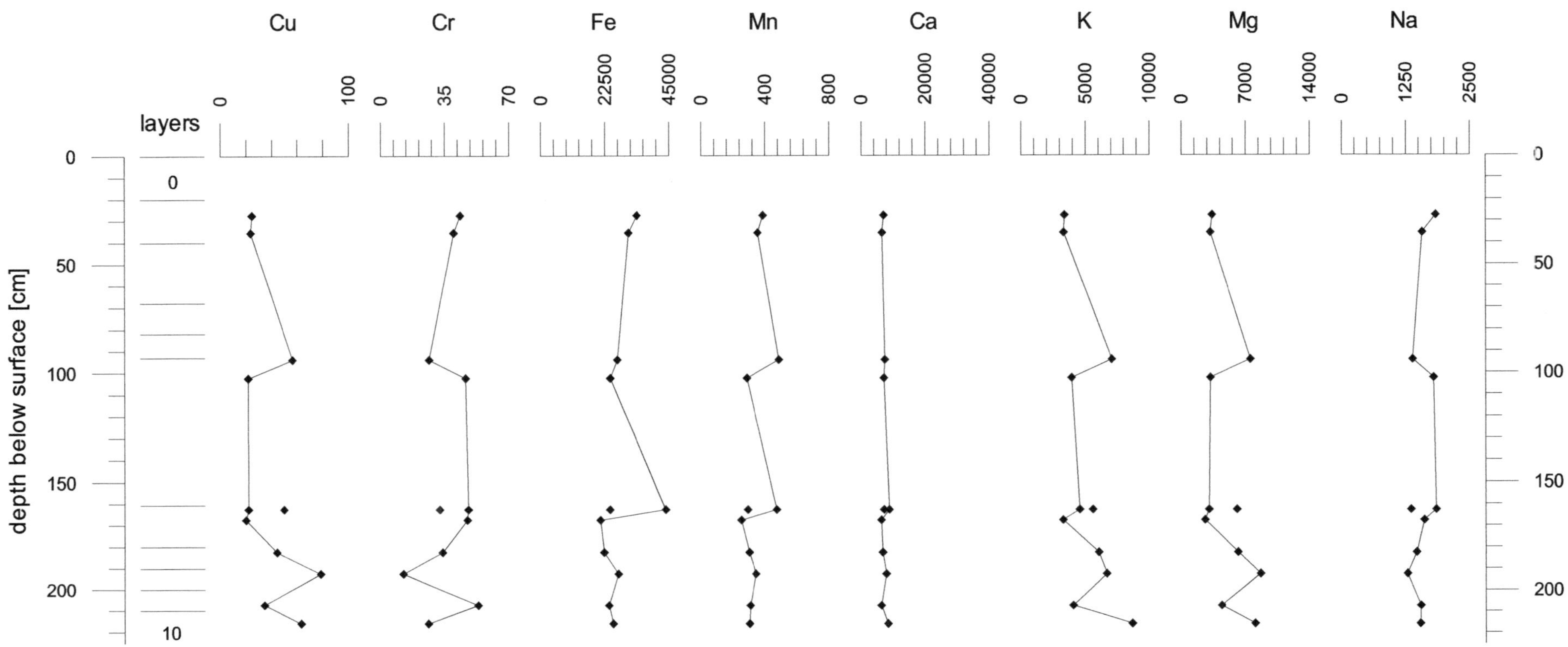

Fig. A-49: AAS measurement results of samples from profile RPQ-3 (vertical scale 1: 30). Note that the lines connecting data points do not imply values between data points but serve solely to aid visualisation.

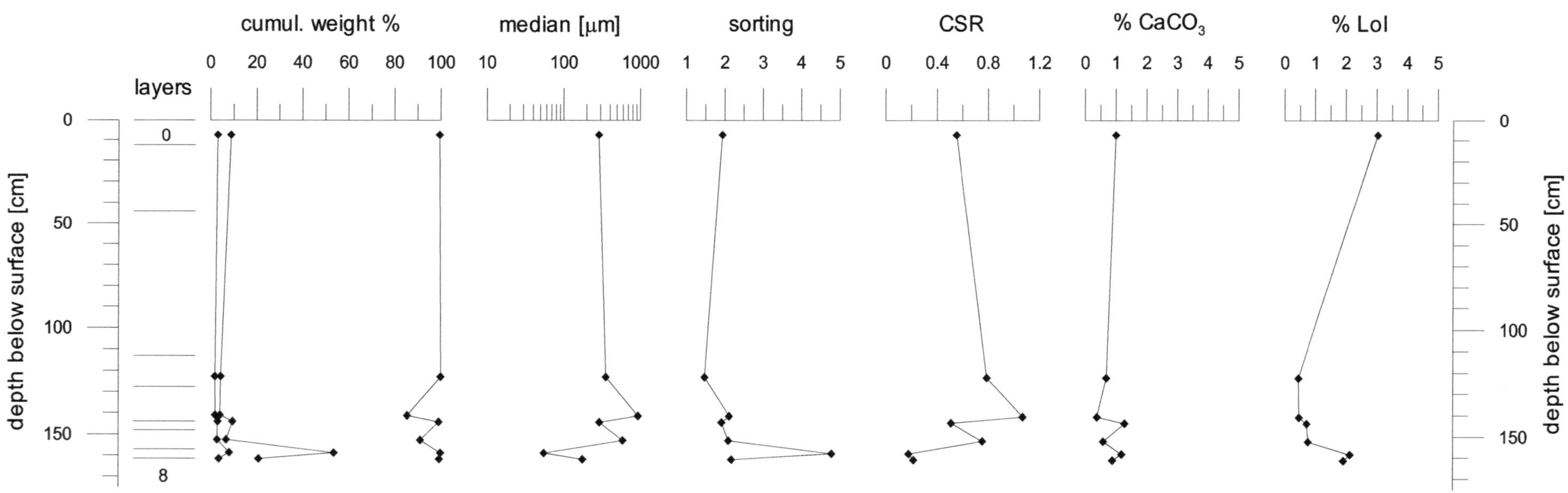

Fig. A-50: Laboratory results of samples from profile RPQ-4 (vertical scale 1: 30). Cumulative weight percent graph shows clay, silt and sand from left to right. Note that the lines connecting data points do not imply values between data points but serve solely to aid visualisation.

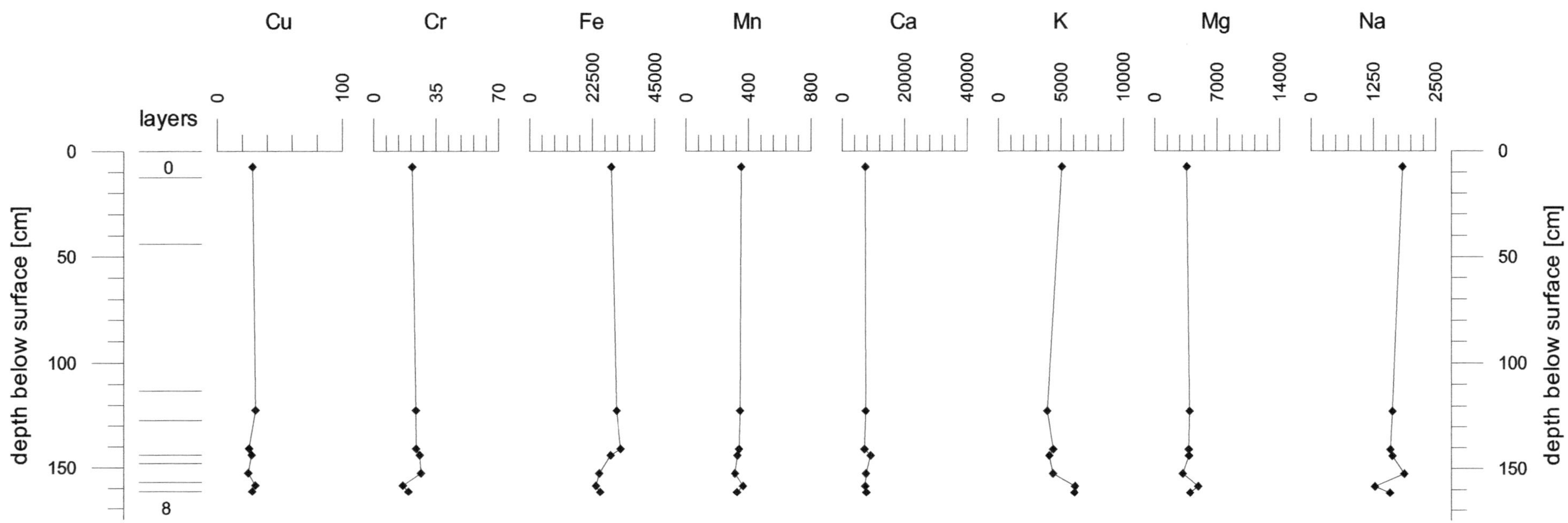

Fig. A-51: AAS measurement results of samples from profile RPQ-4 (vertical scale 1: 30). Note that the lines connecting data points do not imply values between data points but serve solely to aid visualisation.

Profile RPW

Profile RPW (Figure A-52, A-53 and A-54) was recorded in a deep trench close to an erosional natural river bank on the right side of the Rio Palpa, ca. 110 m south-south-west of profile RPE. Aerial photographs show that in 1944 the profile was near the edge of a terrace, to the east of which there had been a 50–100 m wide field. Erosion of the field and creation of the terrace edge occurred in at least two stages between 1944 and 2005 as shown by aerial photographs of 1944 and 1970 and a Quickbird image of 2005.

Layer 0 is structureless, slightly compacted loamy sand covered by organic debris. Layer 1 is undulating coarse sand to fine gravel with a generally horizontal but locally pocketed upper boundary. Layer 2 is compacted, slightly crumbly loamy silt with some gravel. Its upper boundary is undulating. Layer 3 is structureless silty loam. At its base, structure is crumbly. Layer 4 is compacted, structureless loamy sand with some gravel. The upper boundary is undulating. Layer 5 is loamy sand with faint bedding and a horizontal upper boundary. Charcoal is present. Layer 6 is compacted, structureless loamy sand with an undulating, pocketed upper boundary. Layer 7 is loamy sand with charcoal-rich lenses and a horizontal upper boundary. Layer 8 is crumbly sandy loam with some gravel and a horizontal upper boundary. Layer 9 is structureless, well-sorted silty fine sand with a high mica content and a strongly undulating upper boundary. Layer 10 is blackish silty fine sand with disperse charcoal. Layer 11 is poorly sorted loamy sand with ca. 30% gravel and a diffuse, pocketed upper boundary. Charcoal and ceramic shards are present. Layer 12 consists of well-rounded coarse gravel to boulders (up to 25 cm) with some strongly weathered clasts and a pocketed upper boundary. Charcoal from layer 12 (depth 2.58 m) was radiocarbon-dated to 1,690–1,410 cal BP (Poz-8026, 1,705 ±30 BP), i.e. the Middle Nasca period (1,690/1,620–1,420 cal BP, cf. Unkel 2006: 111). LoI values are relatively high (>2.0%) for layers 2 through 8 as well as layer 10. Sorting is poor to very poor except for layer 9. Carbonate content is low (0.5–1.7%) for layers 2 through 10 but somewhat higher (2.3 and 2.7%) for layers 10 and 11, respectively and peaking near the top of the profile in layer 0 (3.2%) and layer 1 (6.8%). Carbonate accumulation is interpreted to be caused by evaporation of irrigation water.

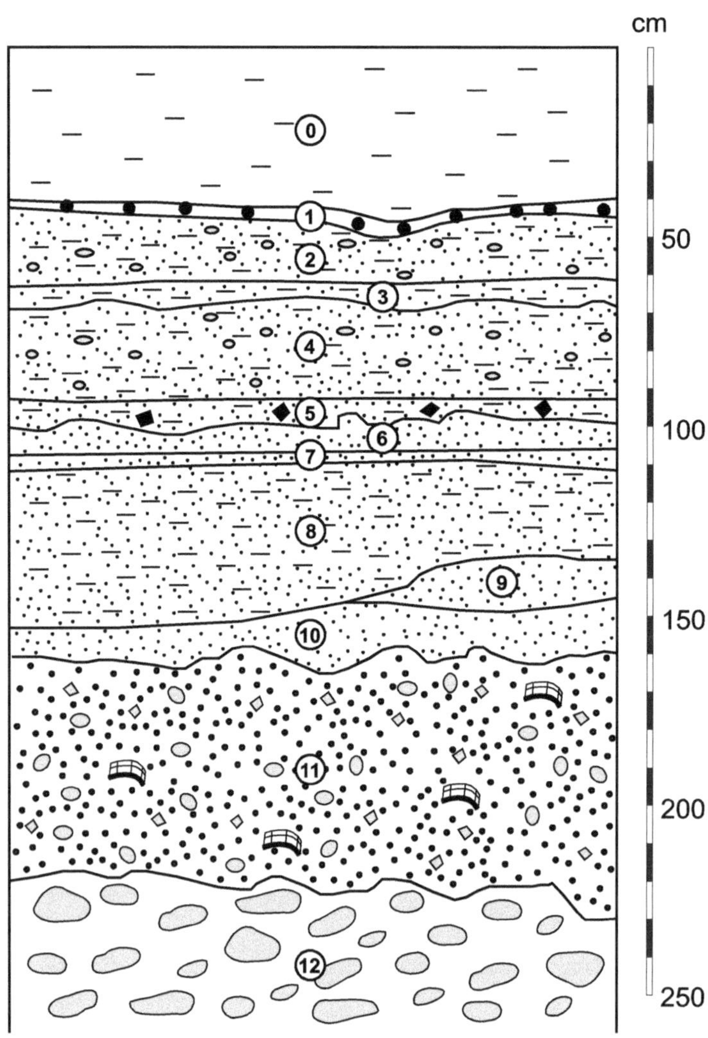

Fig. A-52: Sketch of profile RPW (drawing: R. Mendler, vertical scale 1: 20).

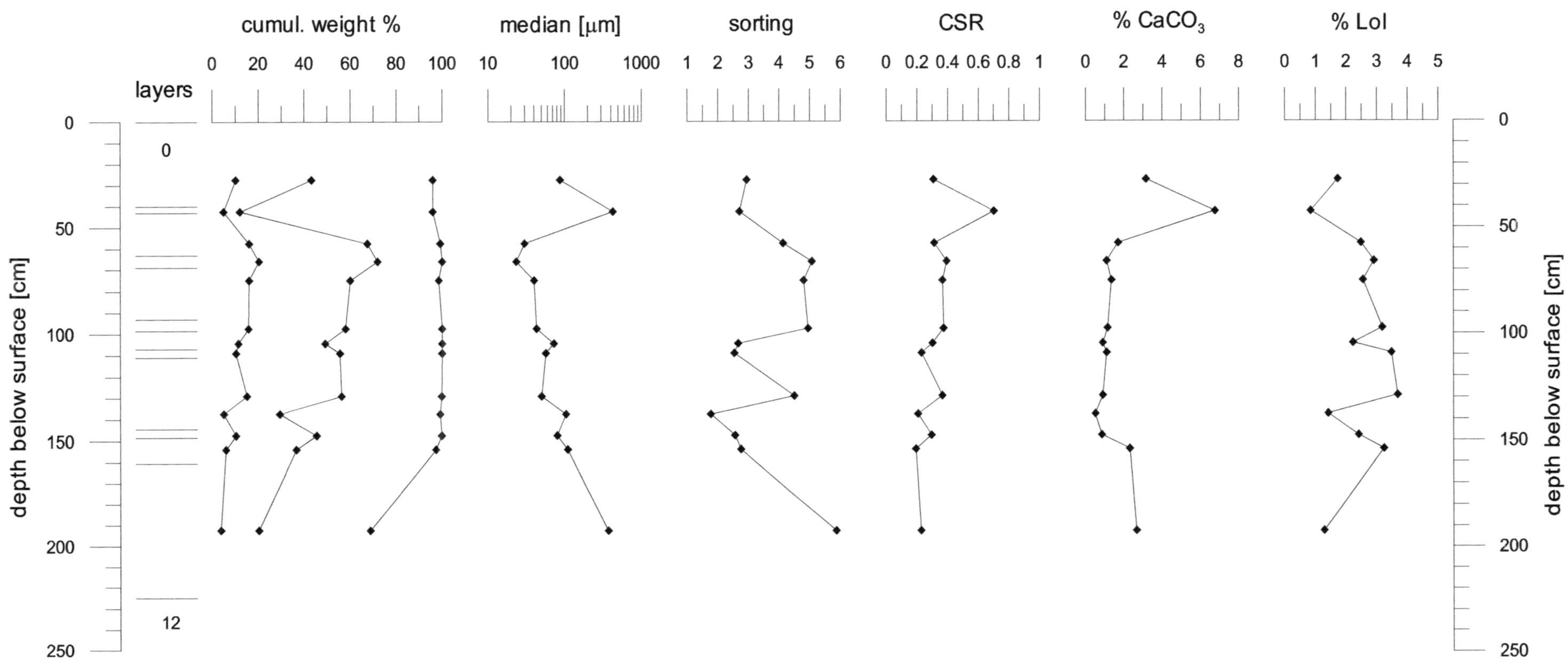

Fig. A-53: Laboratory results of samples from profile RPW (vertical scale 1: 30). Cumulative weight percent graph shows clay, silt and sand from left to right. Note that the lines connecting data points do not imply values between data points but serve solely to aid visualisation.

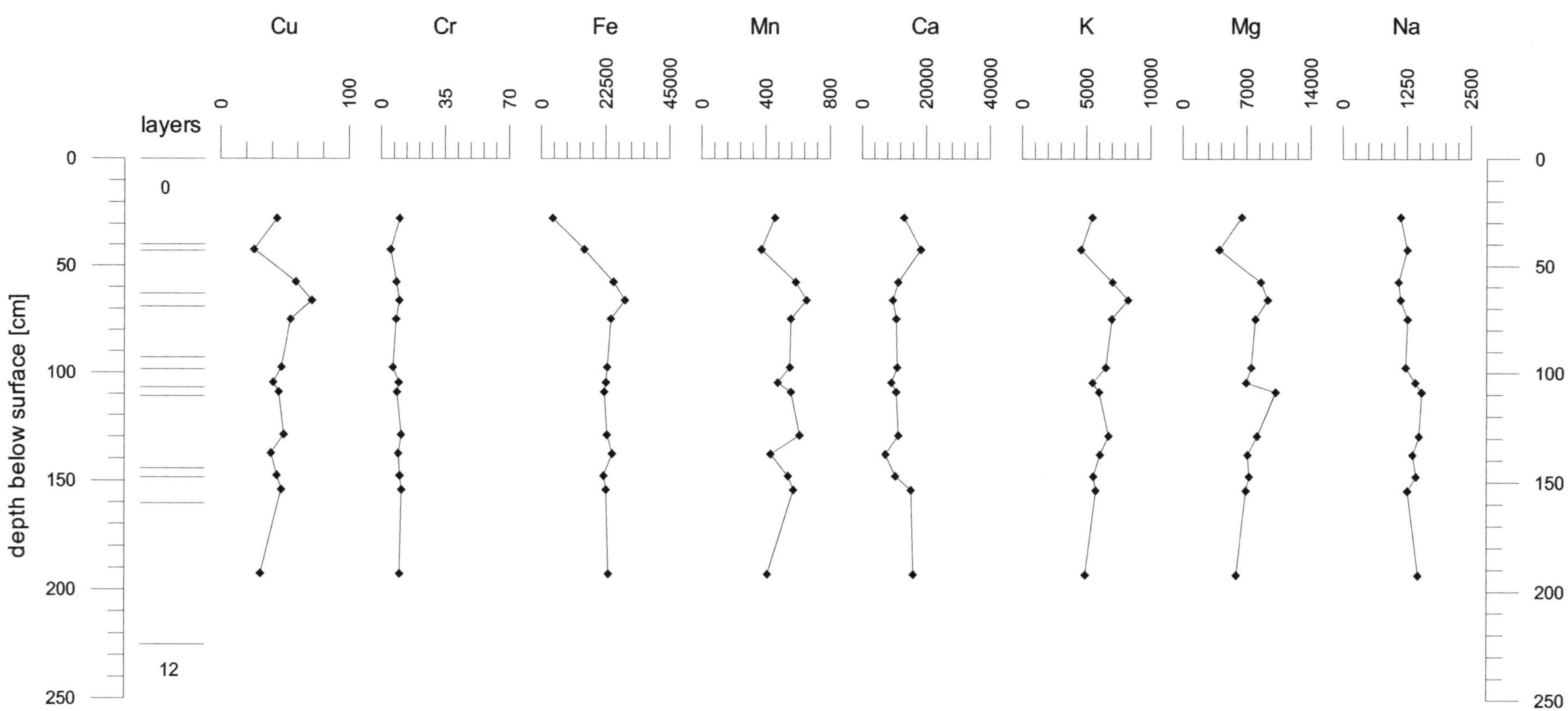

Fig. A-54: AAS measurement results of samples from profile RPW (vertical scale 1: 30). Note that the lines connecting data points do not imply values between data points but serve solely to aid visualisation.

Profile RPW records mainly anthropogenic impact, i.e. the accumulation of irragric anthrosol (all layers except 5, 9, 11 and 12), turbation of fluvial sediment and/or a combination of irragric anthrosol and fluvial sediment (layers 5 and 11) and the incorporation of charcoal and ceramic shards into the sediment. The only undisturbed fluvial layer is layer 9; it is characterised by very good sorting, low LoI and the presence of bedding. The presence of charcoal dated to 1,690–1,410 cal BP almost 0.2 m below the upper boundary of the coarse and partially well-weathered (and therefore presumably very old) basal gravel is somewhat enigmatic. It is tentatively assumed that either the charcoal may have been turbated into the gravel from the overlying, very-poorly sorted and ceramic-bearing stratum (which seems likely because material from the basal gravel also appears to have been turbated into the overlying layer) or that old, weathered gravel may have been remobilised and may have incorporated the charcoal at that time. In any case, the Middle Nasca date of 1,690–1,410 cal BP is a maximum age for the deposition of the overlying sedimentary column. Average sedimentation rate deduced by assuming that this age reflects the begin of fine-grained sedimentation is 0.93–1.11 mm/a.

Profile YN-1

Profile YN-1 (Figures A-55, A-56 and A-57) was a pit in the lower terrace on the right side of the Rio Palpa (ca. 3.5 m above the riverbed) at a distance of 35 m from the river. It is situated 87 m south-east of profile YN-2. The sketch shows the west side of the pit.

Layer 0 (the recent A_p / irrigation horizon) is crumbly fine sandy silt. Layer 1 is fine sandy silt with a sharp, undulating upper boundary. Locally, thin sand bands dipping at various angles as well as pocket-like traces of turbation are present; these are interpreted as plough marks. Layer 2 is structureless fine loamy sand with high porosity and some cohesive clods of silt. Its upper boundary is sharp and horizontal. Layer 3 is a thin (ca. 3 cm) band of fine to

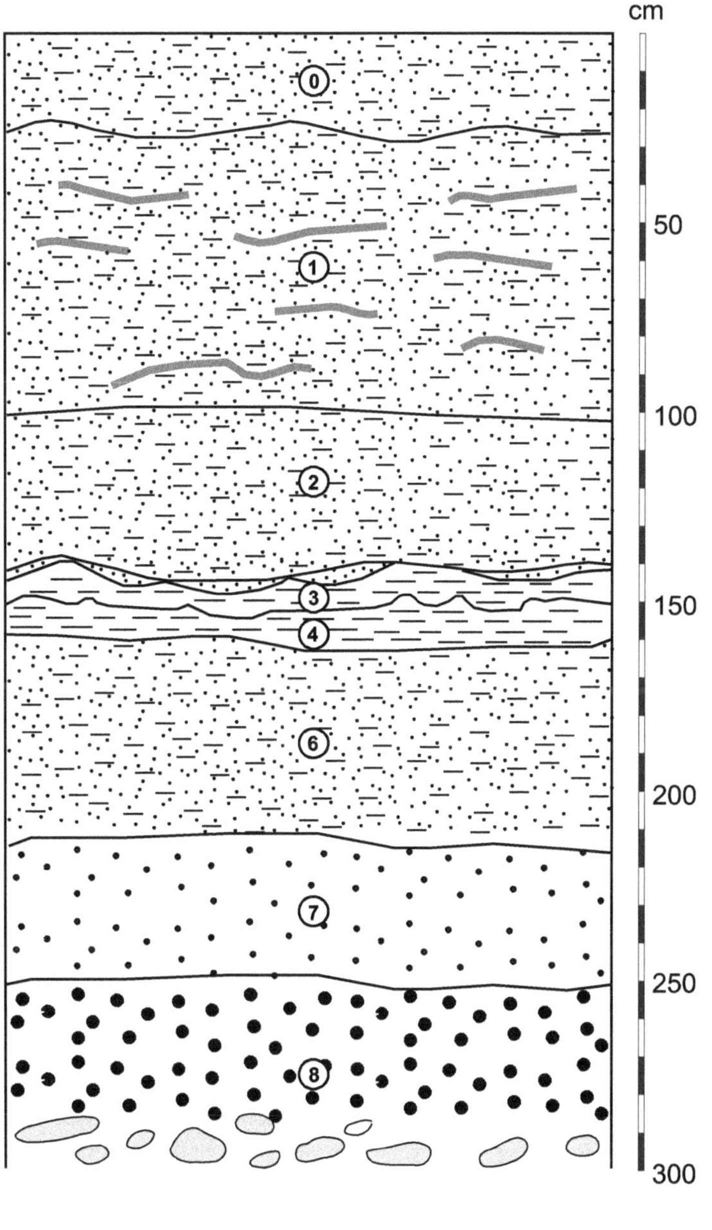

Fig. A-55: Sketch of profile YN-1 (drawing: R. Mendler, vertical scale 1: 20).

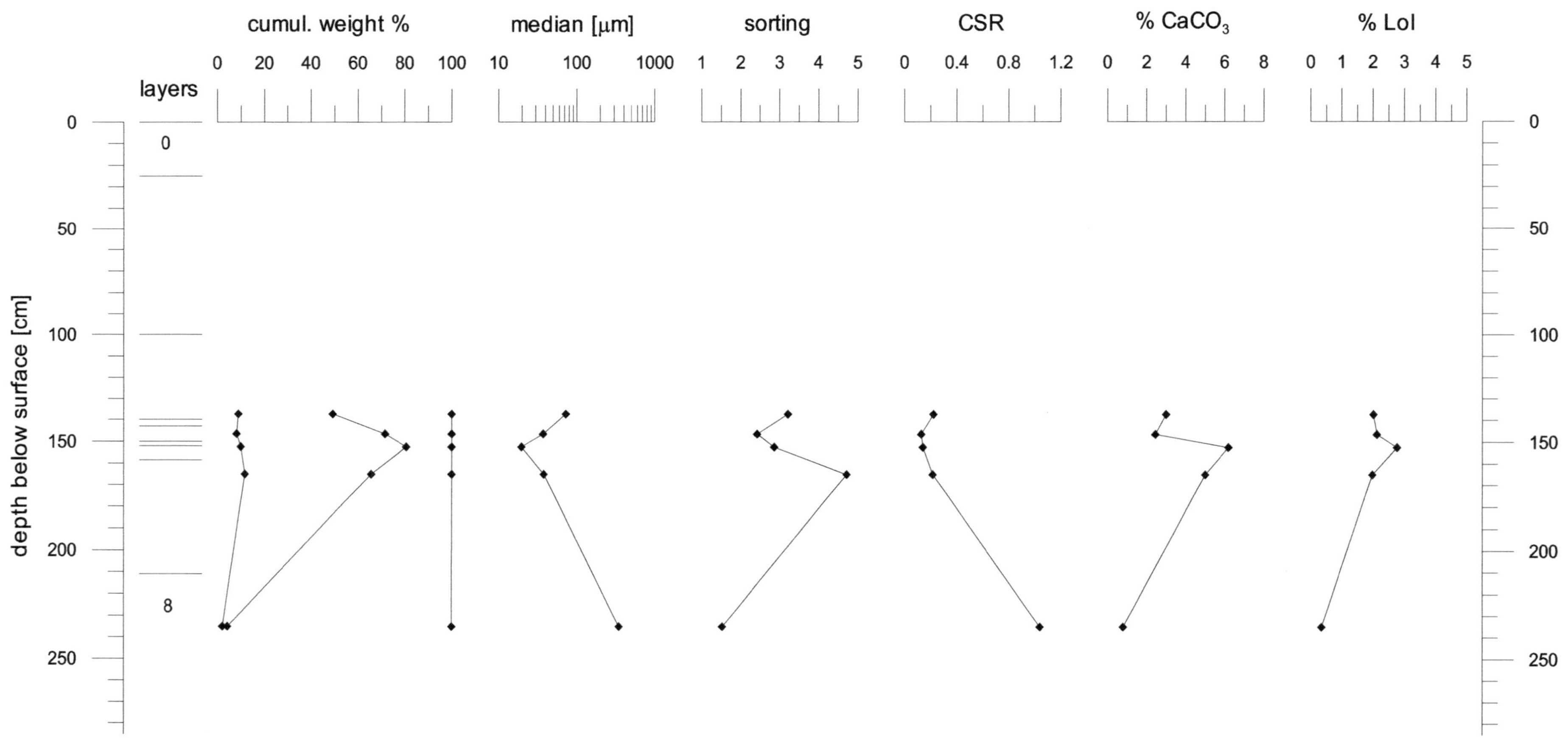

Fig. A-56: Laboratory results of samples from profile YN-1 (vertical scale 1: 30). Cumulative weight percent graph shows clay, silt and sand from left to right. Note that the lines connecting data points do not imply values between data points but serve solely to aid visualisation.

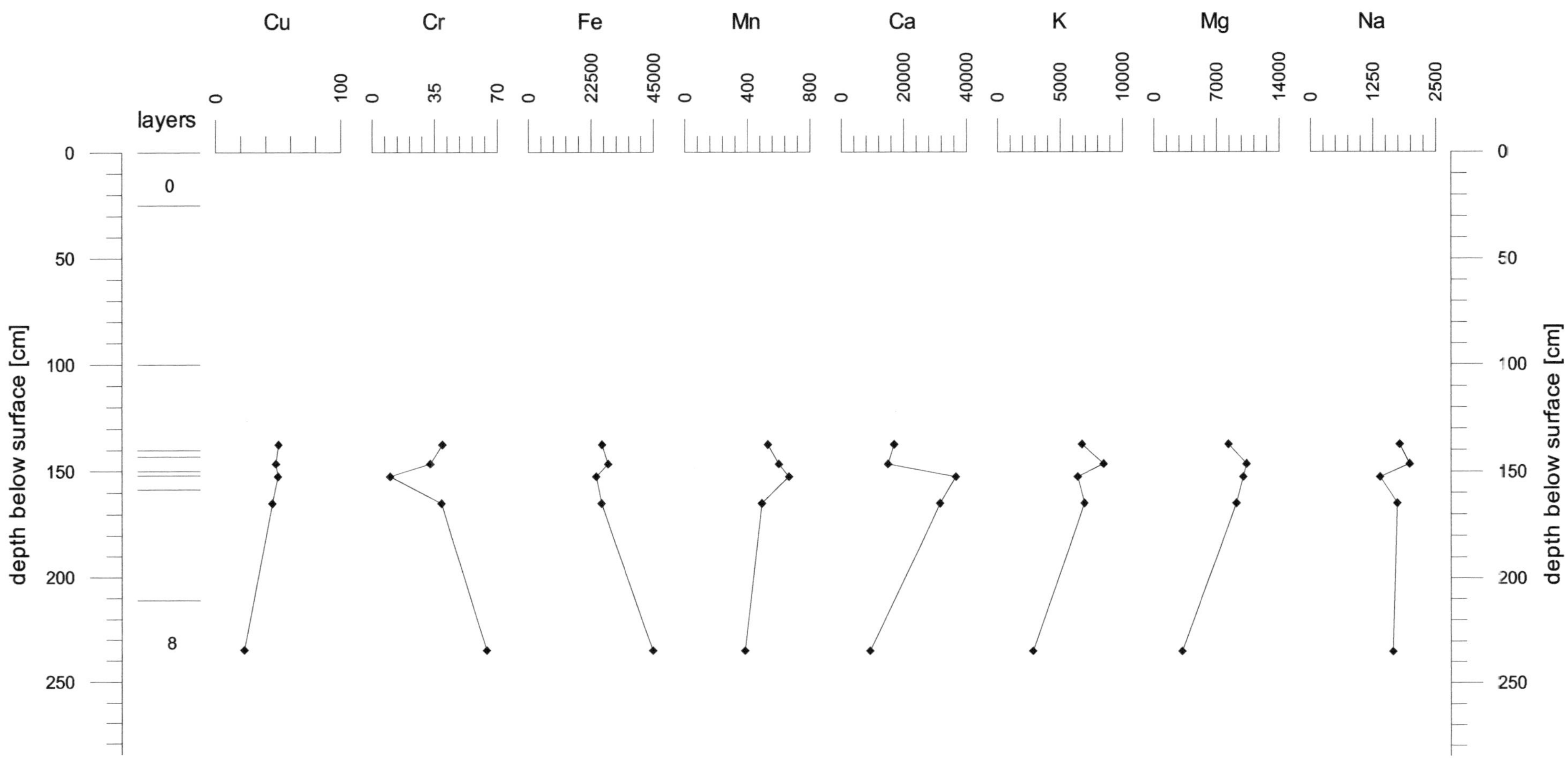

Fig. A-57: AAS measurement results of samples from profile YN-1 (vertical scale 1: 30). Note that the lines connecting data points do not imply values between data points but serve solely to aid visualisation.

medium sand with high mica content and a sharp, undulating upper boundary. Layer 4 is structureless loamy silt with a discontinuous, pocketed/cone-shaped upper boundary. Charcoal is present.

Layer 4 is interpreted as irragric anthrosol which was covered by a flood deposit (layer 4) which filled plough grooves, creating the observed pattern. Layer 5 is a 5–15 cm large lens of dark material at the base of layer 4 in the north-western corner of the pit. Layer 6 is grey clayey silt with faint lamination and blackish bands. Its upper boundary is undulating and pocketed. Layer 7 is structureless, porous loamy silt with a few (ca. 1%) stones and a diffuse upper boundary. A radiocarbon assay of charcoal from the base of layer 7 (depth 2.05 m) gave an age of 750–665 cal BP (Poz-8020, 830 ±30 BP) which pertains to the Late Intermediate Period (950–550 cal BP, cf. Unkel 2006: 111). Layer 8 is very well-sorted medium sand. At a depth of ca. 2.70 m, layers of coarse sand to gravel overlie well-bedded, imbricated coarse gravel.

Profile YN-1 records river-proximal sedimentation on the right side of the Rio Palpa since the Late Intermediate Period. After the retreat of the course of the river (indicated by riverbed gravel) from the site, ca. 0.60 m of well-sorted fluvial sand have been deposited. The maximum age for this deposition is 750–665 cal BP. After this time, mainly irrigation sediments were accumulated, interspersed by two minor flood sediments at depths of 1.60–1.50m and 1.43–1.40 m. Given the proximity of the Rio Palpa, further fine-grained flood deposits may have affected the site; however, if this did occur the associated sediments have been completely incorporated into the irragric anthrosol by tillage. The average sedimentation rate has been 2.55–2.85 mm/a since 750–665 cal BP.

Profile YN-2

Profile YN-2 (Figure A-58) was a pit dug into the main terrace on the right side of the Rio Palpa (ca. 6 m above the riverbed) at a distance of 120 m from the river. It is situated 87 m north-west of profile YN-1. Three sides of the pit (west, north and east) were recorded.

On the western side, layer 0 (the present irrigation / A_p horizon) is crumbly fine sandy silt with ca. 5% clasts. Layer 1 is structureless fine sandy silt with ca. 1% gravel and an undulating upper boundary. Layer 2 is dark-grey medium sand with 40% gravel and a slightly undulating and diffuse upper boundary. Layer 3 is blackish, poorly-sorted silt to coarse sand with 50–70% gravel and a somewhat diffuse upper boundary. Layer 4 is coarse sand to fine gravel (50%) with a sharp upper boundary. Layer 5 is a horizontally bedded sequence of well-sorted silt to coarse sand (from top to bottom fine sand, silt, fine sand, coarse sand). Its upper boundary is sharp and undulating. Layer 6 is poorly sorted fine sand to gravel with a sharp, slightly undulating upper boundary. Layer 7 is poorly sorted coarse gravel in a matrix of coarse sand to fine gravel.

On the northern side of the pit, layer 0 is silty fine sand with a crumbly structure and ca. 5% gravel. Layer 1 is structureless fine sandy silt with an undulating upper boundary. Some stones are present. Layer 2 is a filled-in depression and consists of structureless fine sandy silt with ca. 5% gravel. Some foundation-like, lined-up large clasts are present. The layer has a diffuse upper boundary. Layer 3 is medium sand with 40% gravel and a diffuse upper boundary. Layer 4 is poorly-sorted blackish silt to coarse sand with 10–70% gravel. Its upper boundary is diffuse. Layer 5 is poorly sorted silt to gravel with a sharp upper boundary. Layer 6 is poorly sorted gravel in a silty matrix. Layer 7 is coarse gravel in a gravely matrix with some clods of silt and lenses of medium sand. Imbrication is present. The upper boundary is sharp. At the base of the profile, strongly weathered clasts are present.

On the eastern side of the profile, layer 0 is silty fine sand with ca. 5% gravel and a crumbly structure. Layer 1 is structureless fine sandy silt with ca. 1% gravel and a diffuse, slightly undulating upper boundary. A few lenses of coarse sand are present at the base of the layer. Layer 2 is structureless fine sandy silt with ca. 5% gravel and fills a depression in layer 3. Layer 3 is medium sand with ca. 40% gravel. Layer 4 is blackish, poorly sorted silt to coarse sand with 50–70% gravel and a diffuse upper boundary. Layer 5 is gravel (ca. 60%) in a matrix of fine sand. The upper boundary is sharp. Layer 6 is compact silt with an undulating upper boundary. Layer 7 is slightly imbricated gravel (ca. 70%) in a matrix of coarse sand. Its upper boundary is undulating and pocketed.

Profile YN-2 records high- and low-energy fluvial sedimentation (gravel and bedded silt to sand) as well as anthropogenic influences (strongly turbated coarse- and fine-grained material, a charcoal-rich layer and a filled-in depression into which large stones appeared to have been intentionally set. The somewhat complex lower portion of the profile (depth between 0.60 and 2.20 m) is overlain by 0.60 m of irragric anthrosol.

Profile Ma-T

Profile Ma-T (Figure A-59) was a pit dug into a large, ca. 4 m high upper terrace (ca. 6 m above the riverbed) on the right side of the Rio Palpa at a distance of 240 m south-west from the current river course and ca. 15 m from the terrace edge. Laguna La Maquina, a small spring-fed water body, is 190 m to the west-north-west. The sketch shows the north-western side of the pit.

Layer 1 is crumbly sandy silt to silty sand. Layer 2 is structureless sandy silt with a diffuse, horizontal upper boundary. Layer 3 is structureless sandy silt with a diffuse, horizontal upper boundary. Broken-up pieces of

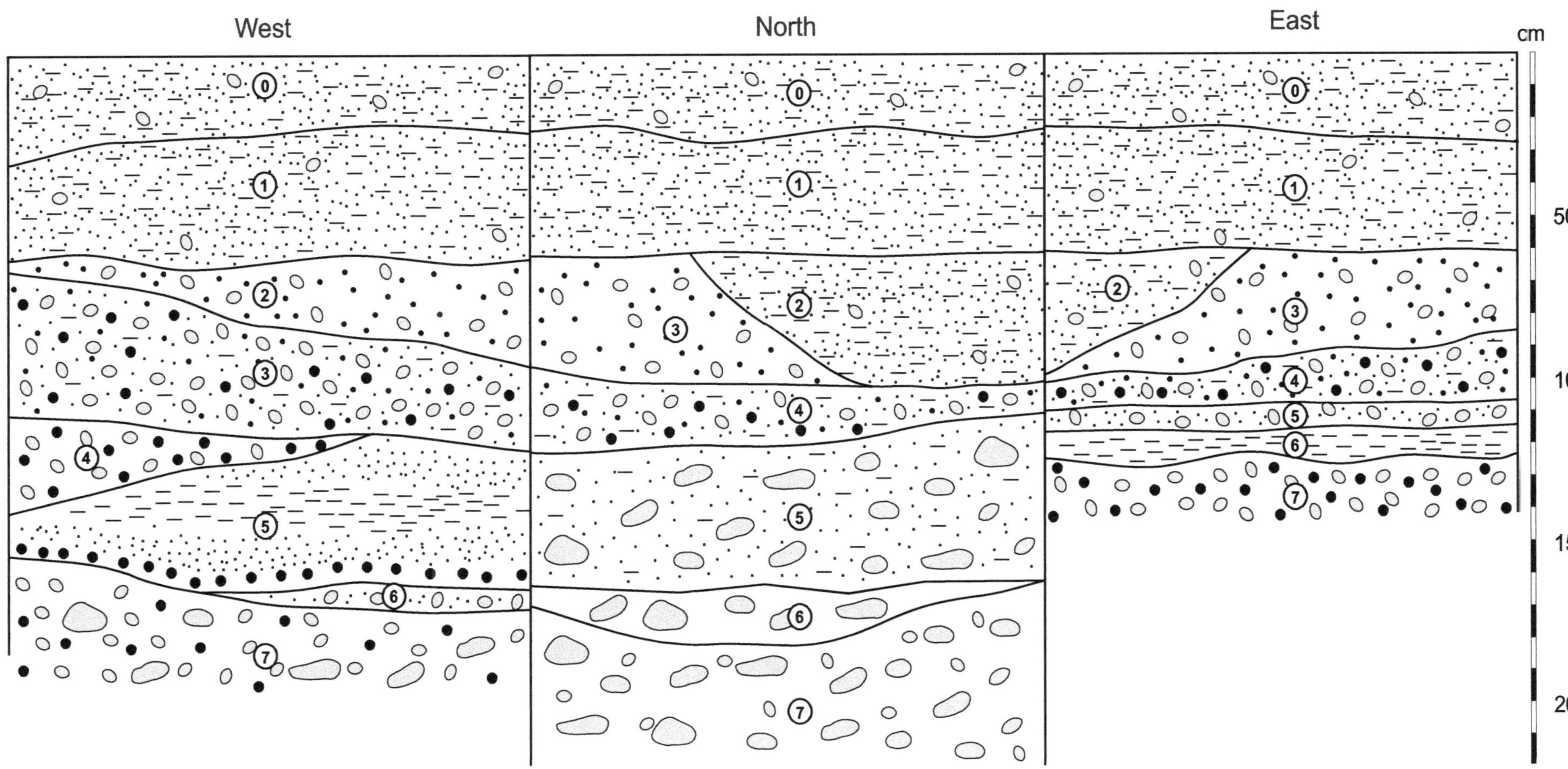

Fig. A-58: Sketch of profile YN-2 (drawing: R. Mendler, vertical scale 1: 20).

material from layer 4 appear to be turbated into layer 3. Layer 4 is whitish, structureless sandy silt with a horizontal, sharp but broken-up upper boundary. Some carbonate nodules are present. Layer 5 is structureless silty sand with a diffuse, horizontal upper boundary which is marked by caliche. Charcoal and slag are present in layer 5. Layer 6 is caliche which is cemented to such a degree as to make grain size determination impossible. Layer 7 is dark grey, cemented material (caliche-like) with a foliar structure. Layer 8 is silty sand with carbonate nodules. At a depth of 1.15 m, pieces of slag have been found. Layer 9 is well-cemented caliche. Layer 10 is silty sand with a sharp, horizontal upper boundary. Layer 11 is sandy silt with a sharp, undulating upper boundary and some carbonate nodules. Layer 12 is sand with reddish and blackish oxidation stains (precipitated iron and manganese oxides). The upper boundary is diffuse and sloping to the north-east. Lenses of fine and coarse sand as well as carbonate nodules are present. Layer 13 is well-cemented caliche with a diffuse, undulating upper boundary. Layer 14 is sand with strong oxidation stains and a diffuse, undulating upper boundary. Layer 15 is a one-clast-thick layer of well-rounded and unweathered stones (3–15 cm) in a sandy matrix and is sloping to the north-east. Layer 16 is sand reddened by oxidation stains and has an undulating upper boundary. Layer 17 is well-rounded coarse gravel to boulders with oxidation stains. The upper boundary is undulating. Some clasts are strongly weathered.

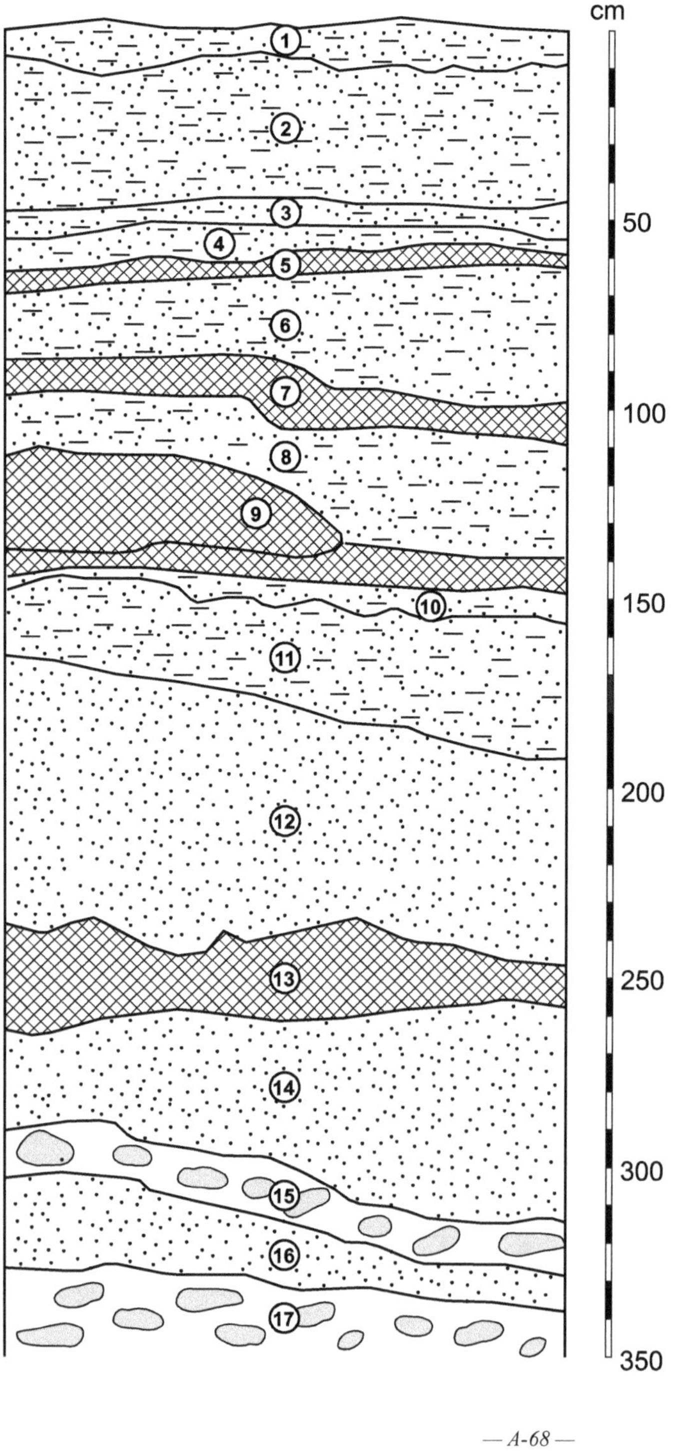

Fig. A-59: Sketch of profile Ma-T (drawing: R. Mendler, vertical scale 1: 20).

Profile Ma-T shows that the large upper irrigated terrace in this part of the Palpa Valley is constructed exclusively of fine-grained material. There are no dating constraints except for the assumption that the strongly-weathered basal gravel (below 3.20 m) is of pre-Holocene age while the unweathered stone layer at a depth between 2.80 and 3.20 is of Holocene age. Given the proximity of Laguna La Maquina, a source of water of high importance in prehistoric times, the lack of artefacts in this excavation is surprising and may have to be interpreted in terms of an early deposition of much of the terrace before human impact. The lack of clearly identifiable irrigation sediment can be explained with the use of sediment-free water from the natural spring, allowing agricultural use without sediment accumulation. This can also explain the presence of five conspicuous layers of well-cemented caliche. The erosional activity of the Rio Palpa creating the edge of the terrace apparently stripped away the fine-grained material and was halted by the underlying gravel base. The thin (a few decimetres) irrigation sediment with agriculturally turbated underlaying gravel on the adjacent lower terrace hints at a young age (on the order of a few hundred years) of this erosional event.

Profiles between Rio Palpa and Rio Vizcas

Tab. A-7: Positions of profiles between Rio Palpa and Rio Vizcas.

Profile	Depth [m]	Northing [m]	Easting [m]	Altitude [m]
AR	2.60	8,390,324.54	476,647.03	286.37
R-2	5.12	8,390,897.46	477,235.78	296.70
R-3	6.10	8,390,925.58	477,143.27	296.42
R-4	3.45	8,390,854.33	477,166.69	296.40
ES	2.58	8,391,019.66	476,569.71	289.59
DO	3.35	8,391,417.24	477,024.68	297.03
RPS	2.00	8,391,774.25	477,006.41	299.18
Jau-1	4.00	8,391,895.52	477,510.55	305.92
FL	2.00	8,391,368.43	477,698.99	303.28
H-1	1.80	8,391,862.95	478,652.75	320.84
Jau-Co	1.70	8,391,878.13	477,992.58	310.69
RP-06	3.60	8,392,816.58	478,596.23	322.87
SF-1	4.50	8,392,561.70	478,964.67	328.64
R-1	4.20	8,392,314.94	479,128.08	329.57
SF-2	3.40	8,392,213.06	479,255.53	331.69
SF-3	3.00	8,392,170.74	479,324.21	332.42
SF-4	2.85	8,392,158.28	479,359.58	332.93
SR	2.55	8,391,011.65	477,115.08	296.41
RVU	3.20	8,391,729.61 – 8,391,638.48	478,589.13 – 478,488.54	316.30 – 317.10

Profile AR

Profile AR (Figures A-60 and A-61) is a 2.60 m high natural bank exposure in the south-western part of the area enclosed by Rio Palpa and Rio Vizcas, 704 m south of profile ES and 745 m south-west of profile R-4. Although the present course of the Rio Grande lies at a distance of 610 m to the west, aerial photographs and the high-resolution DTM of the Palpa Valley (made available by ETH Zurich) clearly show that the bank was cut by the Rio Grande. The arcuate cut bank can be traced for a length of 670 m. Due to large amounts of slope debris and disturbance by tree roots, profile AR had to be recorded in two sections, the lower part (below 2.10 m depth) being located at a distance of 15 metres north-west along the exposure from the upper part. The two sections were connected using a prominent gravel layer and overlying bedded silt to sand (layers 9 and 8, respectively) which were present at similar depths as marker layers.

Layer 1 is very loose, crumbly silt with abundant roots and organic material. Layer 2 is crumbly sandy silt with a few stones (up to 15 cm) and a diffuse, horizontal upper boundary. The layer has been strongly affected by nesting insects (wasps). Layer 3 is sandy silt with faint bedding and a horizontal upper boundary which is partially broken up. Charcoal is present.

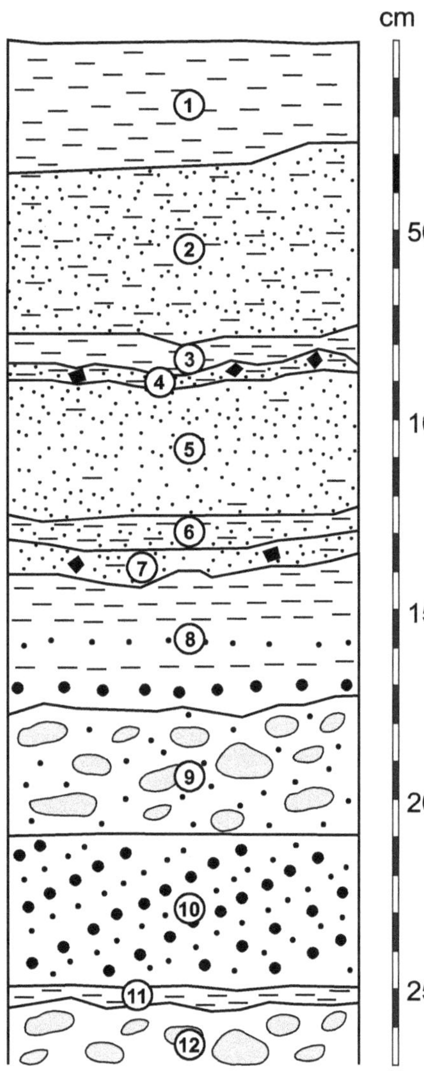

Fig. A-60: Sketch of profile AR (drawing: R. Mendler, vertical scale 1: 20).

Layer 4 is a band of charcoal-rich sandy silt with a sharp, slightly undulating upper boundary. Layer 5 is structureless sandy silt without larger clasts and with a sharp, slightly undulating upper boundary. Layer 6 is a bedded layer of silt to fine sand with a sharp, slightly undulating upper boundary. Layer 7 is charcoal-rich silt to fine sand with a slightly diffuse, horizontal upper boundary. A radiocarbon assay of charcoal from layer 7 (depth 1.37 m) gave an age of 530–330 cal BP (Poz-16829, 460 ±30 BP). Omitting a minor peak in the radiocarbon calibration probability distribution (7.6%) constrains the age to 530–440 cal BP at a probability of 87.8%. The charcoal layer therefore probably was deposited during the Late Intermediate Period (550–418 cal BP, cf. UNKEL 2006: 111). Layer 8 is bedded silt to coarse sand (bedding somewhat disturbed by wasps) with oxidation stains and a diffuse, slightly undulating upper boundary. Layer 9 is well-rounded coarse gravel (up to 15 cm) in a matrix of medium to coarse sand with strong oxidation stains and blackish mineral precipitates (probably manganese oxide). Layer 9 is between 5 and more than 40 cm thick; its upper boundary is sharp and horizontal. Layer 10 is medium sand with strongly (ca. 25°) north-west-dipping bedding planes. The upper boundary is sharp, horizontal and clearly erosional as evidenced by the truncated bedding. Layer 11 is a 2–7 cm thick band of sandy silt with a sharp, horizontal upper boundary. Layer 12 is strongly weathered, well-rounded gravel to boulders with a matrix of silty sand and a sharp, undulating upper boundary.

Profile AR records a sequence from basal, strongly-weathered (likely pre-Holocene) coarse gravel, five low-energy (silt to sand) and one high-energy (gravel) fluvial sediments and five likely anthropogenically dominated layers (irragric anthrosol and charcoal-rich layers). Carbonate content peaks in layer 5 (depth 0.88–1.25 m) with 9.2%, suggesting preferential evaporative enrichment from irrigation-induced groundwater seepage from the field above the exposure. The situation of this large (670 m long) fluvially cut bank in combination with the horizontally layered and bedded sediments in the upper part of the profile and an age of 530–330 cal BP from a depth of 1.37 m indicates that a strong erosional event of the Rio Grande de Nazca has taken place probably during the historic period. Taking into account the accumulation of the top 1.37 m of the profile at an assumed rate of 3.0–5.0 mm/a (cf. profiles ES and DO) constrains the age of this event to ≤260 cal BP.

Profile R-2

Profile R-2 (Figure A-62 and A-63) is part of a group of three profiles closely spaced in the central part of the Palpa Valley (R-2, R-3 and R-4) at a distance of 97 m from profile R3 and 82 m from profile R4. Profile R-2 was a 5.12 m deep pit in the eastern part of a fallow piece of land covered by a few bushes and trees. The Rio Vizcas runs at a distance of 260 m, the Rio Palpa at a distance of 690 m. Layer 1 is a 22–25 cm thick layer of well-bedded sand likely deposited by the 1999 flood event of the Rio Vizcas. Layer 2 is silty loam with faint polyedric structure and a very sharp, slightly undulating upper boundary. Layer 3 is structureless loamy silt with a sharp, horizontal upper boundary. Layer 4 is structureless loamy sand with a diffuse, horizontal upper boundary. Layer 5 is dark loam with a very diffuse, horizontal upper boundary. An unclassifiable ceramic shard was found at a depth of 1.38 m. A radiocarbon assay on charcoal from near the base of layer 5 (depth 1.70 m) gave an age of 1,720–1,540 cal BP (Poz-13608, 1,780 ±30 BP), i.e. the Early to Middle Nasca period (1,870–1,520 cal BP, cf. UNKEL 2006: 111). Layer 6 is structureless, cemented, poorly sorted loamy sand with 6% clasts >2 mm and a diffuse, undulating upper boundary. Layer 7 is structureless loamy sand with a sharp, undulating upper boundary. At the top of the layer there is a midden-like accumulation of abundant Paracas period ceramic fragments, broken fragments of rounded stones,

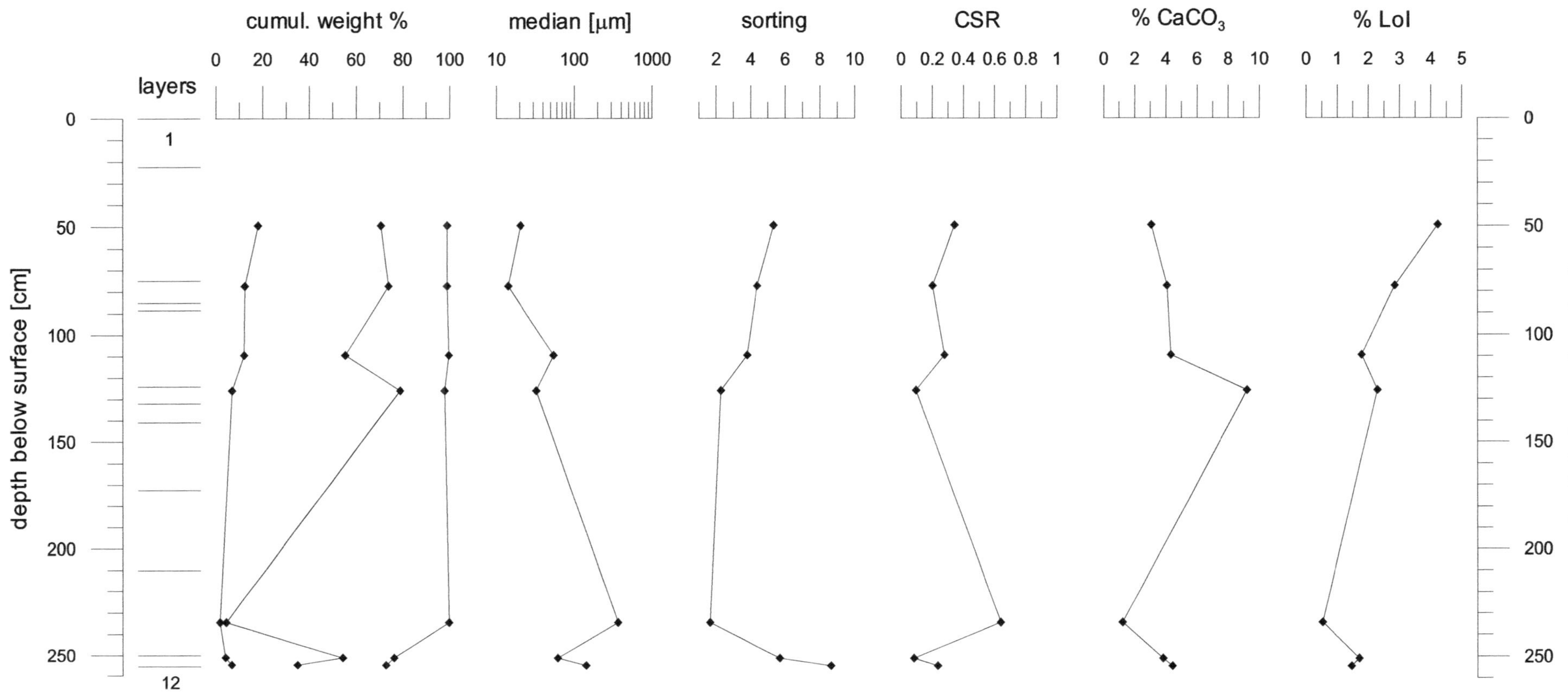

Fig. A-61: Laboratory results of samples from profile AR (vertical scale 1: 30). Cumulative weight percent graph shows clay, silt and sand from left to right. Note that the lines connecting data points do not imply values between data points but serve solely to aid visualisation.

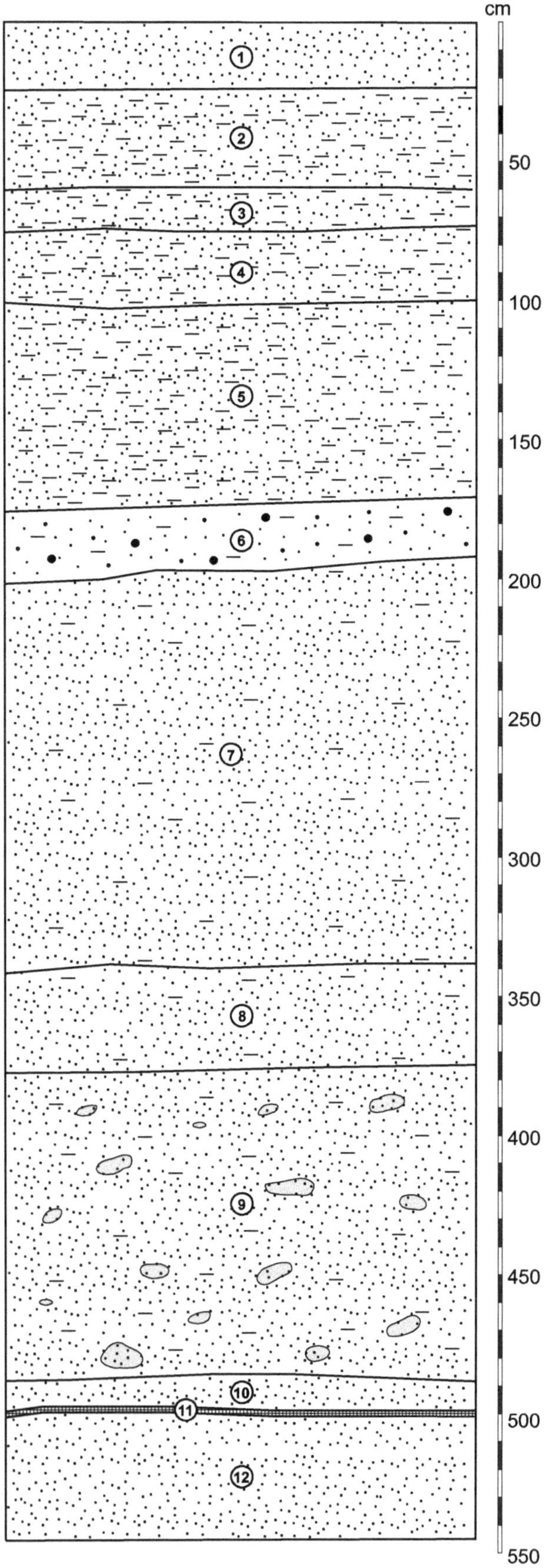

Fig. A-62: Sketch of profile R-2 (drawing: R. Mendler, vertical scale 1: 25).

marine mollusc shells and disperse charcoal. The midden has a thickness of up to 15 cm. Below this accumulation, stone content in this layer is ca. 10% (up to 2 cm). Charcoal and ceramic shards are present, albeit at lower concentration, for at least 30 cm into the layer. A radiocarbon date of charcoal from layer 7 (depth 2.28 m) is 2,700–2,340 cal BP (Poz-13609, 2,460 ±30 BP) and therefore pertains to the Early to Middle Paracas period (2,750–2,350 cal BP, cf. Unkel 2006: 111). Layer 8 is structureless loamy sand similar to layer 7 but with a higher quantity of clasts >2 mm (10–20%, up to 2 cm). The upper boundary is diffuse and horizontal. Layer 9 is cemented loamy sand with ca. 50% gravel-sized clasts (up to 4 cm) with a diffuse, horizontal upper boundary. Layer 10 is well-bedded loamy silt with a sharp, horizontal upper boundary. Layer 11 is a 2 cm thick band of dark loam with a sharp, horizontal upper boundary. Layer 12 is bedded fine to medium sand with oxidation stains and a sharp, horizontal upper boundary. Below the base of the profile (5.12 m), at least 0.30 m of sand are present. Basal gravels were not reached due to time constraints, profile R-2 being the last recorded during the 2005 field season. Layers 2 through 5 as well as 7 are characterised by high LoI values (2.35–3.44% and are interpreted as irragric anthrosol. Layers 6 and 8 consist of cemented, poorly-sorted and very-densely packed material with low LoI values (1.82 and 1.57%, respectively) and are interpreted as hyperconcentrated flow deposits.

Profile R-2 is characterised by long-term human impact since at least the Paracas period, following the deposition of fine-grained fluvial overbank sediments and a massive (depth 4.87–3.76 m, 1.11 m thick) hyperconcentrated flow deposit. A minor hyperconcentrated flow deposit (depth 1.80–1.69, 0.11 m thick) separates a Paracas midden (likely indicating nearby settlement activity) and irragric anthrosol from a later, Early to Middle Nasca irragric anthrosol which is overlain by three further anthrosols separated by diffuse boundaries, recording an apparently almost continuous agricultural use of the site since Early to Middle Nasca times. The top of the profile is marked by the 1999 flood deposit. No sediment comparable with this deposit is present within the 3.67 m of the profile. Sedimentation rates at this river-distal site are 0.50–0.94 mm/a for the period from 2,700–2,340 cal BP (depth 2.28 m) to 1,720–1,540 cal BP (depth 1.70 m) and 0.82–0.91 mm/a for the period from 1,720–1,540 cal BP (depth 1.70 m) to AD 1999 (depth 0.25 m).

Profile R-3

Profile R-3 (Figures A-64 and A-65) is part of a group of three profiles closely spaced in the central part of the Palpa Valley at a distance of 97 m from profile R2 and 75 m from profile R4. Profile R-3 was a 6.10 m deep pit in the eastern part of a fallow piece of land covered by a few bushes and trees. Rio Palpa and Rio Vizcas run at a dis-

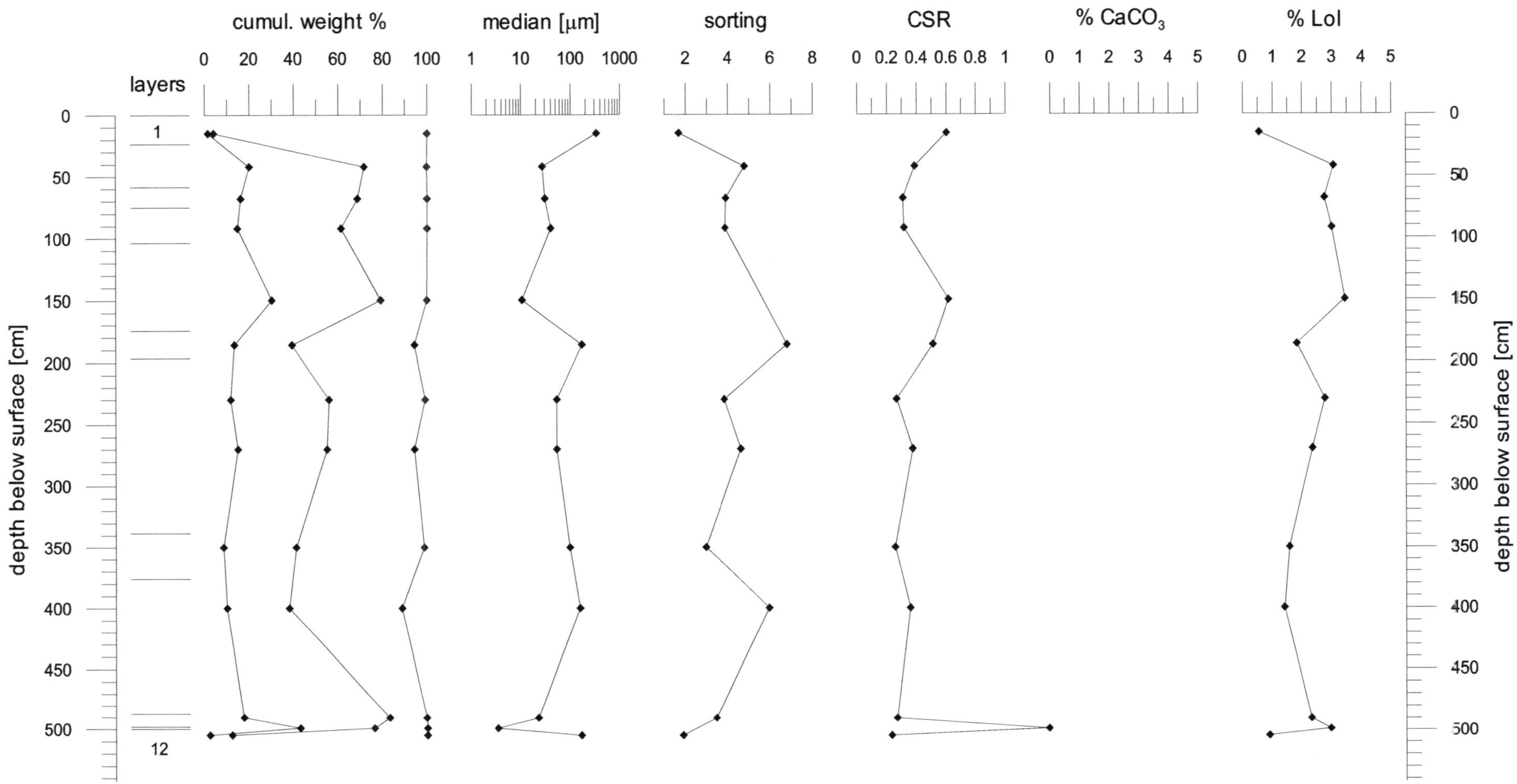

Fig. A-63: Laboratory results of samples from profile R-2 (vertical scale 1: 30). Cumulative weight percent graph shows clay, silt and sand from left to right. Note that the lines connecting data points do not imply values between data points but serve solely to aid visualisation.

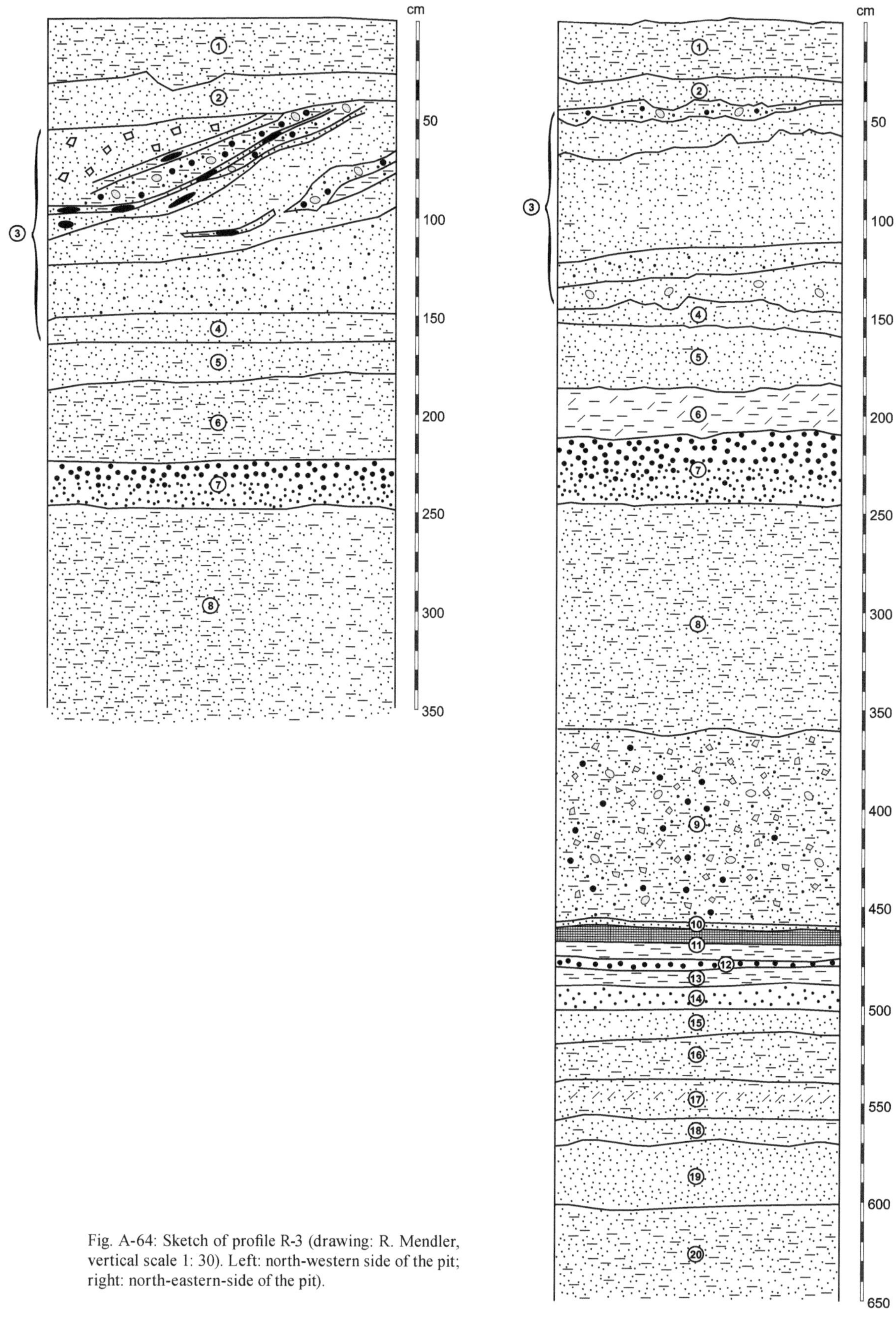

Fig. A-64: Sketch of profile R-3 (drawing: R. Mendler, vertical scale 1: 30). Left: north-western side of the pit; right: north-eastern-side of the pit).

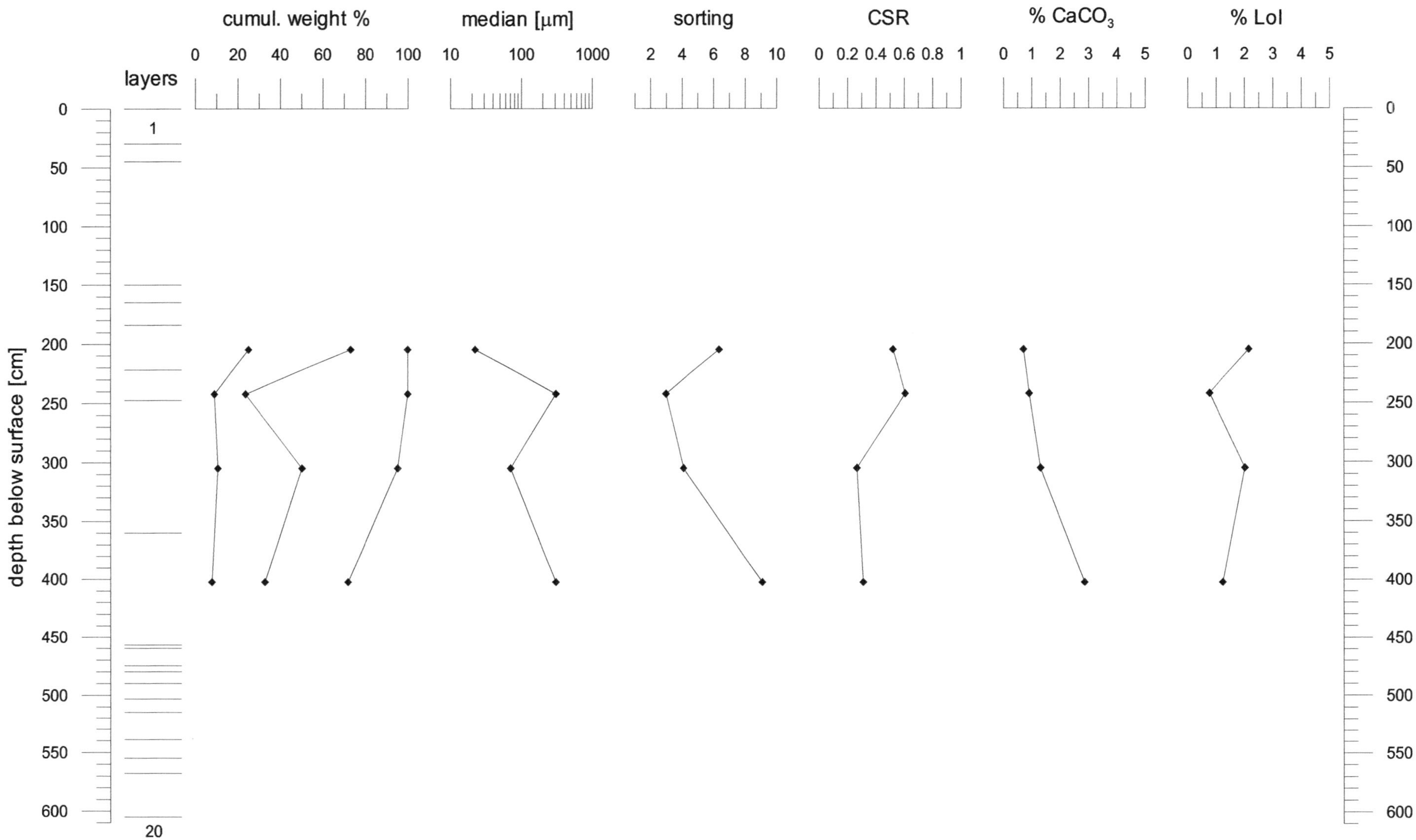

Fig. A-65: Laboratory results of samples from profile R-3 (vertical scale 1: 30). Cumulative weight percent graph shows clay, silt and sand from left to right. Note that the lines connecting data points do not imply values between data points but serve solely to aid visualisation.

tance of 590 and 350 m, respectively. An irrigation channel runs at a distance of ca. 10 m north-east of the pit.

Layer 1 is sandy silt with a crumbly structure and contains clods of cohesive, laminated silt (up to 4 cm in size). Layer 2 is silty sand with cohesive clods of silt and a slightly undulating upper boundary. Locally, faint bedding is discernible. Layers 1 and 2 are interpreted as fine-grained flood deposit (possibly related to the 1999 event albeit with finer grain size than in other profiles) broken up by agriculture. Layer 3 shows a very complex pattern of sediments. The material on the south-western side of the pit consists largely of crumbly silty sand with clods of loam and clay. On the north-western side a conspicuous sequence of south-west-dipping sediments is present. From top to bottom it consists of sand with clods of silt, a charcoal-rich sandy band, a poorly sorted band of silt to fine gravel, a charcoal-rich band, a band of poorly sorted silt to fine gravel with abundant silt clods, a sandy charcoal band, a wedge of structureless silty sand increasing in thickness to the north-east, a lens of sand to gravel, a wedge of silty fine sand, a lens of charcoal-rich silty sand and a structureless wedge of fine to medium sand.

On the north-eastern side, the wedge-shaped sediment bodies grade into beds which appear to have more or less constant thickness along the profile side. The material on the south-western side is interpreted as an irragric anthrosol in the immediate vicinity of the irrigation channel the excavation spoil (due to cleaning of accumulated sediment) of which is seen on the north-western and north-eastern side. Layer 4 is loamy sand with a slightly undulating upper boundary. At its base, the layer contains a charcoal rich band overlain by white material (ash). Layer 5 is structureless silty sand with a few stones (up to 3 cm) and a diffuse horizontal upper boundary. Grain size shows a upward-coarsening tendency. Late Intermediate Period ceramic shards (950–555 cal BP, cf. Unkel 2006: 111) were found at a depth of 1.70 m in layer 5. Layer 6 is loamy silt with faint polyedric structure and an undulating upper boundary. At the base of layer 6, a 3 cm thick silty band is faintly discernible. Layer 6 is interpreted as a paleosol (irragric anthrosol). Layer 7 is medium to coarse sand with an upwards-coarsening tendency and a slightly undulating upper boundary. Layer 8 is structureless silty sand to sandy silt with a few stones (up to 2 cm) and abundant root molds. The upper boundary is sharp and horizontal; charcoal and unclassifiable ceramic shards are present at the upper boundary. Minor amounts of charcoal are present throughout the upper part of layer 8. The bottom part of the layer is characterised by increased content of gravel and carbonate nodules. The result of a radiocarbon assay of charcoal from the top of layer 8 (depth 2.47 m) is 2,300–1,950 cal BP (Poz-16825, 2,150 ±35 BP). Omitting a minor peak in the radiocarbon calibration probability distribution (1.7%) constrains the age to 2,160–1,950 cal BP with a probability of 93.8%. The paleosurface at the top of layer 8 therefore probably pertains to the Initial Nasca period (2,150–1,890 cal BP, cf. Unkel 2006: 111).

Layer 9 is structureless, very poorly-sorted silt to coarse gravel (10–20% gravel-sized clasts up to 12 cm, mostly angular to subangular). The material is compacted and cemented and contains carbonate nodules. Layer 9 is interpreted as a massive (depth 3.60–4.57 m, 0.97 m thick) hyperconcentrated flow deposit. The upper boundary is diffuse, therefore the increased abundance of coarse clasts near the bottom of the overlying layer 8 is interpreted as being caused by agricultural activity. Layer 10 is structureless, poorly-sorted sand with a slightly undulating upper boundary. Layer 11 is a sequence of (top to bottom) a 5 mm thick band of clay, structureless silt with carbonate nodules up to 2 cm in size and a lens of coarse sand. The upper boundary is sharp and horizontal. Layer 12 is structureless coarse sand with oxidation stains and a sharp, horizontal upper boundary. Layer 13 is silt with a sharp, horizontal upper boundary, carbonate nodules and locally conspicuous laminations. Layer 14 is structureless, well-sorted medium sand with oxidation stains transgressing into the layer from its diffuse horizontal upper boundary. Layer 15 is structureless, well-sorted fine sand with faint oxidation stains and a slightly diffuse, horizontal upper boundary. Layer 16 is structureless silt with some fine sand and clay and clay clods up to 1 cm in size. Its upper boundary is diffuse and horizontal. Layer 17 is bedded silty to loamy sand with a diffuse, horizontal upper boundary. Layer 18 is structureless silt to coarse sand with a downward tendency to higher sand content. Its upper boundary is horizontal and slightly diffuse. Layer 19 is structureless, upwards-coarsening medium to coarse sand with a sharp, slightly undulating upper boundary. On the north-western side of the pit, a grey, funnel-shaped structure (possibly due to leaching by groundwater) is present. Layer 20 is structureless, well-sorted fine sandy silt with some oxidation stains and a sharp, slightly undulating upper boundary. Basal gravels were not reached in profile R-3 because excavation had to be terminated due to groundwater inflow.

Profile R-3 records a sequence of sediments from both fluvially and anthropogenically dominated environments. All eleven layers at the bottom of the profile (depth 6.10–4.57 m) are interpreted as low-energy overbank deposits due to their grain size, sorting and bedding characteristics. Layer 9 (depth 4.57–3.60 m) is interpreted as a massive (0.97 m thick) hyperconcentrated flow deposit. Incipient irrigation and agricultural use of the site are indicated by the diffuse upper boundary of the hyperconcentrated flow deposit and the apparent turbation of coarse-grained material from this deposit into the overlying fine-grained irrigation sediment. The sharply defined top of this irragric anthrosol is marked by ceramic shards and abundant charcoal, possibly indicating nearby settlement activity at 2,300–1,950 cal BP; probably during the Initial Nasca period (2,150–1,890 cal BP, cf. Unkel 2006:

111). A ca. 25 cm thick fluvial layer separates this usage horizon from a second likely irragric anthrosol at a depth between 1.80 and 2.24 m. Another fluvial layer with a turbated upper boundary containing Late Intermediate Period ceramic shards (950–555 cal BP, cf. UNKEL 2006: 111) is overlain by a third irragric anthrosol which in turn is capped by a complex sequence of irrigation sediment and material excavated from the nearby irrigation channel. The two top layers of profile R-3 are interpreted as agriculturally reworked flood deposit possibly deposited by the 1999 event. Average sedimentation rate for the upper portion of profile R-3 is 1.05–1.23 mm/a, however, the complex situation near the top of the profile has to taken into account.

Profile R-4

Profile R-4 (Figures A-66 and A-67) is part of a group of three profiles closely spaced in the central part of the Palpa Valley at a distance of 82 m from profile R2 and 75 m from profile R3. Profile R-4 was a 3.45 m deep pit in the eastern part of a fallow piece of land covered by a few bushes and trees. Rio Palpa and Rio Vizcas run at a distances of 640 and 280 m, respectively. The sketch shows the north-eastern side of the pit.

Layer 0 is well-bedded and well-sorted coarse silt and fine sand deposited by the 1999 flood event. Layer 1 is loamy sand with a crumbly structure and a sharp, horizontal upper boundary. It is the pre-flood A_p irrigation horizon. Layer 2 is loamy silt with crumbly to polyedric structure and a sharp, slightly undulating upper boundary. The upper boundary appears to be characterised by plough marks. Layer 3 is silty loam with columnar structure (conspicuous vertical cracks). Layer 4 is silty sand with a sharp, horizontal upper boundary . In its lower portion, layer 4 is bedded; however, in its upper portion bedding is broken up and cohesive clods of silt are dispersed in the matrix. Layers 4 and 3 are interpreted as a succession from a fluvially deposited sediment to irrigation sediment with beginning agricultural activity disturbing the upper boundary of the underlying sediment. Layer 5 is sandy loam with a sharp, horizontal upper boundary. Like layer 4, layer 5 is characterised by the presence of bedding in its lower part and a turbated upper part; however, the low layer thickness and depth of turbation in this case support the interpretation as natural turbation (e.g. by animals walking across the surface of the fluvial sediment). At the base of layer 5 there is a 5 mm thick, well-defined band of silt. Layer 6 is loamy sand with weak columnar structure and a sharp, horizontal upper boundary. A single small, unidentifiable ceramic fragment was found at a depth of 1.10 m. Layer 7 is structureless silty sand with a horizontal upper boundary. Locally, an up to 2 cm thick band of silt marks the upper boundary; where it is not present, the layer boundary is diffuse.

Layer 8 is silty sand with a high concentration of disperse charcoal and a sharp, horizontal upper boundary. Layer 9 is loam to loamy sand with polyedric structure down to a depth of ca. 2.25 m from where on the material appears structureless without a visible layer boundary. The upper boundary is horizontal and slightly diffuse. Clay and silt content peak in the middle of the layer. Charcoal is present throughout the layer. Several unclassifiable ceramic shards were present at a depth of 2.20 m. A few larger, sand-filled root molds are present; smaller root molds are characterised by incipient carbonate precipitation.

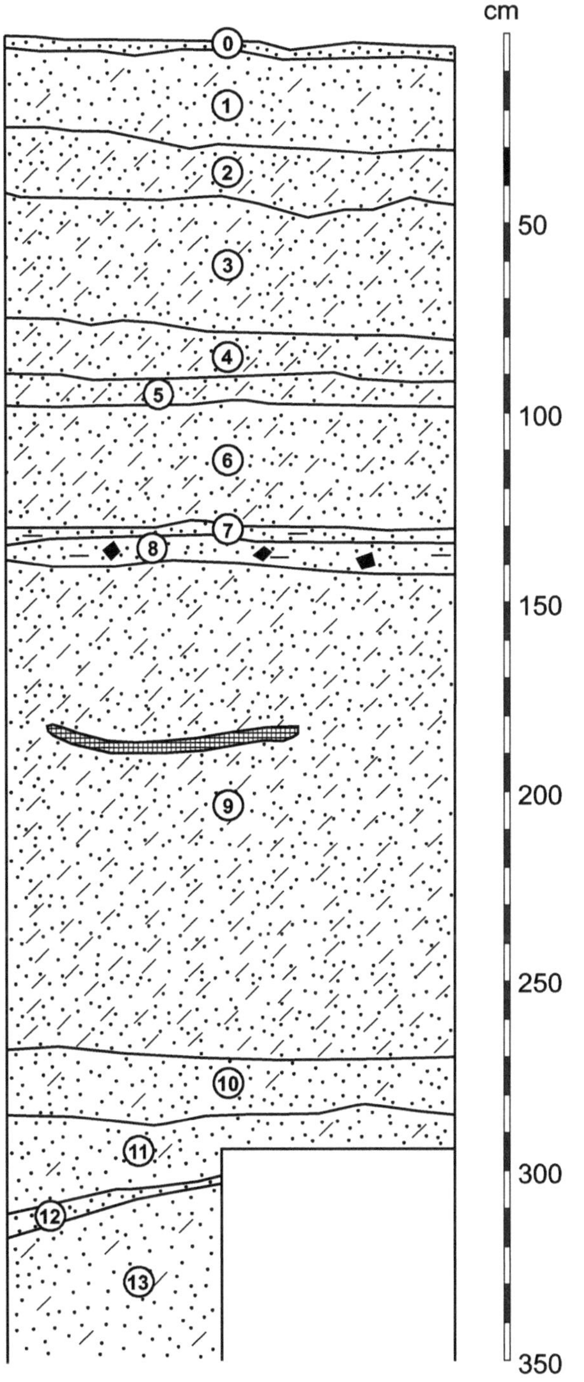

Fig. A-66: Sketch of profile R-4 (drawing: R. Mendler, vertical scale 1: 20).

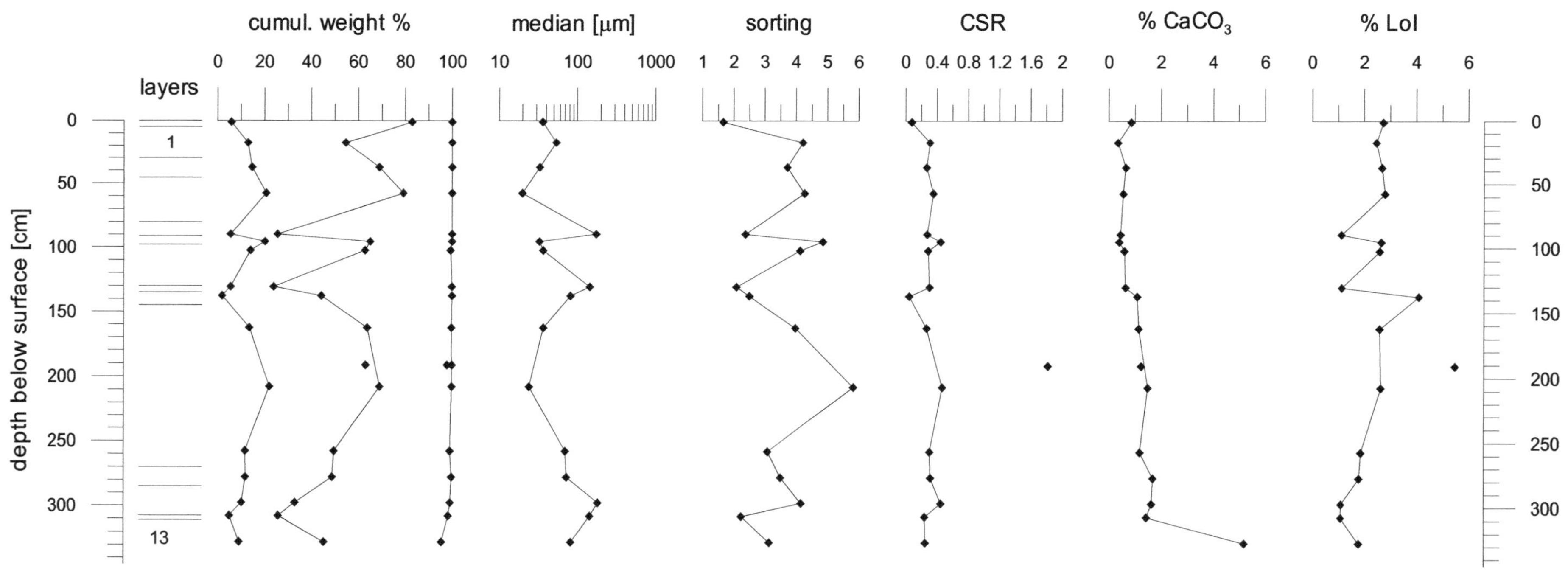

Fig. A-67: Laboratory results of samples from profile R-4 (vertical scale 1: 30). Cumulative weight percent graph shows clay, silt and sand from left to right. Note that the lines connecting data points do not imply values between data points but serve solely to aid visualisation.

A ca. 4 cm thick and 80 cm wide, slightly through-shaped band of silty clay is present at a depth of 1.85–1.95 m. Layer 10 is structureless loamy sand very similar to layer 9. The upper boundary is horizontal and diffuse except for some patches of a ca. 1 cm thick band of sand where it appears sharp. Some carbonate nodules (up to 1 cm) are present. Layer 11 is structureless loamy sand with a higher proportion of medium to coarse sand than layers 9 and 10. Silt content diminishes from top to base of the layer. The upper boundary is diffuse and horizontal. Layer 12 is a band of upwards-coarsening, silty fine to medium sand which is dipping towards the north-west. The upper boundary is sharp. Layer 13 is structureless loamy sand with a sharp, north-west sloping upper boundary. A few angular stones (up to 5 cm) and a few carbonate nodules and disperse precipitation carbonate are present. Some unclassifiable ceramic shards were present at a depth of 3.05 m and appeared densely packed together with some stones. Basal gravels were not reached due to time constraints (profile R-4 being the last recorded during the 2006 field season).

Profile R-4 records a sequence of mainly anthropogenic influence interspersed by a few fluvial events. A lower irragric anthrosol (depth below 3.07 m) containing ceramic shards of unknown age was capped by two layers of fine-grained fluvial sediment following erosional truncation. Re-initiation of agriculture took place by irrigation sedimentation and turbation of the upper portion of the fluvial sediment. A sediment interpreted as irragric anthrosol with poor sorting and high LoI values is present between depths of 2.85 and 0.05 m interrupted only by three fluvial layers (layers 4, 5 and 7) which all show turbation of their top with the base of the overlying layers. The profile is capped by a thin fluvial deposit attributed to the 1999 flood event.

Profile ES

Profile ES (Figures A-68 and A-69) is a 2.58 m high natural bank exposure on the left bank of the Rio Palpa. It is situated 606 m south-west of profile DO and 546 m west of profile SR. Shortly before the profile was recorded (04.04.2006), the bank had been actively eroded by the Rio Palpa by ca. 10 m as shown by a satellite image taken close to the end of the discharge period of the preceding year (Quickbird, 25.03.2005). The river bank had been at a distance of ca. 40 m in 1970 and ca. 45 m in 1944 as shown by aerial photographs. In 1944 the site had been situated in a field and in 1970 as well as in 2006 at the edge of a plantation.

Layer 1 is crumbly loamy silt. Layer 2 is structureless sandy loam with a diffuse, horizontal upper boundary. Layer 3 is sandy silt with a diffuse, horizontal upper boundary. Layers 1 through 3 contain abundant roots. Layer 4 is a sequence of silt (top, diffuse horizontal upper boundary), clay (very thin, turbated), fine sand (ca. 2 cm) and organic material (0.5–1.0 cm). Layer 5 is bedded silty sand to sandy silt with an upwards-coarsening tendency, cross-bedding and a sharp, slightly undulating upper boundary. Layer 6 is a ca. 2 cm thick band of loam with a sharp, horizontal upper boundary.

Layer 7 is bedded fine sandy silt with a sharp, horizontal upper boundary. Layer 8 is poorly sorted material ranging from silt to gravel (up to 5 cm). Charcoal and cohesive clods of silt are present. Within the top 10 cm of the layer, faint bedding is discernible and silt content is higher. The upper boundary is sharp and slightly undulating. Layer 9 is bedded sandy silt with lenses of medium sand and a sharp, horizontal upper boundary. Layer 10 is a charcoal-rich band of sandy silt with a very sharp and slightly undulating upper boundary which in places is marked by ash. Oxidation stains are present along root molds. Layer 11 is a thin layer of loam with a somewhat diffuse, slightly undulating upper boundary. Layer 12 is silty sand with faint bedding and a sharp, slightly undulating upper boundary. Root molds are marked by oxidation stains and

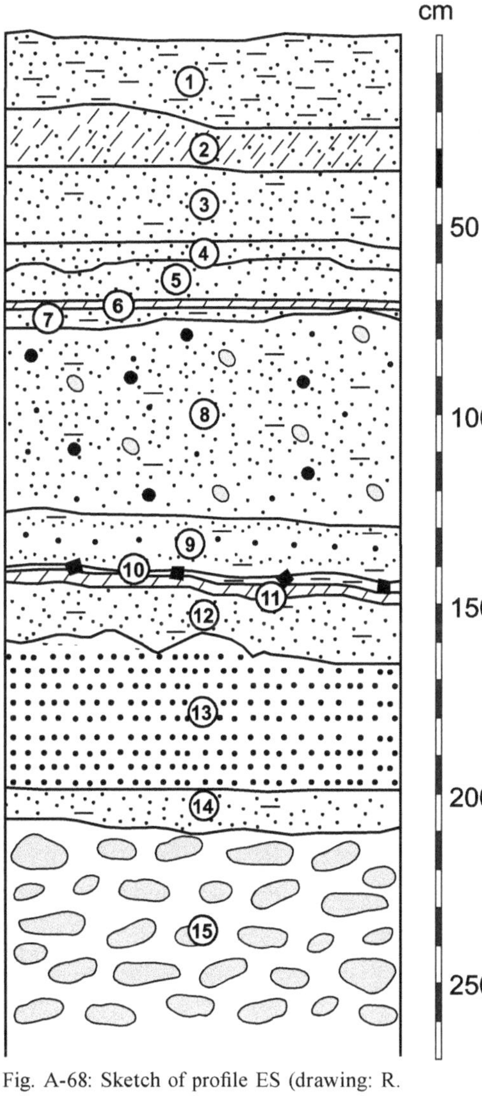

Fig. A-68: Sketch of profile ES (drawing: R. Mendler, vertical scale 1: 20).

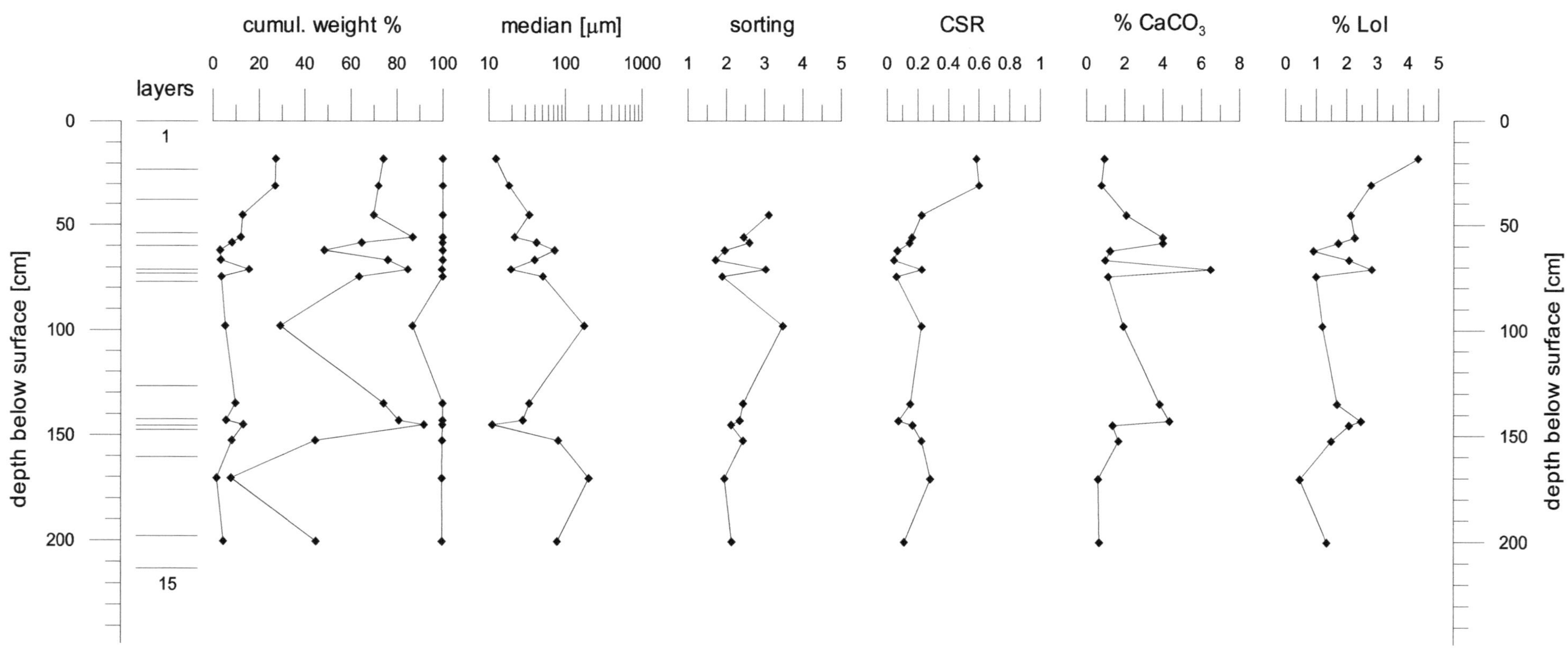

Fig. A-69: Laboratory results of samples from profile ES (vertical scale 1: 30). Cumulative weight percent graph shows clay, silt and sand from left to right. Note that the lines connecting data points do not imply values between data points but serve solely to aid visualisation.

charcoal is present. A radiocarbon assay of charcoal from the base of layer 12 (depth 1.58 m) gave an age ≤270 cal BP (Poz-16828, 130 ±30 BP), thus placing the sedimentation of the top 1.58 m of the profile into the historic period. Layer 13 is well-bedded (including cross-bedding) fine to medium sand with a sharp, undulating upper boundary and oxidation stains along root molds. Layer 14 is structureless silty sand with a diffuse, undulating upper boundary. Layer 15 is well-rounded, unweathered coarse gravel (up to 15 cm) with a sharp, undulating upper boundary.

CSR is low throughout the profile except for layers 1 and 2. LoI reaches a high value of 4.3% in layer 1 and modest values between 2.0 and 3.0% in layers 2–4, 5–6 and 10–11. Carbonate content varies widely between 0.7% (layers 13–14) and 6.5% (layer 6); the higher values are attributed to evaporative precipitation from preferential groundwater seepage at these depths.

Profile ES records repeated fine-grained overbank deposition in a river-proximal situation during the historic period. At least 6 separate events can be discerned, 5 of which occurred since 270 cal BP, i.e. at an average rate of ≥2.2 events per century. Anthropogenic influence can be assumed for the top 0.54 m of the profile (irragric anthrosol and/or agriculturally turbated overbank deposit) and for the charcoal-rich layer at a depth of ca. 1.45 m. A massive layer of poorly sorted material at a depth between 0.76 and 1.27 m is interpreted as a hyperconcentrated flow deposit.

Average sedimentation rate since ≤270 cal BP has been ≥4.86 mm/a and has to be attributed mainly to fluvial deposition, with irrigation sediment accounting for ≤34% of the stratigraphy above the datum. Assuming that the land use change from field to plantation between 1944 and 1970 (as seen in aerial photographs) has led to changes visible in the sedimentary record, the boundary between layers 1 and 2 may be attributed to this land use change, suggesting a sedimentation rate by late historic plantation irrigation of 3.03–3.71 mm/a.

Profile DO

Profile DO (Figures A-70 and A-71) was a pit in a fallow field 200 m from the left bank of the Rio Palpa and 360 m south of profile RPS.

Layer 26 is the recent A_p / irrigation horizon and consists of silty sand. Layer 25 is structureless silty sand with disperse cohesive clods of silt and a diffuse horizontal upper boundary. Layer 24 is structureless silty sand with a diffuse, undulating upper boundary, cohesive clods of silt and disperse charcoal. Layer 23 is medium sand with a diffuse, undulating upper boundary. Layer 22 is silty fine sand with a sharp and slightly undulating upper boundary. Layer 21 is fine to medium sand with a sharp and slightly undulating upper boundary. Layer 20 is sandy silt with a sharp and slightly undulating upper boundary. Layer 19 is medium to coarse sand with a sharp and slightly undulating upper boundary. Layer 18 is sandy silt with a sharp and slightly undulating upper boundary. Layer 17 is fine to medium sand with a sharp and slightly undulating upper boundary. Layer 16 is sandy silt with a sharp and slightly undulating upper boundary. Layer 15 is bedded fine to coarse sand with a sharp and slightly undulating upper boundary. Layer 14 is fine sandy silt. The sharp and undulating upper boundary is partially broken-up. Layer 13 is fine to coarse sand with a sharp, undulating upper boundary. Layer 12 is fine sandy silt with a sharp, slightly undulating upper boundary. A charcoal sample from the base of layer 12 (depth 1.64 m) was radiocarbon-dated to 490–320 cal BP (Poz-13546, 380 ±30 BP), i.e. the late prehistoric (Late Horizon, 550–418 cal BP, cf. Unkel 2006: 111) to early historic period.

Layer 11 is fine to coarse sand with a sharp, slightly undulating upper boundary. Oxidation stains (50%) are present; at the top and bottom of layer 11 there are cohesive clods and lenses of silty fine sand. Layer 10 is fine sandy silt with faint bedding and a diffuse, strongly undulating upper boundary. In the top of the layer there is a bedded band of ash and charcoal underlain. Layer 9 is a clearly delimited, undulating band of fine sandy silt of even thickness (ca. 2 cm). Layer 8 is silty fine sand with a sharp, undulating upper boundary and some oxidation stains (10%). Layer 7 is fine sand with a sharp, horizontal upper boundary and 50% oxidation stains. Layer 6 is sandy silt with a sharp, slightly undulating upper boundary. Layer 5 is silty fine to medium sand (thin beds of fine sand, silt with disperse charcoal and medium sand from top to bottom) with a sharp, horizontal upper boundary. There are 50% oxidation stains. A charcoal sample from layer 5 (depth 2.39 m) was radiocarbon-dated to 900–670 cal BP (Poz-15235, 880 ±40 BP), i.e. Late Intermediate Period (950–550 cal BP, cf. Unkel 2006: 111).

Layer 4 is fine to medium sand with a diffuse, strongly undulating upper boundary and 80% oxidation stains. Layer 3 is well-rounded, coarse gravel (up to 10 cm) with a sharp, undulating upper boundary. Layer 2 is well-sorted and well-bedded medium sand with a sharp, horizontal upper boundary. A lens of gravel (up to 7 cm) is present and a thin (1 mm) black band (possibly disperse charcoal) are present. Oxidation stains cover 20–80% of the layer. Layer 1 is coarse, well-rounded gravel to boulders (up to 20 cm) in a slightly cemented matrix of coarse sand. The upper boundary is sharp and strongly undulating.

Profile DO records a sequence of initial deposition of coarse fluvial gravel followed by river-proximal fluvial events and late anthropogenic influence. Twenty-two low- to medium-energy fluvial layers alternating between sand and silt have been identified, five of them with internal layering/bedding. Eight of the fluvial events occurred between 900–670 cal BP and to 490–320 cal BP, i.e. at an average rate of 1.4 to 4.4 events per century. Twelve events occurred after 490–320 cal BP, i.e. at an average rate of 2.4 to 3.8 events per century. From the bottom of the profile upwards, there is a tendency towards thinner layers as would be expected for overbank deposits as the bank accumulates and thereby rises relative to the riverbed.

Sorting is generally good in the lower, fluvially dominated portion of the profile; however, LoI reaches ca. 2.0% for several of the fluvial layers (generally in the more silty sediments). Since charcoal is present only in very small amounts throughout the fluvial portion, these relatively high LoI values are best explained by accumulation of organic material through plant growth during periods between flood events. The top 0.95 m (layers 26 through 24) of profile DO differ significantly from the lower part in that they are characterised by poor sorting, relatively high CSR and LoI, diffuse layer boundaries and lack of bedding. These differences are attributed to both a different mode of sedimentation (irrigation instead of floods) and strong turbation by agriculture. These irragric anthrosol layers have been deposited since to 490–320 cal BP. Average sedimentation rates are 1.01–4.17 mm/a for the period from 900–670 cal BP to 490–320 cal BP and 3.00–4.37 mm/a for the period since 490–320 cal BP.

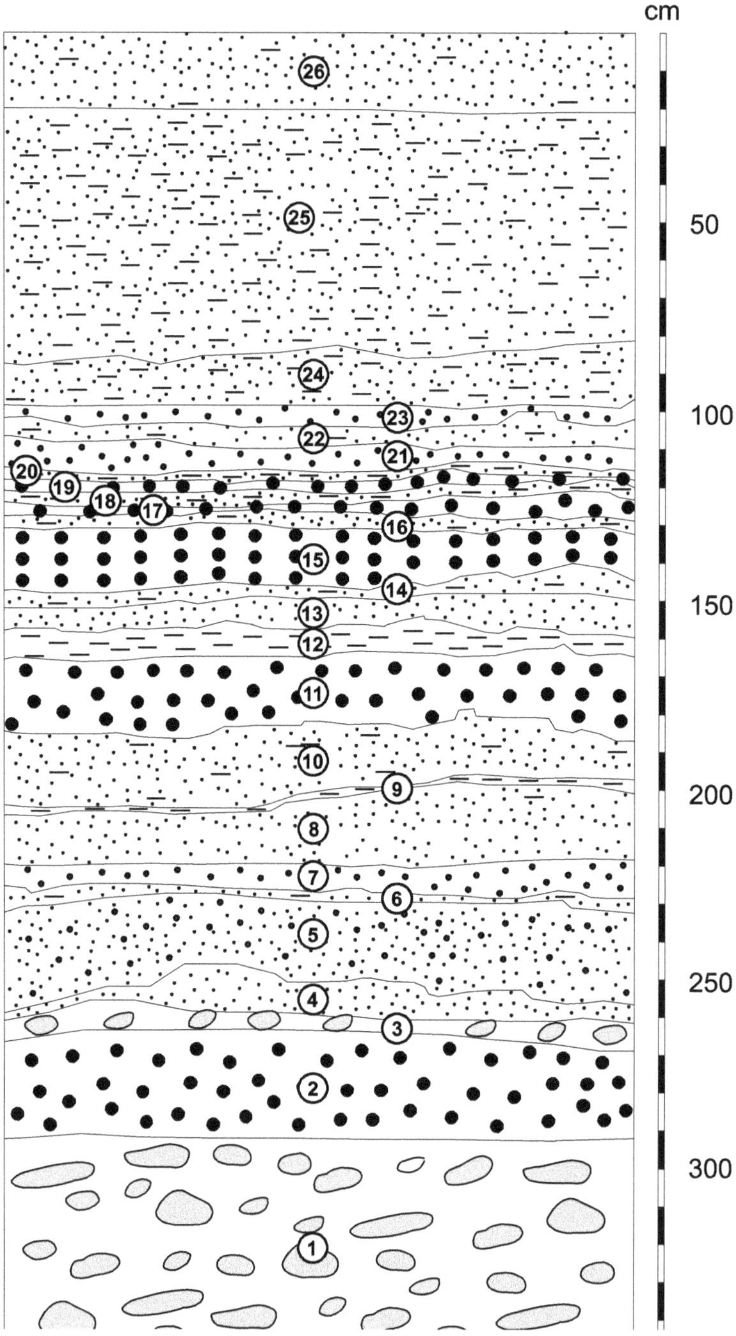

Fig. A-70: Sketch of profile DO (drawing: R. Mendler, vertical scale 1: 20).

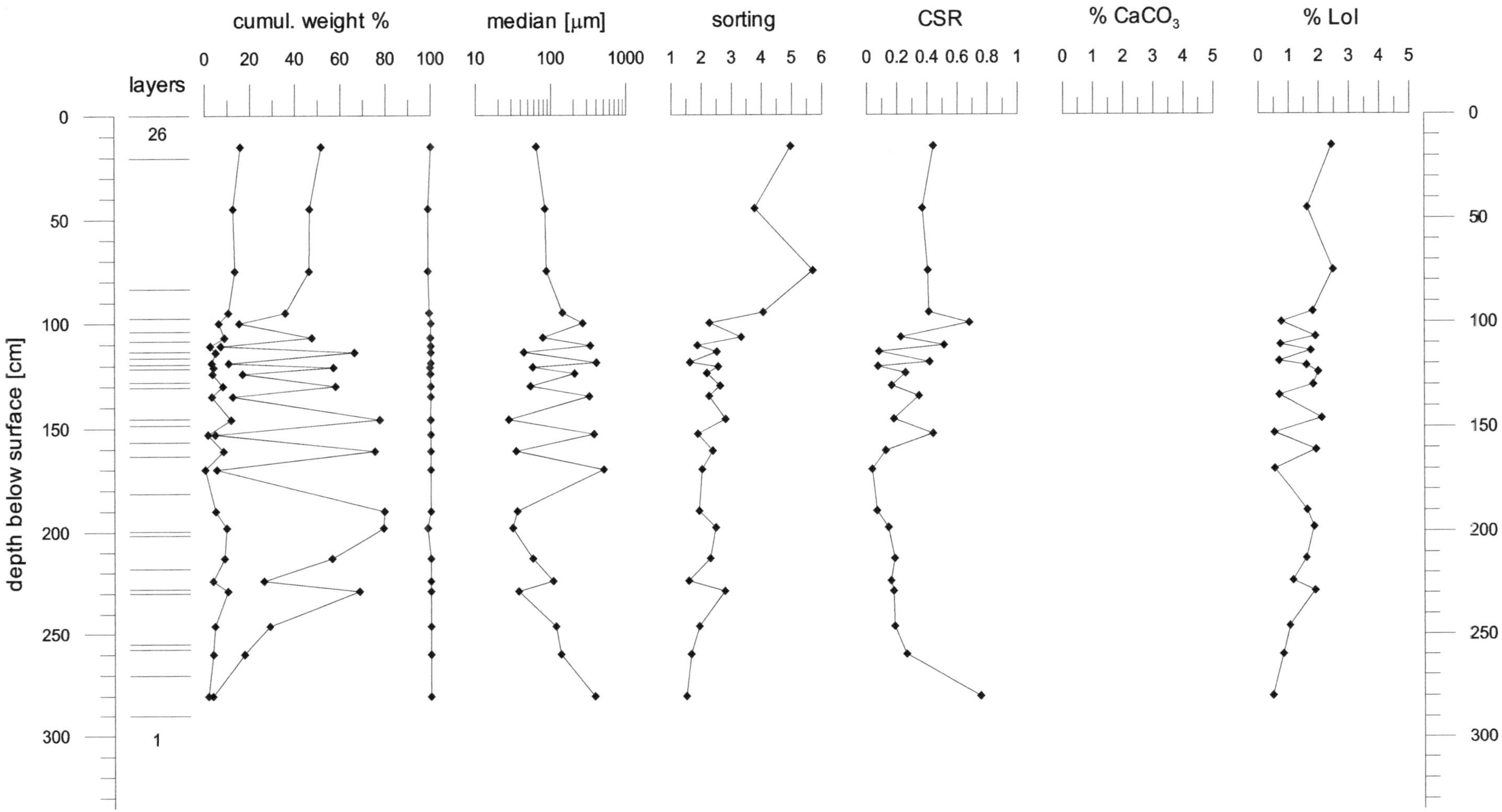

Fig. A-71: Laboratory results of samples from profile DO (vertical scale 1: 30). Cumulative weight percent graph shows clay, silt and sand from left to right. Note that the lines connecting data points do not imply values between data points but serve solely to aid visualisation.

Profile RPS

Profile RPS (Figure A-72) was recorded in a natural bank exposure on the left (south-western) side of the Rio Palpa near the hacienda of Don Oscar Tijero Ríos. It is situated 519 m west-south-west from profile Jau-1 and 360 m north of profile DO. No samples were analysed from profile RPS.

Layer 0 is silty sand with high content of organic material. Layer 1 is silty sand with angular gravel. Layer 2 is medium to coarse sand with a diffuse upper boundary. Turbation is obvious through the presence of cohesive clods of silt throughout the layer. Layer 3 is laminated silt with a sharp upper boundary and some traces of turbation. Layer 4 consists of 2 bands of structureless silt with a sharp upper boundary. Layer 5 is poorly sorted sand with a sharp upper boundary. Layer 6 is well-sorted silty fine sand with a sharp upper boundary marked by a thin band of silt. Faint bedding is present. Layer 7 is silty fine sand with varying thickness (4–10 cm) and a sharp upper boundary. Disperse charcoal is present. The top of layer 7 is marked by subpolyedric structure. Layer 8 consists of alternating bands of coarse sand to gravel with a upwards-fining tendency. The upper boundary of layer 8 appears pocketed. At the base of layer 8, some oxidation stains are present. Layer 9 is coarse sand to fine gravel overlying coarse gravel with a sandy matrix. There are abundant oxidation stains.

Profile RPS records alternating high- and low-energy fluvial sedimentation and some human impact (as indicated by the presence of abundant charcoal in the layer 7) in a river-proximal environment.

Profile Jau-1

Profile Jau-1 (Figures A-73, A-74 and A-75) was a four-metre-deep pit in a field near the hacienda of Don Oscar Tijero Ríos in the direction of the present-day settlement of Jauranga. It is situated 240 m south-east of the left bank of the Rio Palpa and 482 m west of the profile Jau-Co. The sketch shows the northern side of the pit.

Layer 0 (the present A_p horizon) is crumbly, poorly sorted silty sand with some fine gravel. Layer 1 consists of bedded fine sand to fine gravel. Bedding is marked by thin bands of fine sand interspersed in thicker bands of coarse sand to fine gravel. Crossbedding and ripple marks are present. The upper boundary of layer 1 is undulating. Layer 2 is well-sorted , bedded silty fine sand with an undulating upper boundary. Beds are 5–15 cm thick; some lenses of coarse sand are present. Layer 3 is porous, structureless silt with an undulating upper boundary and contains abundant charcoal. Layer 4 is bedded (with ripple marks) fine sandy silt with an undulating to diffuse upper boundary. The layer dips slightly towards the east. Layer 5 well-bedded fine sand to gravel. Cross-bedding is present. At the top of the layer, there is a 2–10 cm thick band of fine to medium sand, followed by three bands of gravel (each 2–3 cm thick) interspersed with sand. In the lower part of the layer, bedding consists exclusively of fine to medium sand. The upper boundary of layer 4 is undulating.

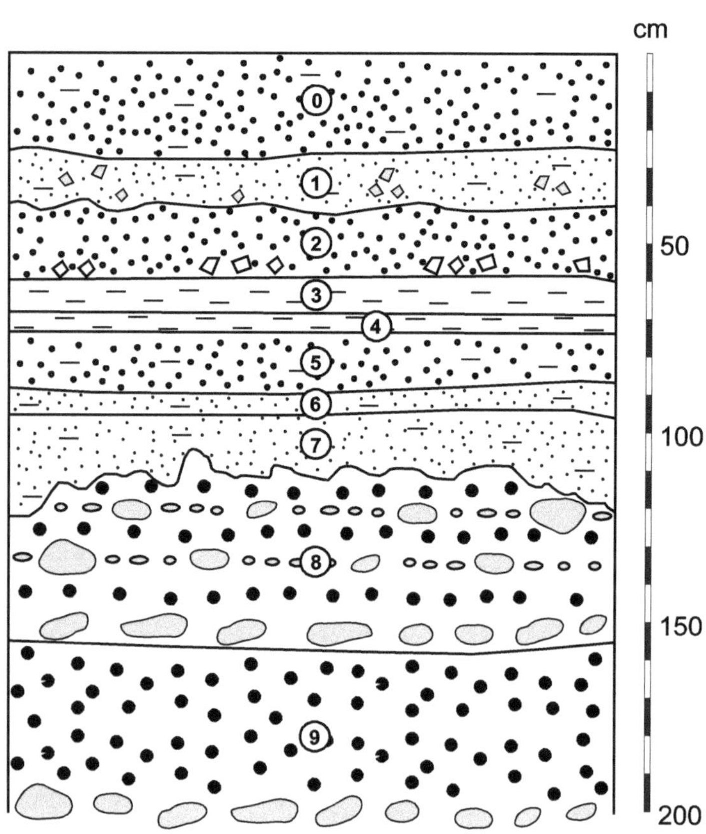

Fig. A-72: Sketch of profile RPS (drawing: R. Mendler, vertical scale 1: 20).

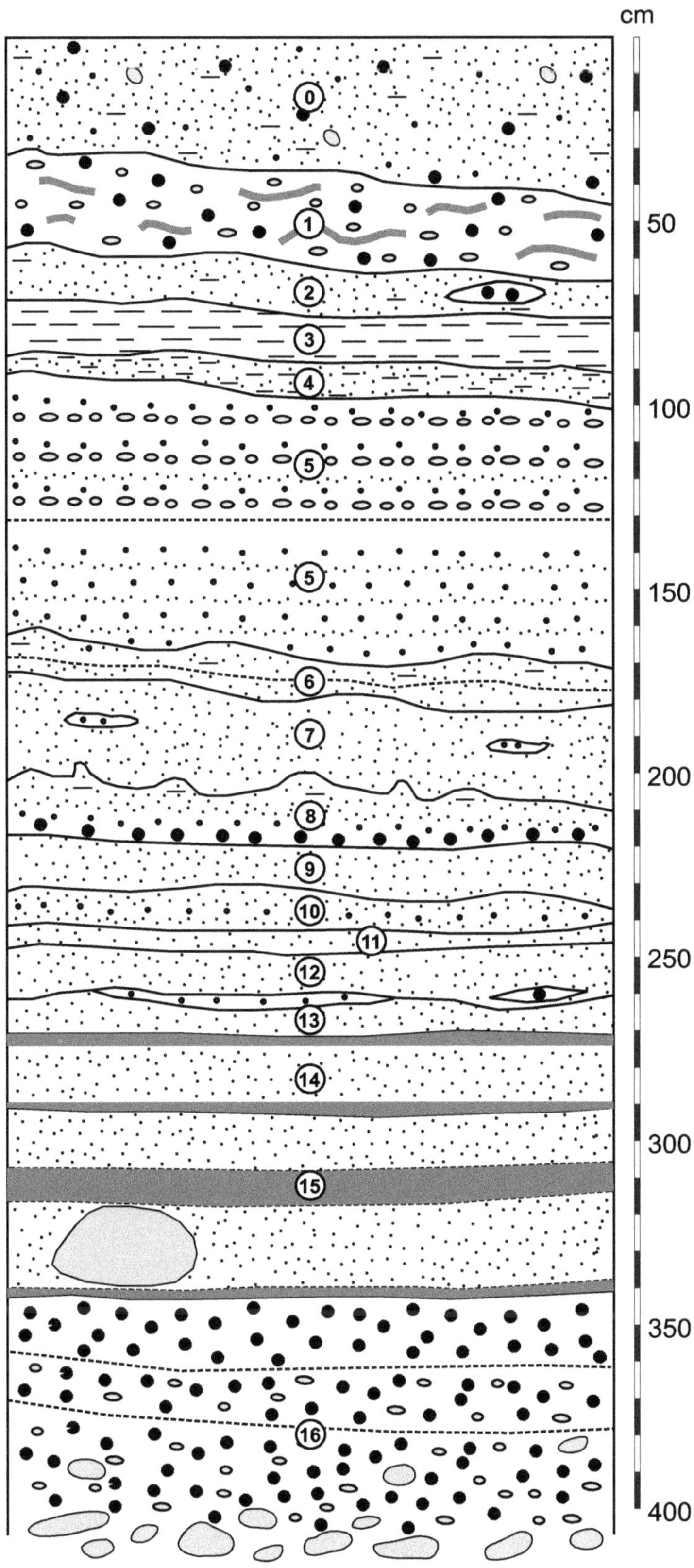

Fig. A-73: Sketch of profile Jau-1 (drawing: R. Mendler, vertical scale 1: 20).

Fig. A-74 (following page): Laboratory results of samples from profile Jau-1 (vertical scale 1: 30). Cumulative weight percent graph shows clay, silt and sand from left to right. Note that the lines connecting data points do not imply values between data points but serve solely to aid visualisation.

Fig. A-75 (page A-87): AAS measurement results of samples from profile Jau-1 (vertical scale 1: 30). Note that the lines connecting data points do not imply values between data points but serve solely to aid visualisation.

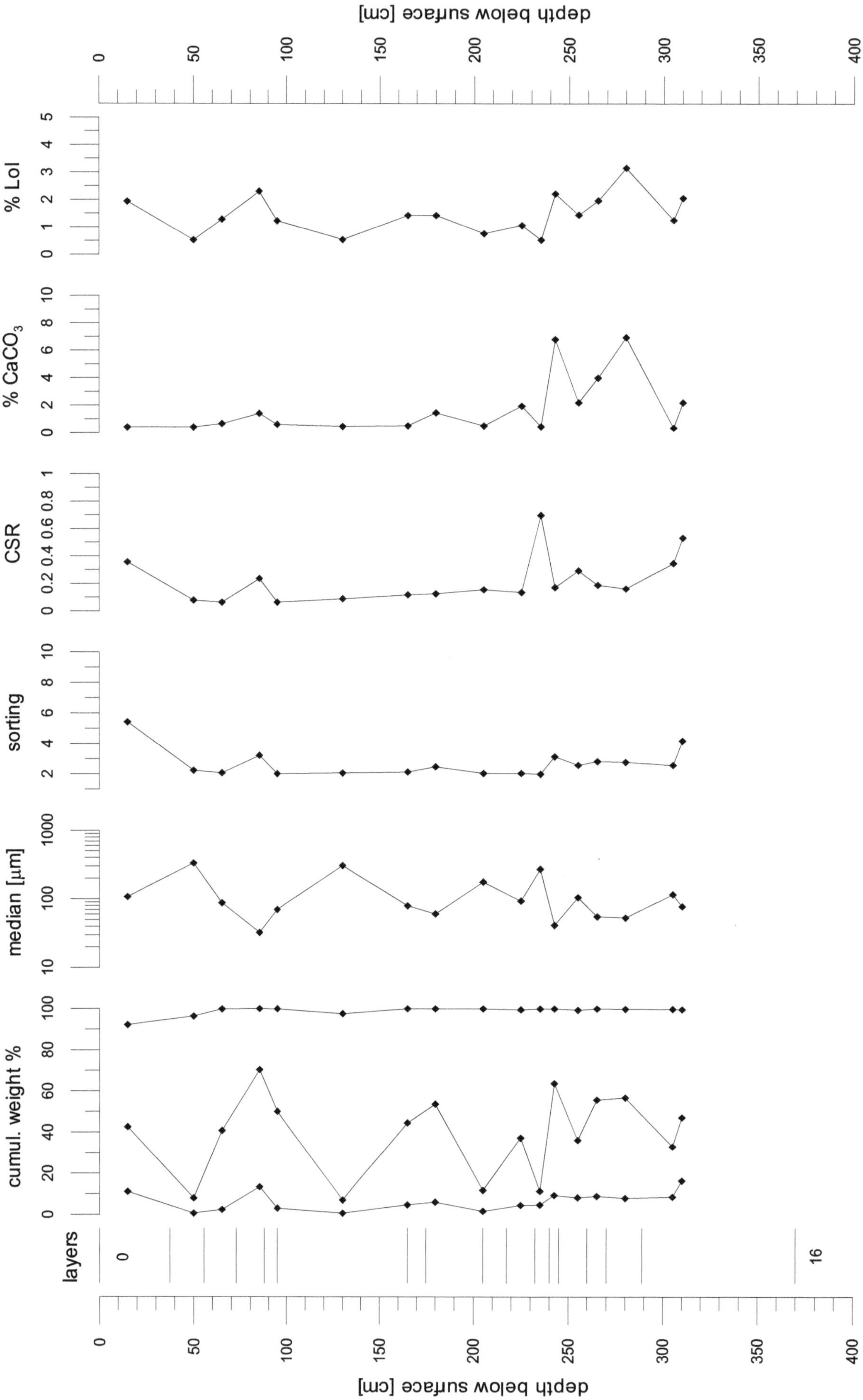
depth below surface [cm]
% LoI
% $CaCO_3$
CSR
sorting
median [µm]
cumul. weight %
layers
0
16
depth below surface [cm]

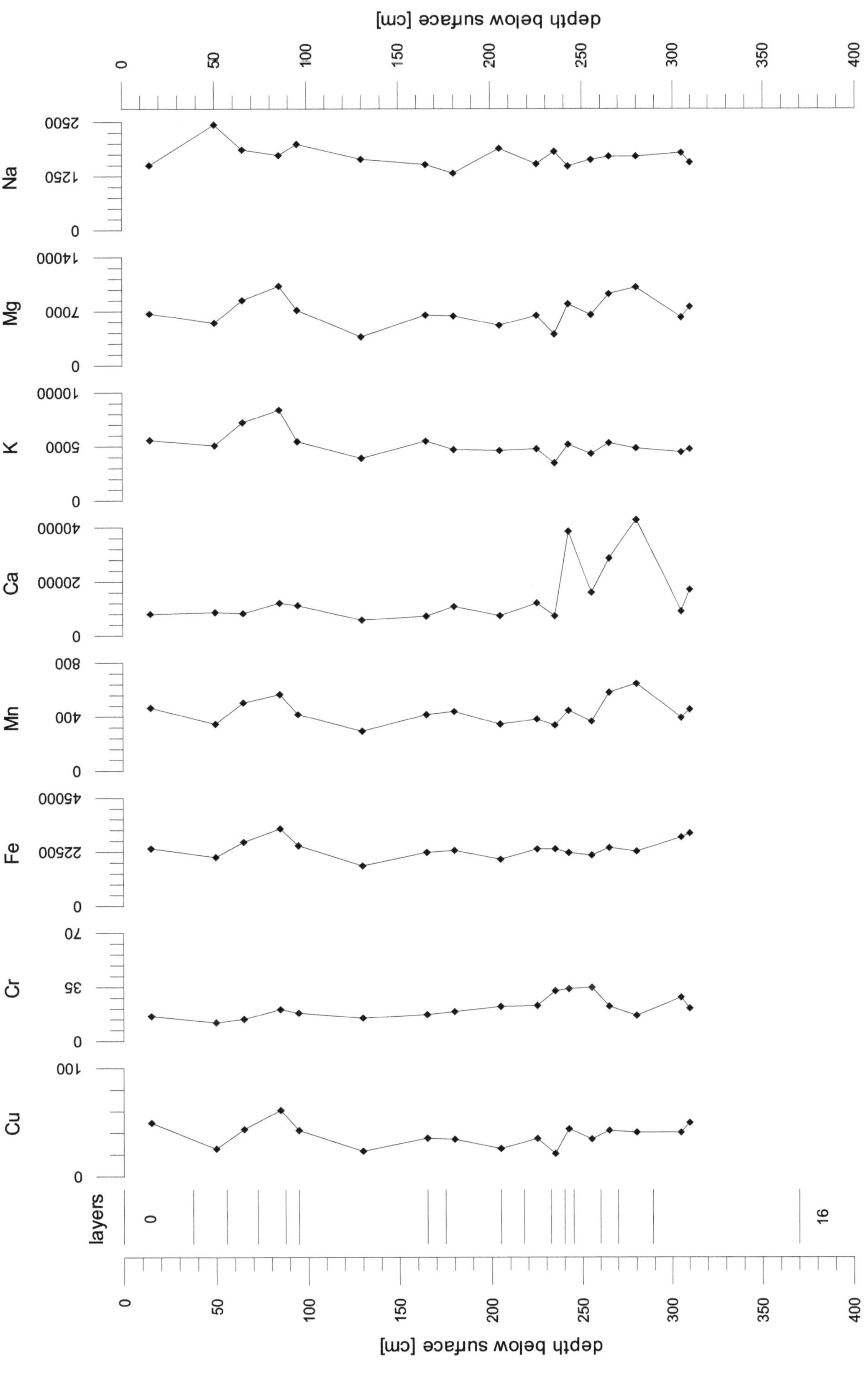

depth below surface [cm]
Na
Mg
K
Ca
Mn
Fe
Cr
Cu
layers

A radiocarbon assay of charcoal from the top of layer 4 (depth 1.00 m) gave an age of 500–320 cal BP (Poz-8006, 395 ±25 BP) and places the upper section of profile Jau-1 into the early historic or late prehistoric period. Layer 6 is an undulating band (both regarding its upper and lower boundary) of bedded silty fine sand and fine sand. Oxidation stains are present. Layer 7 is structureless silty fine sand with an undulating upper boundary which is marked by a thin black band. Charcoal and lenses of medium to coarse sand are present. Layer 8 is bedded (with cross-bedding) silty fine sand to coarse sand with a sharp, pocketed upper boundary. On the north-western side of the pit, a 2 cm thick band of fine sand is present at the base of layer 8. Charcoal and oxidation stains are present. Layer 9 is structureless silty fine sand with oxidation stains and a slightly undulating upper boundary. Layer 10 is well-sorted and bedded fine to medium sand with oxidation stains and an undulating upper boundary. Charcoal from this layer (depth 2.30 m) was radiocarbon-dated to 550–505 cal BP (Poz-8008, 545 ±25 BP), indicating a late prehistoric (Late Horizon, 550–418 cal BP, cf. Unkel 2006: 111) age.

Layer 11 is a 0–5 cm thick band of silty fine sand with a sharp and slightly undulating upper boundary. Pieces of charcoal are present. Layer 12 is porous silty fine sand with oxidation stains, abundant charcoal and basal lenses of medium to coarse sand (up to 10 cm thick and 100 cm long). Layer 13 is porous silty fine sand with an undulating upper boundary. Charcoal is present, oxidation stains are lacking. Layer 14 is silty sand with a well-defined, horizontal upper boundary. The layer dips slightly to the north-east. The top 10–15 cm of layer 14 have a strongly disturbed appearance which may be due to tillage of the then-surface. A thin, blackish band is present at the base of the layer and also seems to have been affected by turbation. Layer 15 consists of silty fine sand with two thin blackish bands (at depths 3.05–3.15 m and 3.35–3.40 m) containing slightly more silt and clay and increased LoI. These bands are interpreted as humus- or charcoal-rich. A single large rounded stone (20 cm) lies exactly between the two blackish bands. The lower of the two bands marks the lower boundary of layer 15. Charcoal from near the base of layer 15 (depth 3.20 m, eastern side) was radiocarbon dated to 1,490–1,300 cal BP (Poz-8007, 1,545 ±25 BP), i.e. the Late Nasca period (1,520–1,300 cal BP, cf. Unkel 2006: 111).

Layer 16 is coarse gravel in a matrix of coarse sand to fine gravel with oxidation stains. Gravels are imbricated and a strong downwards-coarsening trend is present. The upper boundary of layer 16 is sharp. Grain size parameters change repeatedly throughout the stratigraphy, with unambiguous fluvial layers (identified on the basis of bedding characteristics) characterised by higher sand content. The material is well-sorted except for layers 0, with layers 3, 11 and 15 having slightly higher values than the rest of the profile. CSR is very low except for layers 0, 3, 10 and 15. Carbonate content is very low except at depth below 2.30 (layers 10 through 15), LoI values are low except for layers 0, 3 and 10 through 15.

Profile Jau-1 shows an alternation of fluvial and anthropogenic influences. Basal fluvial gravels are present at a depth of 3.40 m. Since 1,490–1,300 cal BP (depth of 3.20 m) and until historic times, the site was repeatedly impacted by fluvial events which deposited well-sorted silt, sand and fine gravel. Some layers (1, 2, 4–6, 8 and 10) are of clearly fluvial origin as indicated by bedding characteristics and were deposited in low- to medium energy flow environments (ripple marks and cross-bedding in silty sand to fine gravel). It is likely that there are hiatus in the stratigraphy since several of the layer boundaries probably have erosional character. Some fluvial layers (layers 4 and 8) are characterised by diffuse or pocketed upper boundaries overlain by well-turbated, structureless fine-grained sediment (layers 3 and 7) containing disperse charcoal. These situations are interpreted as incipient agricultural use and irrigation on fluvial sediment. Other anthropogenically dominated layers are the surface A_p irrigation horizon and layers 12 to 15 which are characterised by increased LoI values and the presence of charcoal and of a prominent boulder in a matrix of silty sand. Sedimentation rates are 0.91–1.20 mm/a for the period 1,490–1,300 cal BP (depth 3.20 m) to 550–505 cal BP (depth 2.30 m), 0.57–26.0 mm/a for the period 550–505 cal BP (depth 2.30 m) to 500–320 cal BP (depth 1.00 m) and 1.80–2.67 mm/a for the period since 500–320 cal BP (depth 1.00 m). Possible hiatus and the strongly episodic nature of deposition in profile Jau-1 have to be taken into account when interpreting these sedimentation rates.

Profile FL

Profile FL (Figures A-76 and A-77) was a pit on a fallow field between Rio Palpa (760 m) and Rio Vizcas (290 m), 591 m south-west of profile Jau-Co.

Layer 16, the recent A_p horizon, consists of loamy silt with a crumbly structure and contains abundant roots. Layer 15 is loamy silt with polyedric structure (edge lengths 5–10 cm) and a sharp, horizontal upper boundary. Layer 14 is loamy silt with a polyedric to columnar structure (edge lengths 5–10 cm) and a diffuse, horizontal upper boundary which is in places marked by a well-defined horizontal crack possibly related to ploughing. Layer 13 is structureless sandy loam with a diffuse horizontal upper boundary. Layer 13 is more compacted than the overlying layer 14; the vertical cracks related to the columnar structure of layer 14 penetrate into the top of layer 13. Layer 12 is structureless silty sand with a diffuse, undulating upper boundary. A radiocarbon assay of charcoal from layer 12 (depth 1.04 m) gave an age of 510–320 cal BP (Poz-15238, 415 ±30 BP).

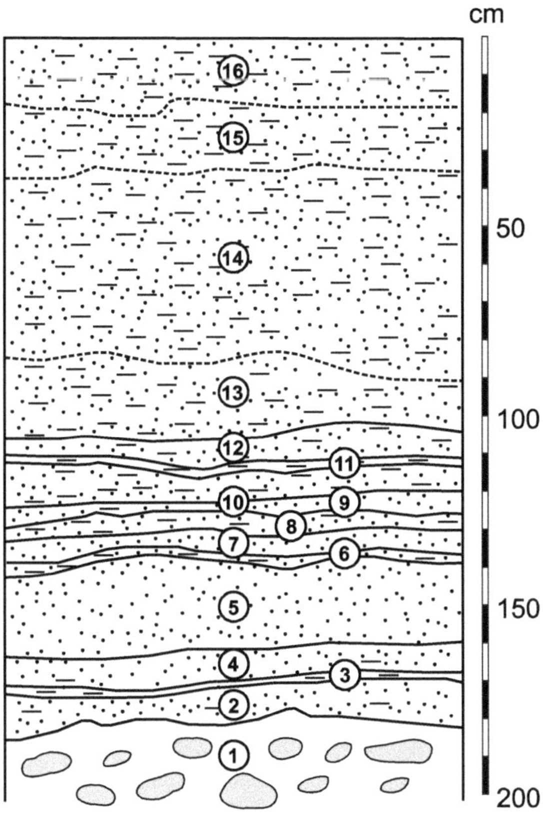

Fig. A-76: Sketch of profile FL (drawing: R. Mendler, vertical scale 1: 20).

Layer 11 is a well-defined, 1–3 cm thick band of silt with a sharp and slightly undulating upper boundary. Layer 10 is sandy silt with a sharp and slightly undulating upper boundary. Layer 9 is silty sand with a sharp and slightly undulating upper boundary. It is well-defined on the western and northern and also present on the southern side of the pit but not present on the eastern side of the pit where layer 10 directly overlies layer 8 with a very diffuse boundary. Layer 8 is structureless sandy silt with a sharp and slightly undulating upper boundary. In the lower part of layer 8, sand lenses (layer 7) are present. These sand lenses are 5 to 8 cm thick, have faint bedding and sharp boundaries and intrude wedge-like from the southern side (from the Rio Vizcas) into the profile. Layer 6 is a 1–5 cm thick, well-defined band of silty fine sand to sandy silt. Layer 5 is structureless silty fine sand with ca. 5 % cohesive clods of silt and a few (<1%) small stones. Its upper boundary is sharp and slightly undulating. Layer 4 is a 3–10 cm thick band of structureless silty fine sand with a diffuse, undulating upper boundary. Layer 3 is a 1.5–3.0 cm thick band of loamy silt with a sharp and slightly undulating upper boundary.

Layer 2 is structureless, poorly sorted, loamy sand with some stones (up to 4 cm). A radiocarbon assay on charcoal from the base of layer 2 (depth 1.80 m) gave an age of 1,060–920 cal BP (Poz-13547, 1,125 ±35 BP). The deposition of this layer therefore probably occurred during the later part of the Middle Horizon (1,300–950 cal BP, cf. Unkel 2006: 111) or the early part of the Late Intermediate Period (950–550 cal BP, cf. Unkel 2006: 111). This date furthermore provides a minimum age for the deposition of the underlying layer 1, a coarse, well-rounded gravel (up to 15 cm) with a matrix of coarse sand and a sharp, undulating upper boundary. A few of the gravel clasts are weathered. Layer 1 is interpreted as high-energy fluvial river bed sediment. Given the distance from the recent bed of the Rio Vizcas and the relatively young age of the overlying sediments as compared to other profiles in the central Palpa Valley, an erosional event cutting a new river bed and leaving channel deposits is assumed to have affected the site.

Profile FL records the results of a high energy event of erosion and deposition followed by a sequence of fine-grained flood deposits (layers 2 through 11) and anthropogenically induced sedimentation (irragric anthrosol, layers 12 through 16). Fluvial overbank deposits are clearly distinguishable from irrigation sediments on the basis of sorting, LoI and bedding characteristics. The fine-grained fluvial deposits possibly represent three events (layers 2 and 3, layers 4 through 6, layers 7 through 11) with respective internal stratigraphic sequences as suggested by the recurrent capping of these three sedimentary packages by well-defined silty layers. Average sedimentation rates are 1.0–1.9 mm/a for the lower, fluvially dominated part of the profile (depth 1.04–1.80 m, 1,060–920 cal BP) and 1.8–2.8 mm/a for the upper part of the profile (above 1.04 m, since 510–320 cal BP).

Profile H-1

Profile H-1 (Figure A-78) was a pit on a fallow field near the Rio Vizcas. The field lies lower than the next field towards the west and north in a terrace-like manner and is sloping towards the river. The erosional bank of the Rio Vizcas related to the 1999 flood event lies at a distance of 150 m, the artificial embankments of the river lie at a distance of 350 m. Profile H-1 is situated at a distance of 150 to 280 m north-east of profile RVU and 767 m south-west of profile SF-4. No samples were taken from profile H-1. A large boulder (>1.2 m) is present in the north-western corner of the profile extending from a depth of 0.5 m below the surface down beyond the limit of excavation.

Layer 7 is the recent ploughing horizon and consists of very loose, crumbly sandy silt with a high concentration of plant parts. Layer 6 is sandy silt with a broken-up polyedric structure and an undulating upper boundary. Layer 5 is structureless sandy silt with an undulating horizontal upper boundary. Layer 4 is a ca. 9 cm thick, laminated band of silt which is best recognisable on the eastern side of the profile and only diffuse on the northern and western sides. Its upper boundary is horizontal. Layer 3 is very heterogeneous throughout the pit. It consists mainly of silty sand and has a diffuse horizontal upper boundary. On the eastern side, a laminated, north-sloping band of

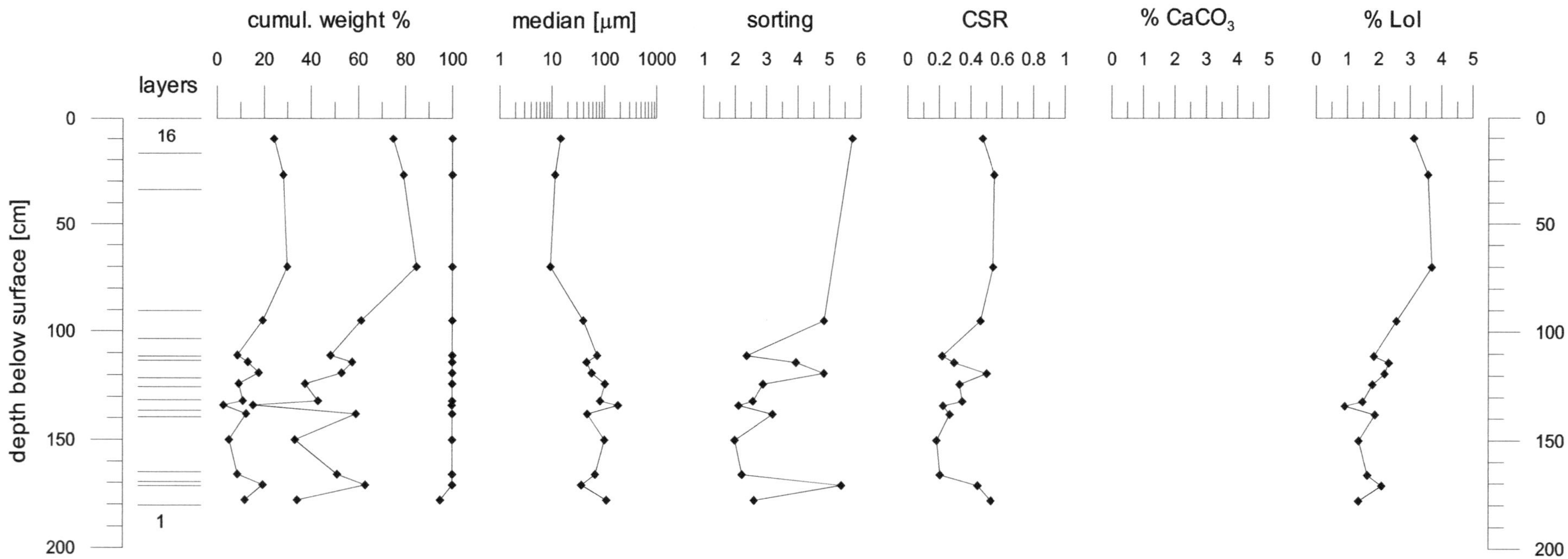

Fig. A-77: Laboratory results of samples from profile FL (vertical scale 1: 30). Cumulative weight percent graph shows clay, silt and sand from left to right. Note that the lines connecting data points do not imply values etween data points but serve solely to aid visualisation.

Fig. A-78: Photograph of profile H-1 (vertical scale 1: 20, photograph by J. Baade).

silt to fine sand, a disturbed loamy band and a 70 cm wide lens of coarse sand with a basal stone and a sharp lower and diffuse upper boundary are present. On the northern side, small (25 cm wide) lenses of coarse sand were recorded.

On the western side, a channel-like lens (35 cm wide, 20 cm thick) of coarse sand with a sharp lower and somewhat diffuse upper boundary is present. The sand lenses in the eastern and western side are interpreted as remnants of an irrigation channel. Layer 2 is coarse gravel (30%, up to 10 cm) in a matrix of sand. It has a sharp, undulating upper boundary. Layer 1 is densely packed coarse gravel (80%) with a sandy matrix. Granites are strongly weathered. This layer is interpreted as pre-Holocene river bed gravel due to the strong weathering.

Profile H-1 is characterised by basal gravels of probably pre-Holocene age as judged from their strong weathering, overlain by poorly sorted, younger (Holocene) gravels which in turn are overlain by 1.25–1.40 m of irrigation sediment and/or turbated fine-grained overbank flood sediment. A very large boulder indicates the occurrence of an extreme flood event of unknown age. An irrigation channel filled with coarse sand and capped by silty sand was identified at a depth of 1.05 m.

Profile Jau-Co

Profile Jau-Co (Figure A-79) was a pit on a fallow near the present-day settlement of Jauranga in the centre of the Palpa valley. It is situated between Rio Palpa (470 m) and Rio Vizcas (480 m), 1.19 km south-west of profile SF-1. The sketch shows the north-west side of the pit. In the top 100 cm, a few Nasca shards were found. The 1944 aerial photograph shows at that time there was a plantation at this site.

Layer 1 is a loose, crumbly layer of silty sand with a high concentration of roots. Layer 2 is a structureless layer of sandy silt containing ca. 1% gravels with lengths of up to 3 cm as well as some brick fragments and Middle to Late Nasca ceramic shards (1,620–1,300 cal BP, cf. UNKEL 2006: 111; identified by J. Palomino). The upper boundary is undulating. Near the lower boundary, plastic foil and a metal can were found. Due to the presence of 20th century refuse at a depth of 70 cm this layer is interpreted as landfill related to the nearby settlement; the ceramic age is therefore discarded. Beginning near the base of layer 2 and penetrating through layers 3, 4 and 5, layer 9 is a narrow (20–30 cm) and well-delimited column of silty sand. It is interpreted as a post hole.

Layer 3 is an upwards-fining (coarse sand to silt) sedimentary package. The top 4 cm are clearly bedded; the upper boundary is horizontal and slightly undulating. This layer is unambiguously a fluvial sediment related to a flood event showing characteristics of waning flow deposition (decreasing grain size). Layer 4 consists of structureless silt with disperse clods of cohesive silt. These clods indicate a mechanical break-up of the material, likely by cultivation. The upper boundary of this layer is sharp and sloping towards the west. The large round stones (up to 40 cm) set in the top of this layer can not be explained by natural processes and further indicate human activity at the site. Layer 5, consisting of silt to fine sand and containing mica, is a fluvial overbank deposit or a sediment related to incipient irrigation. Layer 6 is a lens of fine sand with a sharp, undulating upper boundary and is interpreted as a fluvial overbank deposit. Layer 7 consists of poorly sorted sand with 25% gravel. The upper boundary is undulating. The layer contains charcoal; radiocarbon dating of a sample from the base (depth 1.55 m, close to the gravel surface) gave an age of 1,700–1,420 cal BP (Poz-16814, 1,730 ±35 BP). Due to the presence of charcoal and disperse gravel the layer is interpreted as anthropogenic. At a depth of 1.43 to 1.60 m, basal gravels (layer 8) were reached. They consist of well-rounded, unweathered and uncemented clasts with lengths of up to 30 cm. The upper boundary of layer 8 is sharp and undulating. This layer is interpreted as a fluvial sediment of a former river bed.

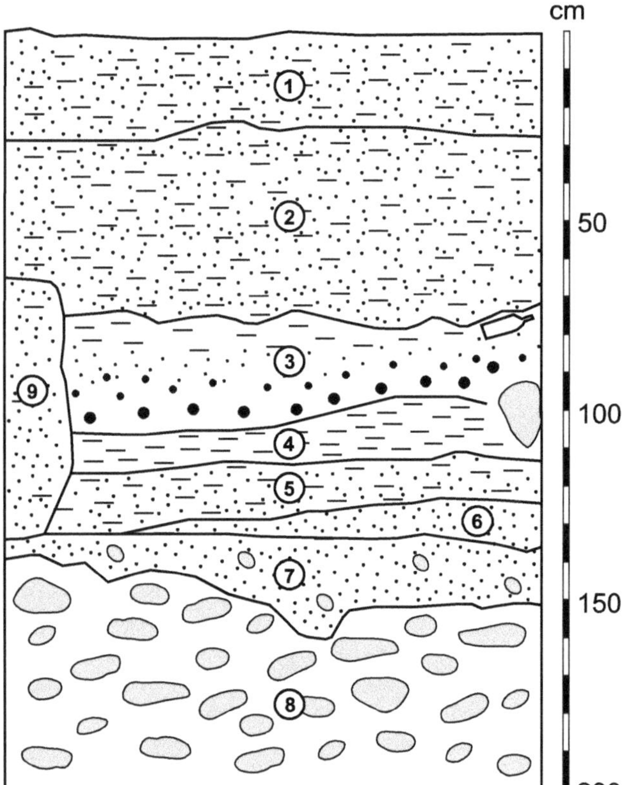

Fig. A-79: Sketch of profile Jau-Co (drawing: R. Mendler, vertical scale 1: 20).

Profile Jau-Co is interpreted as a succession of riverbed (layer 8) and irrigation (layers 7, 4, 2, 1, 9 and possibly 5) sediments with interspersed fluvial overbank deposits (layers 6 and 3). The minimum age of river bed gravel deposition is given by a radiocarbon date of 1,700–1,420 cal BP from the base of the overlying fine-grained sediment. Layer 3 represents a major flood event that deposited 15 to 30 cm of coarse sand to silt. The top 0.75 m of the profile contain Nasca shards as well as 20th century refuse and are therefore considered to represent material related to nearby 20th century settlement activity (landfill). Excluding the top 0.75 m of the profile and assuming a plausible age of AD 1900–2000 (50 to -50 cal BP), fine-grained sedimentation of fluvial and irrigation sediment occurred at an average rate of 0.57–0.58 mm/a before the deposition of refuse.

Profile RP-06

Profile RP-06 (Figures A-80 and A-81) is a 3.60 m high bank exposure on the left side of the Rio Palpa, ca. 1.1 km south-west of the Panamericana and 449 m north-west of profile SF-1. The current artificial embankments of the Rio Palpa are at a distance of only 20 m. These embankments are clearly visible and intact on the satellite images of March 2005 (Google Earth), however, in March 2006 they had been almost completely reworked by fluvial activity even though river discharge was low (maximum 15.25 m^3/s) in the intervening year (according to river discharge data made available by the Junta de Usarios Palpa). Although the exposure is an erosional bank of the Rio Palpa, aerial photographs of 1944 and 1970 show that the river bank at this site has not retreated by more than 20 m since 1944, with most of the erosion occurring between 1944 and 1970.

In 1944, the position of the profile was at the centre of a oval structure 26 by 42 m in size which had apparently been exempted from agricultural use. This may be due to the site surface lying slightly higher than the nearby irrigation channel. The 1970 and 1997 aerial images as well as the 2005 satellite image and 2006 ground evidence all confirm this pattern. Several wall foundations (large rounded stones set into adobe-like material) penetrating the upper 0.60–1.00 m of the profile were identified by J. Palomino.

The top of the profile (layer 16) is very loose sandy silt with a layer of ash containing also stones, ceramic shards and bones at its base. Carbonate content (6.6%) and LoI (5.6%) are high. Layer 15 consists of structureless loamy silt with some stones and ceramic shards. Its upper boundary is undulating. Carbonate content (5.7%) and LoI (2.8%) are lower than in layer 16. Layer 14 consists of structureless, poorly sorted material (sandy silt with ca. 30% gravel-sized clasts). The upper boundary is diffuse and horizontal and contains some stones and ceramic shards. Carbonate (6.0%) and LoI (2.9%) values are similar to those of layer 15. Layer13 is structureless silty sand with some rounded stones, fragments of rounded stones, marine mollusc shells and some dark (fired under reductive conditions) ceramic shards. The upper boundary is diffuse and horizontal.

Layer 12 is structureless silty fine sand; its diffuse horizontal upper boundary is marked by a layer containing stones, marine mollusc shells and ceramic shards. All ceramic shards from layer 12 were consistently assigned Late Paracas ages (Ocucaje 9, 2,300–2,150 cal BP, cf. Unkel 2006: 111) by J. Palomino. Layer 11 is structureless silty fine sand with high porosity and some stones. Its upper boundary is diffuse and horizontal. A radiocarbon assay of charcoal from layer 11 (depth 1.40 m) gave an age of 3,080–2,840 cal BP (Poz-16816, 2,890 ±35 BP), placing the sedimentation of this layer into the late Initial Period (3,750–2,810 cal BP, cf. Unkel 2006: 111). Layer 10 consists of poorly sorted silty sand with 30–50% angular to subrounded clasts (up to 3 cm). The upper boundary is diffuse and horizontal.

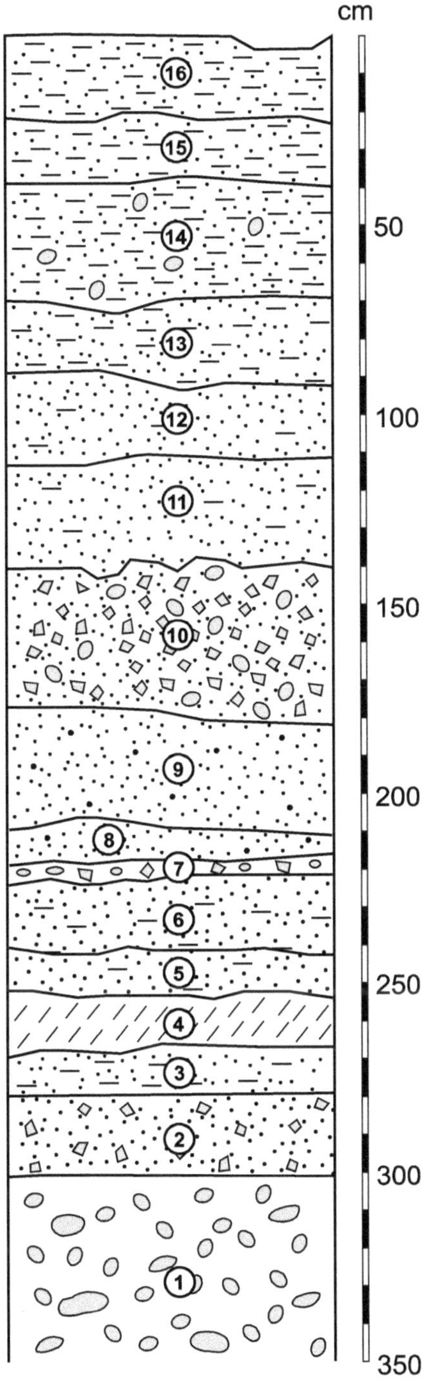

Fig. A-80: Sketch of profile RP-06 (drawing: R. Mendler, vertical scale 1: 20).

The lower 20–25 cm of layer 10 are interpreted as an undisturbed hyperconcentrated flow deposit with an upwards-coarsening tendency while the upper 15 cm are characterised by increased sand content which is interpreted as resulting from turbation from the overlying layer 11. Layer 9 is structureless fine to medium silty sand with a few small stones. The upper boundary is sharp and horizontal with some pocket-like depressions suggesting erosional activity before (and probably by) the hyperconcentrated flow. Some small carbonate concretions are present, but overall carbonate content is low (1.2%). Layer 8 is structureless fine to medium sand with some small carbonate concretions and a few stones. The upper boundary is diffuse. Layer 7 is fine angular to subrounded gravel with a sharp, undulating upper boundary. Locally it is only present as lenses. It is interpreted as a minor hyperconcentrated flow deposit. Layer 6 is dark brown to black silty fine sand. Its upper boundary is sharp and slightly sloping towards north-east. Towards the south-west, layer 6 peters out. Layer 5 is structureless, well-sorted fine to medium sand with a high mica content and a sharp, undulating upper boundary. Layer 4 is dark brown to black silty sand with a higher CSR (up to 0.38), higher carbonate content (3.0–9.5%) and higher LoI (2.2–3.1%) than the overlying layers. Some small stones (up to 3 cm) are present. A faint polyedric structure is present. The upper boundary is sharp and slightly undulating. Layer 4 is interpreted as a fossil A horizon capped by erosion. A radiocarbon assay of charcoal from layer 4 (depth 2.85 m) gave an age of 6,190–5,920 cal BP (Poz-16817, 5,320 ±40 BP). Layer 3 is grey-brown silty sand with faint polyedric structure and a diffuse upper boundary and some stones. It is interpreted as the C horizon (hyperconcentrated flow deposit) of the fossil soil due to the lower LoI values (1.6–1.9%) and similar grain size parameters. Layer 2 consists of poorly sorted sand with some gravel and an upwards-fining tendency. Mica content is high; the upper boundary is diffuse. Layer 1 is coarse gravel and boulders with a diffuse upper boundary. The material is partially strongly weathered and slightly cemented. It is interpreted as the pre-Holocene basal material of the river valley comparable to the material of the bajadas.

Profile RP-06 is a sequence from natural to human-dominated sedimentation. There is an increase in carbonate content and LoI from a depth of 2.00 m to the surface. The increase in carbonate content is explainable by evaporative processes. While pore water was available underground from a nearby irrigation channel, there was no irrigation (and hence no downward percolation flushing out soluble salts) on the site during at least the past 62 years. Carbonate and LoI also peak in layer 4 which is interpreted as a paleosol. The lowest LoI values (0.6–0.8%) of profile RP-06 was measured in a hyperconcentrated flow deposits and the fluvial layers. At the base of the profile, gravels of probably pre-Holocene age are exposed. A shallow soil developed during the mid-Holocene in a hyperconcentrated flow deposit. Fluvial

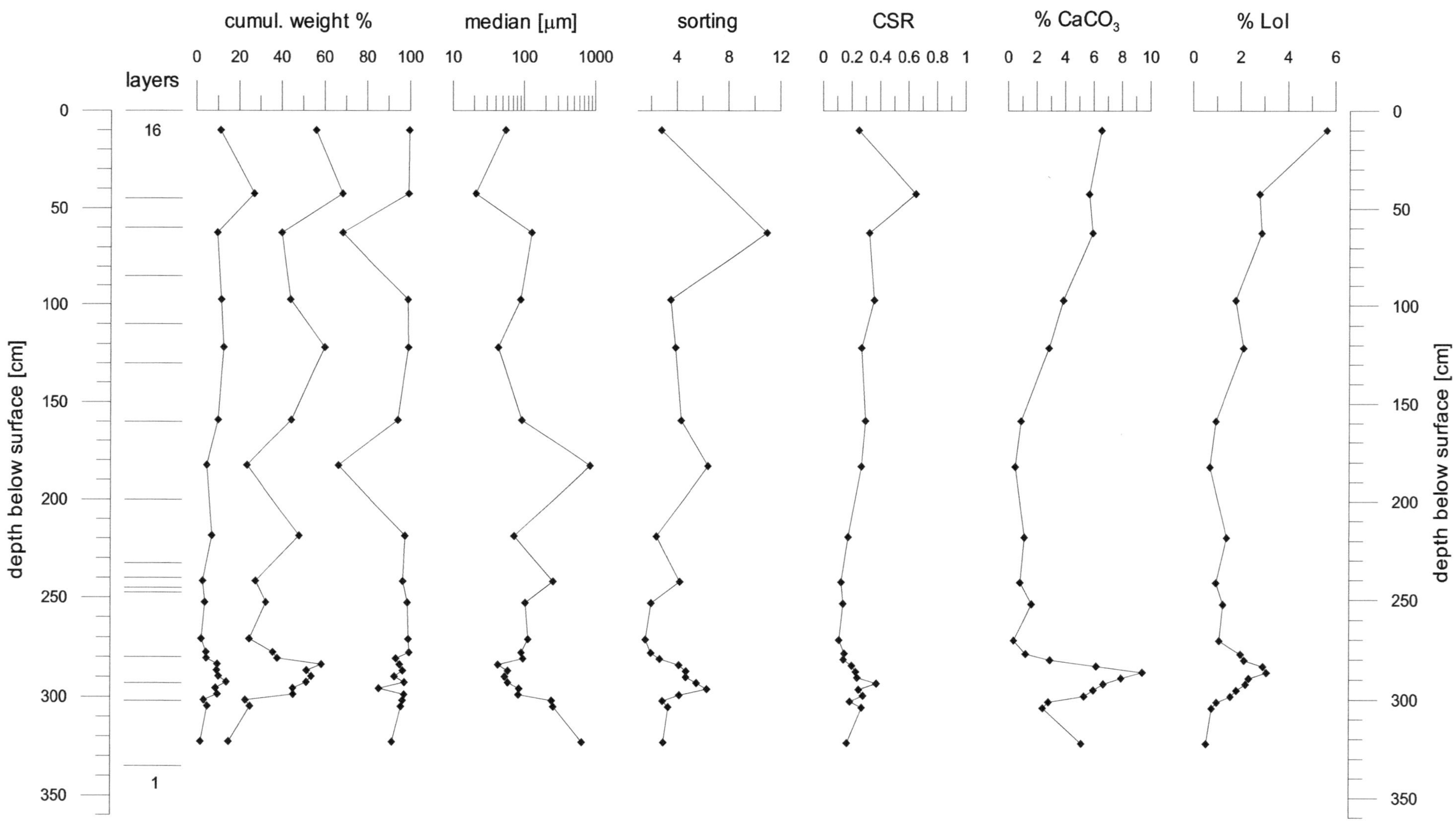

Fig. A-81: Laboratory results of samples from profile RP-06 (vertical scale 1: 30). Cumulative weight percent graph shows clay, silt and sand from left to right. Note that the lines connecting data points do not imply values etween data points but serve solely to aid visualisation.

overbank deposits (layers 5, 6, 8 and 9) and hyperconcentrated flow deposits (layers 7 and 10) characterise the period from 6,190–5,920 cal BP to 3,080–2,840 cal BP, followed by deposition of irrigation sediment (irragric anthrosol) during the late Initial Period (3,080–2,840 cal BP) and human use of the site during late Paracas (Ocucaje 9, 2,300–2,150 cal BP, cf. UNKEL 2006: 111) times.

Possible and actual anthropogenic alteration of the stratigraphy (e.g. by fire, deposition of material and erection of buildings) inhibit interpretation of the as of yet undated upper 0.85 m of the profile. Average sedimentation rates are 0.43–0.51 mm/a for the mid- to late Holocene sequence of fluvial and hyperconcentrated flow deposition and paleosol development (6,190–5,920 to 3,080–2,840 cal BP), 0.32–0.56 mm/a for the incipient irrigation agriculture (3,080–2,840 to 2,300–2,150 cal BP) and 0.47–0.50 mm/a for the anthropogenically dominated upper part of the profile (after 2,300–2,150 cal BP).

Profile SF-1

Profile SF-1 (Figures A-82 and A-83) was a pit dug on a fallow covered by bushes and small trees between Rio Grande and Rio Vizcas, ca. 380 m from the left bank of the Rio Palpa, 610 m from the right bank of the Rio Vizcas and 980 m SW from the Panamericana. It is situated between profiles RP-06 (449 m) and R-1 (297 m). Two irrigation channels run close to the site. The sketch shows the north-western side of the pit.

Layer 18 is sand with clods of silt. Bedding is not present. The disperse broken-up silt clods indicate disturbance through tillage. This layer is interpreted as the disturbed upper part of a flood deposit, likely from the 1999 event. Layer 17 has a sharp, horizontal but slightly undulating upper boundary and consists of well-bedded sand. At the base of the layer there is a 5 mm strong, well-defined band of silt; the bottom 3 cm of layer 17 show an upwards-coarsening grain size. This layer is interpreted as a flood deposit, likely from the 1999 flood which according to locals inundated large parts of the valley between Rio Palpa and Rio Vizcas. Layer 16 has an undulating, sharp upper boundary and consists of sandy silt. Charcoal is present throughout the layer. At its base (mainly on the north-western to northern side of the pit), there is a distinct layer of charcoal, pieces of which allowed identification as burned cane. Layer 15 has a slightly undulating, sharp upper boundary and consists of structureless silt. The layer is free of stones. Layer 14 has a horizontal, sharp upper boundary and consists of sandy silt. It has a loose, crumbly structure. Layer 13 has a slightly undulating, sharp upper boundary. On the northern side of the pit it consists of well-bedded brown fine sand and white silt; on the southern side it is disturbed and mixed with organic material or disperse charcoal. The layer rises slightly to the east. At a depth of 1.25 m there is a single stone (15 cm, rounded but facetted). Layer 13 is interpreted as

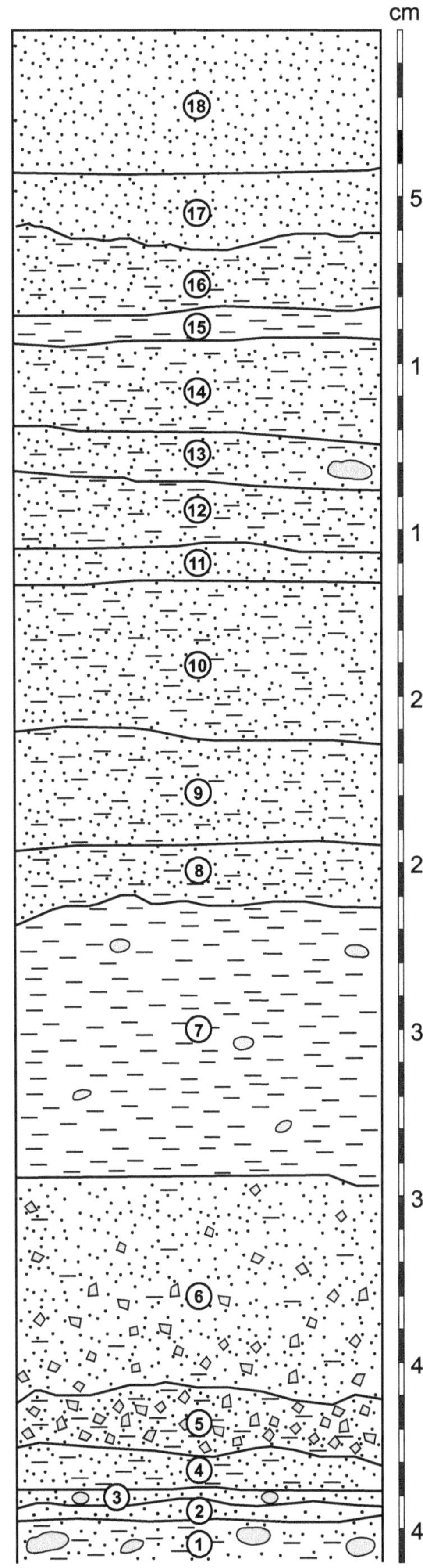

Fig. A-82: Sketch of profile SF-1 (drawing: R. Mendler, vertical scale 1: 20).

a deposit from alternating slowly flowing and standing water, possibly a small pool in an irrigation context. Layer 12 has a sharp, horizontal upper boundary and consists of sandy silt. At its base there is a black (organic material or disperse charcoal) band up to 4 cm strong which is continuous around the pit but partially disturbed. Layer 11 has a sharp horizontal upper boundary and consists of structureless silty sand.

Layer 10 has a diffuse horizontal upper boundary and consists of structureless sandy silt with a few stones (up to 2 cm). A radiocarbon assay of as charcoal sample from a depth of 2.00 m (context of a not classifiable ceramic shard) gave an age of 440–140 cal BP (Poz-16815, 255 ±35 BP). Layer 9 has an undulating, diffuse upper boundary (marked by faint sand lenses) and consists of structureless sandy silt with a few stones and some clay clods. Disperse charcoal is present. Layer 8 has an undulating diffuse upper boundary (defined by a change in colour) and consists of structureless fine sandy silt without stones but with disperse charcoal. Layer 7 has an undulating diffuse upper boundary (defined by a change in colour) and consists of structureless silt with a few stones. In the top of the layer there is a single large stone (15 cm). Carbonate precipitation is present along root molds and cracks; disperse charcoal is present. A radiocarbon assay on a charcoal sample from the base of the layer (depth 3.50 m) gave an age of 6,280–5,990 cal BP (Poz-16821, 5,410 ±40 BP). Layer 6 has a horizontal and slightly diffuse upper boundary and consists of structureless silty sand with a stone content of 20–25% at the base and 10% at the top of the layer. The stones have long axes of up to 5 cm, are angular to subangular and are not imbricated. Disperse charcoal is present. The layer is interpreted as a hyperconcentrated flow deposit with high fine material content. Alternatively, it could be interpreted as the result of turbation and associated inclusion of fine material into the top of a hyperconcentrated flow deposit with low fine material content. However, the necessary depth (>0.5 m) and intensity of turbation and the character of the lower boundary (abrupt increase in stone content) make the interpretation as an independent hyperconcentrated flow deposit more probable.

Layer 5 has a slightly undulating but generally horizontal, diffuse upper boundary and consists of a matrix of silty sand with a stone content of 80% (mainly angular and a few rounded clasts, up to 6 cm). Imbrication is weak. Layer 4 has a sharp horizontal but slightly undulating upper boundary and is a layer of structureless sandy silt between 12 (north-west) and 17 cm (south-west) strong. Layer 3 has a sharp horizontal upper boundary and consists of poorly sorted sand and fine gravel without bedding. Layer 2 consists of medium sand and is only present on the north-western side. Layer 1 has a sharp, undulating upper boundary and consists of coarse, well-rounded gravel (up to 15 cm) in a very loose matrix of sandy silt with a low carbonate content. Some of the gravel clasts (in particular granite clasts) are strongly weathered. A few angular clasts are present. Layers 18 through 11 do not contain carbonate. In layers 10 through 7, carbonate is present as disperse precipitate (white dots of approximately 1 mm), in layer 6, carbonate precipitation is along cracks and root molds.

Profile SF-1 records a sequence of natural sedimentation from riverbed gravels (layer 1) followed by fluvial channel or proximal overbank deposits (sand to fine gravel, layers 2 and 3), distal or waning flow overbank deposits (sandy silt, layer 4) and hyperconcentrated flow deposits (layers 5 and 6) which is capped by 3.55 m of irrigation sediment. An age of 6,280–5,990 cal BP for incipient fine-grained sedimentation places the underlying hyperconcentrated flow deposit into the early to mid-Holocene. The low average sedimentation rate between 6,280–5,990 cal BP and 440–140 cal BP (0.24–0.27 mm/a) and the high average sedimentation rate since 440–140 cal BP (4.0–10.3 mm/a) are anomalous in comparison with other profiles. Possible explanations are reworking of older material, turbation of younger material, a pre- or early historic hiatus or mechanised anthropogenic deposition in late historic times.

Profile R-1

Profile R-1 (Figures A-84 and A-85) was a pit dug in a field between Rio Vizcas and Rio Palpa, 297 m from profile SF-1 and 164 m from SF-2.

Layer 16 (the recent A_p horizon) and consists of broken-up sand and silty sand. Locally, remnant bedding is discernible. Judging from the distribution of the 1999 flood deposit in profiles SF-1 through SF-4, Layer 16 probably contains reworked flood deposit. A piece of textile is present at the base of this layer. Layer 15 is structureless silty sand with ca. 10% silt clods and locally discernible bedding of sand. Its upper boundary is sharp and horizontal. Layer 14 is silty sand containing up to 50% clods of cohesive silt (mainly in the top of the layer). The upper boundary is horizontal and diffuse. Ploughing marks in a north-south direction are present at a depth of 0.42–0.60 m and at a east-west direction at a depth of 0.55–0.70 m. Layers 13 and 12 consist of silty sand. Inclined thin bands of sand on the eastern and western side of the pit are interpreted as ploughing marks. The upper boundaries of layers 13 and 12 are sharp and horizontal and are marked by abundant (ca. 30%) clods of cohesive silt which suggest fine-grained low-energy / still water sedimentation followed by agricultural turbation.

Fig. A-83 (following page): Laboratory results of samples from profile SF-1 (vertical scale 1: 30). Cumulative weight percent graph shows clay, silt and sand from left to right. Note that the lines connecting data points do not imply values etween data points but serve solely to aid visualisation.

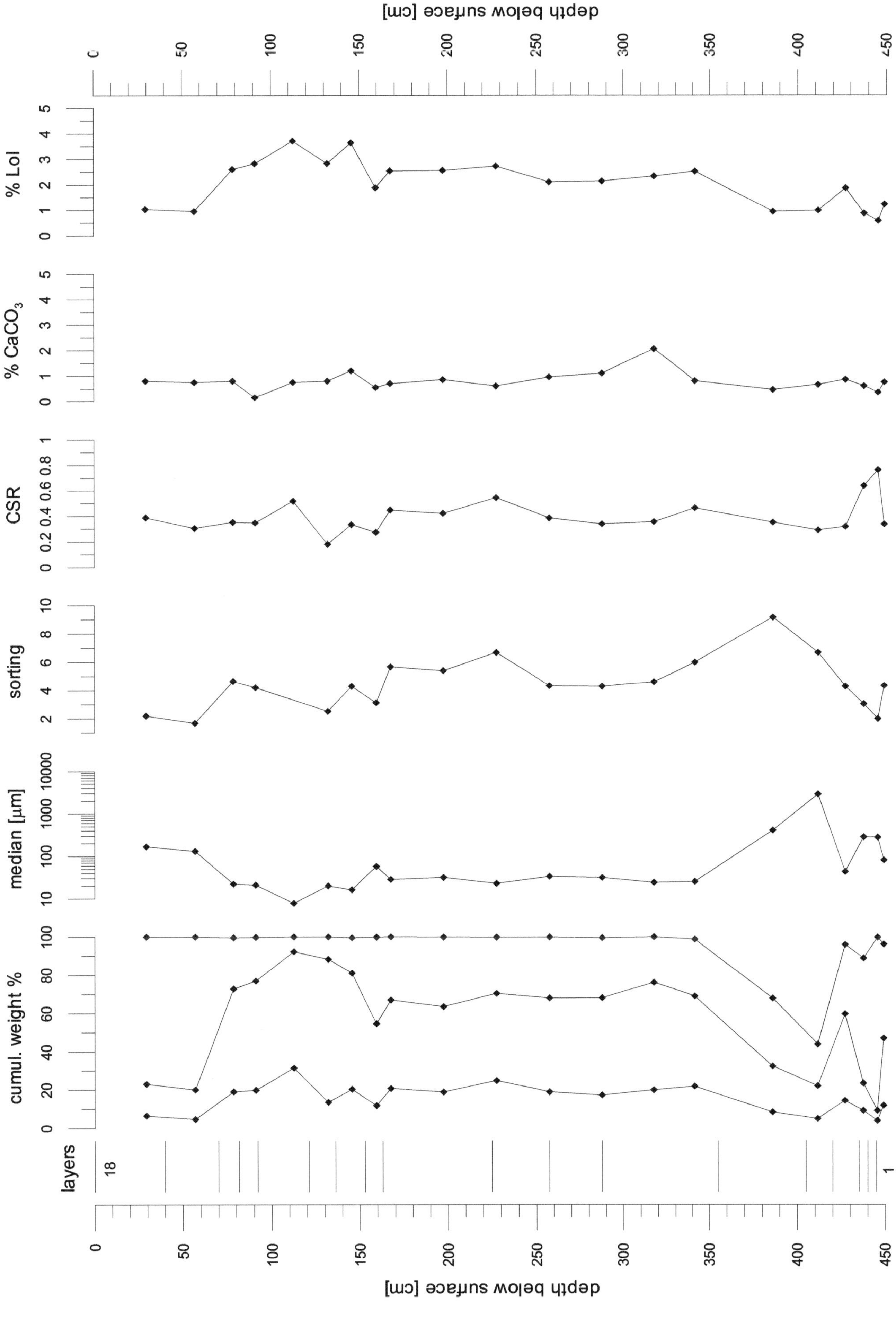
depth below surface [cm]
% LoI
% CaCO$_3$
CSR
sorting
median [µm]
cumul. weight %
layers
18
1
depth below surface [cm]

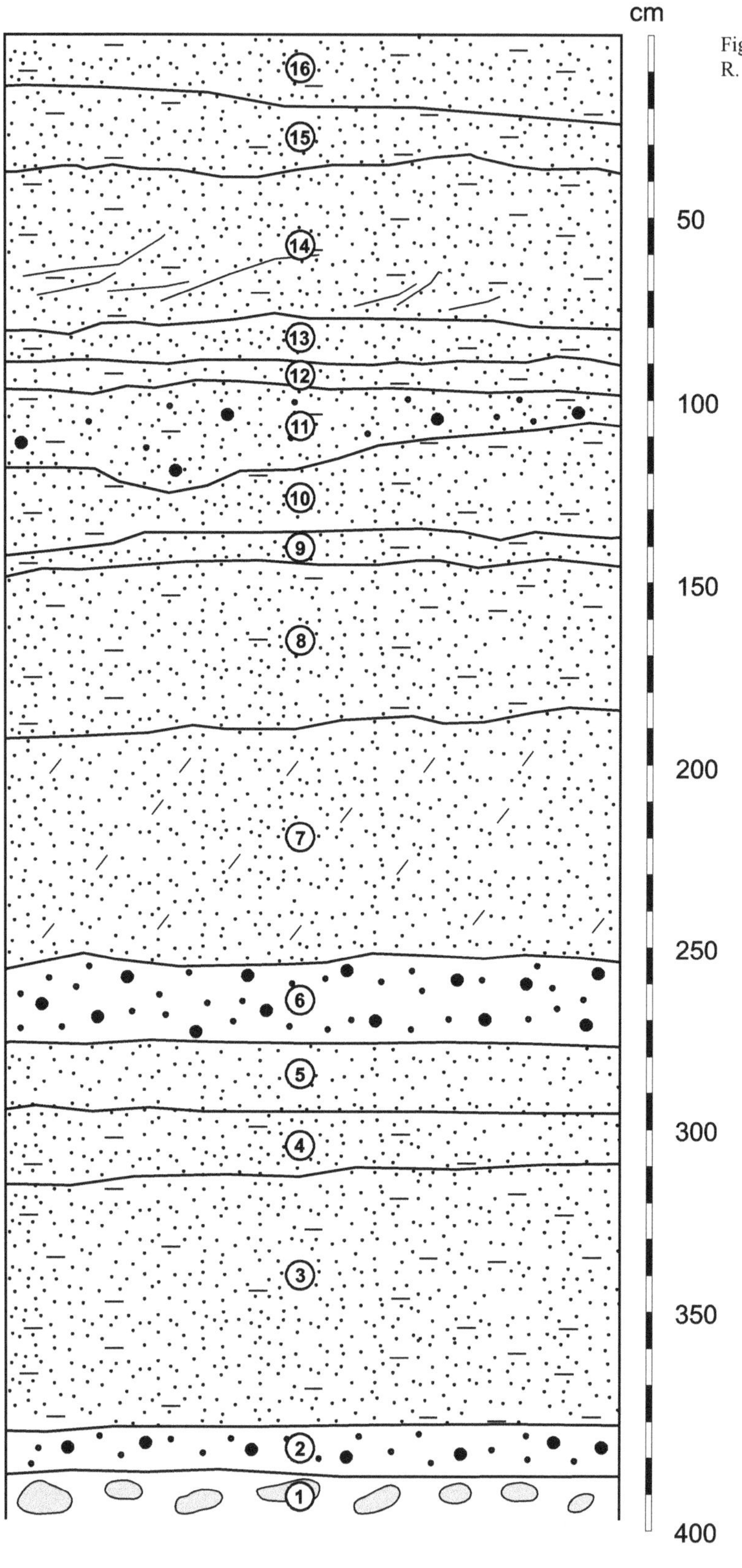

Fig. A-84: Sketch of profile R-1 (drawing: R. Mendler, vertical scale 1: 20).

Fig. A-85 (following page): Laboratory results of samples from profile R-1 (vertical scale 1: 30). Cumulative weight percent graph shows clay, silt and sand from left to right. Note that the lines connecting data points do not imply values etween data points but serve solely to aid visualisation.

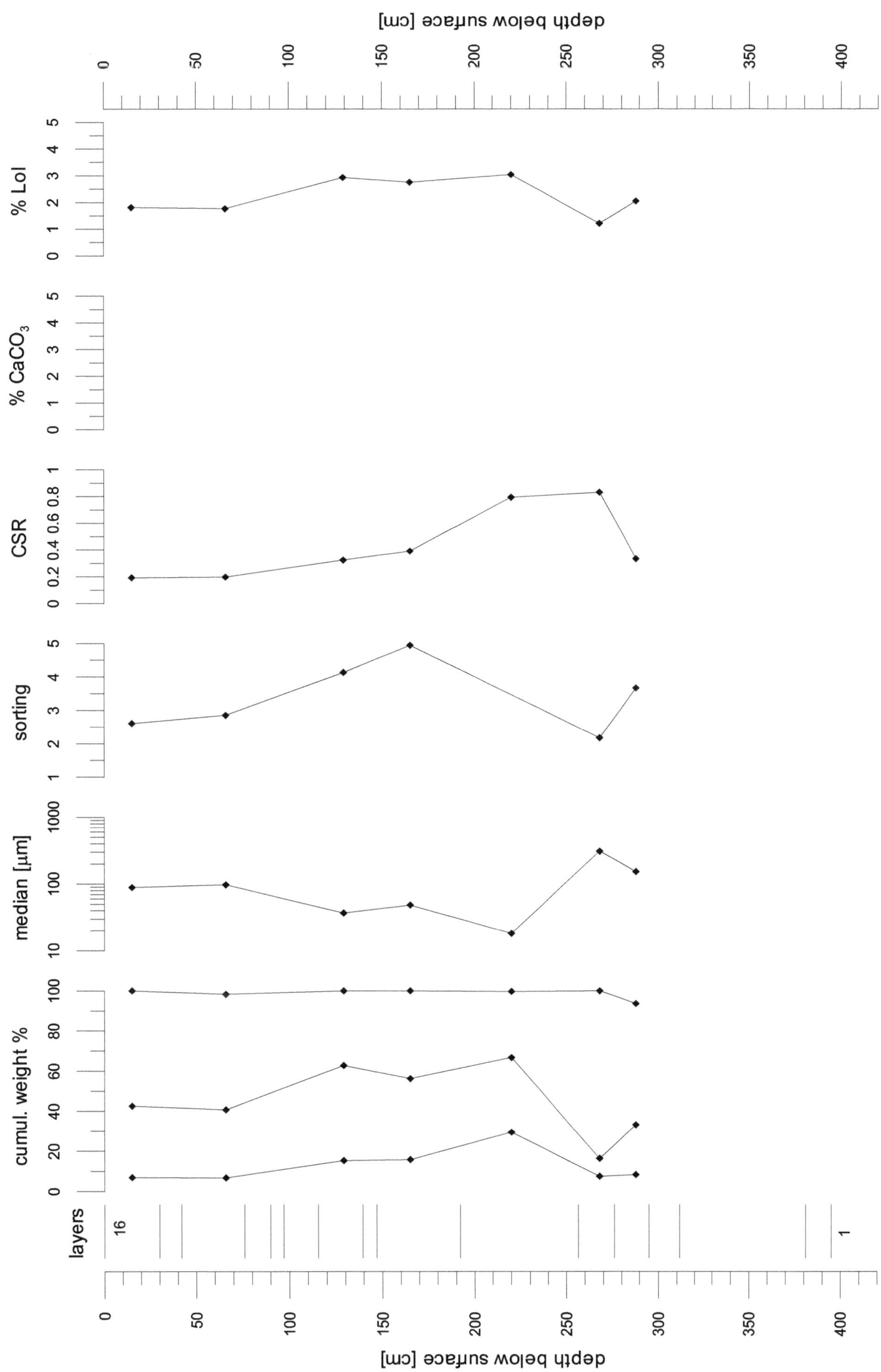
depth below surface [cm]
layers
cumul. weight %
median [µm]
sorting
CSR
% CaCO3
% LoI
depth below surface [cm]

Layer 11 consists of bedded silty fine sand to coarse sand (the beds being between 2 and 5 cm thick). The sharp, horizontal upper boundary is marked by a 3 cm thick band of silty fine sand; the top 15 cm of layer 11 have polyedric structure. Bedding is clear on the western side of the pit while on the northern side it is wedge-shaped and strongly turbated with the underlying layer 10. Layer 10 is structureless silty sand with an undulating and more or less distinct upper boundary. A single larger stone (10 cm) is embedded in the fine-grained matrix. Ceramic shards are present at a depth of 1.20 and 1.25 m.
Layer 9 is charcoal-rich silty sand with an undulating upper boundary. A charred tree root (15 cm thick) is present in the eastern side and can be followed into the underlying layer 8. A radiocarbon assay from layer 9 (depth 1.40 m) gave an age of 490–320 cal BP (Poz-15236, 385 ±30 BP). Layer 8 is structureless silty sand with a sharp, undulating upper boundary. A thin, broken-up band is faintly visible at a depth of 1.68 m. Layer 7 is loam to loamy sand with columnar to polyedric structure (edge length 5–10 cm). The upper boundary is horizontal and diffuse. Some angular stones (<1%, up to 4 cm) are present. An unclassifiable ceramic shard was found at a depth of 2.32 m. Two radiocarbon assays of charcoal from depths of 2.10 and 2.51 m gave ages of 1,530–1,350 cal BP (Poz-14006, 1,595 ±30 BP) and 1,870–1,630 cal BP (Poz-13607, 1,885 ±30 BP), respectively, thus placing the sedimentation of layer 7 into the Nasca period (1,870–1,300 cal BP, cf. Unkel 2006: 111).

Layer 6 is medium to coarse sand with a broken-up upper boundary. Clods of loamy material appear to have been turbated from the overlying layer 7 into layer 6. A rounded clast (13 cm) is present at the base of layer 6. Layer 5 is structureless sand with ca. 5% larger clasts (up to 5 cm) and an undulating, sharp upper boundary. Layer 4 is silty sand with a horizontal and somewhat broken-up upper boundary. Layer 3 is structureless silty sand with an undulating diffuse / broken-up upper boundary. A blackish sand lens is present a the base of layer 3. Layer 2 is structureless medium to coarse sand with an undulating and very diffuse upper boundary. Grain size differs laterally, and material from the overlying layer 3 appears to have been turbated into layer 2. Layer 1 is well-rounded coarse gravel (up to 15 cm) with a matrix of coarse sand. Some clasts are weathered. Layer 1 is interpreted as fluvial riverbed sediment.

Profile R-1 records a sequence of fluvial (layers 1, 2, 5 and 6) and irrigation (layers 7 through 15) sedimentation. LoI is highest for layers 7, 8 and 10 (2.8–3.0%). The highest CSR (0.79, omitting the silt-poor layer 6) also occurs in layer 7. Two radiocarbon dates (1,530–1,350 cal BP and 1,870–1,630 cal BP) place this layer into the Nasca period (1,870–1,300 cal BP, cf. Unkel 2006: 111). Incipient irrigation agriculture in early Nasca times at this site is interpreted from the turbation of clods of irrigation sediment into underlying fluvially deposited material, however, given the unclear origin of the yet lower layers 3 and 4, an earlier period of irrigation agriculture is possible. The recent ploughing horizon likely contains reworked material from the 1999 flood event. Sedimentation rates are 0.79–4.1 mm/a for the Nasca irrigation sediment, 0.58–0.81 mm/a for Late Nasca to historic time and 2.6–3.8 mm/a for historic times.

Profile SF-2

Profile SF-2 (Figures A-86 and A-87) was a pit dug in a fallow between Rio Vizcas and Rio Palpa, ca. 170 m north-west of the right bank of the Rio Vizcas. It was situated between profiles R-1 (164 m) and SF-3 (80.8 m). The sketch shows the north-western side of the pit. The entire profile is free of carbonate.

Layer 12 is laminated medium to coarse sand alternating with fine sand. Cross-bedding is present. The sand is cemented, but carbonate is not present. Layer 11 has an undulating but sharp upper boundary and consists of sandy silt. The structure is crumbly. Layer 10 has an undulating, sharp upper boundary and consists of structureless sandy silt. The upper boundary is defined by plough marks. Layer 9 has an undulating, diffuse upper boundary and consists of structureless, compacted silt. At the base of the layer there are disturbances of fine sand and silt possibly created by ploughing. Layer 8 has a horizontal, diffuse upper boundary and consists of structureless, sandy silt. There are a few stones (up to 15 cm) and lenses of medium to coarse sand. A mollusc shell fragment was found at 1.05 m; a small (5 mm) piece of rusted iron was found at a depth of 0.80 m.

Layer 7 has a sharp to diffuse, horizontal upper boundary. In places the upper boundary is marked by a 1–2 cm strong layer of coarse sand. The layer consists of structureless sand with diffuse gravel (1–2%). At 1.34 m there was a hearth with a diameter of approximately 40 cm and a depth of a few cm. The hearth contained abundant charcoal as well as burned ceramic shards. At the same depth and in direct context of the hearth there were also a few more painted shards which were classified by J. Palomino as belonging to the Late Intermediate Period (ca. 950–550 cal. BP, cf. Unkel 2006: 111). A charcoal sample from the hearth gave a radiocarbon age of 1,080–930 cal BP (Poz-16820, 1,165 ±30 BP) whose lower limit overlaps with the older age limit of the ceramic style epoch. Layer 6 has a horizontal and slightly undulating upper boundary and consists of sand with oxidation stains. The material is upwards-fining from coarse sand with fine gravel to medium sand. There is faint bedding and a single larger stone (10 cm).

The upper boundary contains root molds. The excavated material from a depth between 0.75 and 1.75 m (layers 8 through 6) contained ceramic shards pertaining to the Middle Horizon (Loro and Late Loro / Wari; 1,300–950

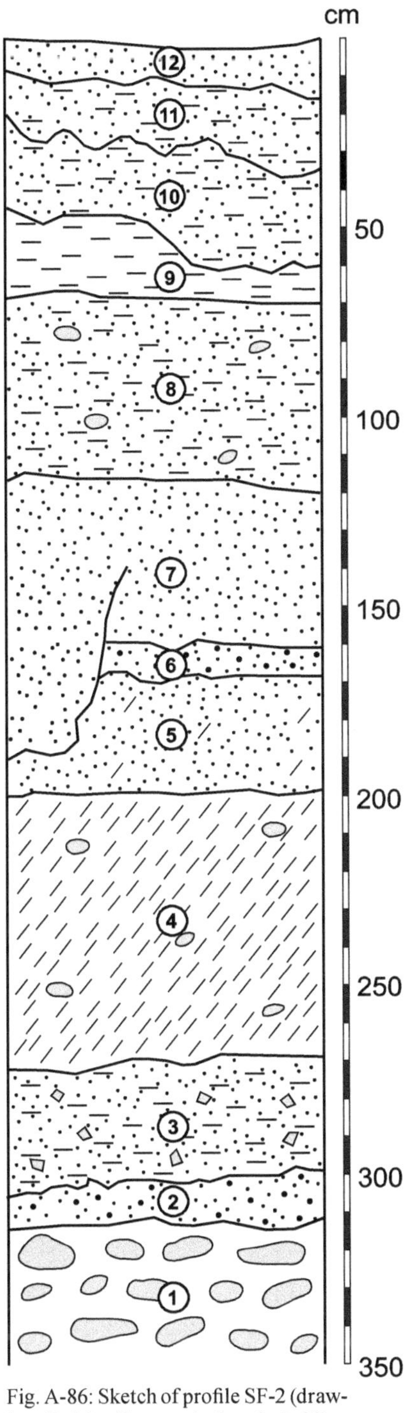

Fig. A-86: Sketch of profile SF-2 (drawing: R. Mendler, vertical scale 1: 20).

cal BP, cf. Unkel 2006: 111; ceramic assignment by J. Palomino). Layer 5 has a horizontal, slightly undulating and sharp upper boundary. It consists of structureless loamy sand. Bedding is not present. On the north-eastern side of the profile there is an up to 5 cm strong lens of coarse sand at the base of the layer. On the south-western side of the profile, there is a structure (7a) of poorly sorted medium sand containing disperse charcoal. The upper boundary was very poorly defined, however, its sides were diffuse but distinct. This structure penetrates from the middle of layer through layer 6 and well into layer 5. It is interpreted as a pit dug and filled-in during the time of deposition of layer 7. A ceramic fragment (handle) was found at 1.70 m and possibly pertains to Paracas period according to J. Palomino, however, since ceramic style assignment was not secure and the find is from a filled-in pit, it is omitted from chronological interpretation.

Layer 4 has a horizontal upper boundary which is in places sharp, in other places diffuse. It consists of loam with a polyedric structure and contains a few stones (up to 5 cm, sub-angular). Layer 3 has a diffuse horizontal upper boundary and consists of structureless, compact, poorly-sorted silty sand with 20% angular to subangular stones (up to 4 cm). The layer is interpreted as a hyperconcentrated flow deposit. Layer 2 has a sharp and slightly undulating upper boundary and consists of structureless, grey, fine to medium sand. Bedding is not present, but there is a slight upwards-fining tendency. A sediment sample from layer 2 (depth 3.20 m) containing disperse charcoal was radiocarbon-dated to 3,220–2,950 cal BP (Poz-16818, 2,980 ±35 BP). Layer 1 has a diffuse undulating upper boundary and consists of poorly sorted gravel (90%, long axes up to 50 cm) with a sandy matrix. Some of the clasts are strongly weathered. The gravel is slightly cemented but free of carbonate.

Profile SF-2 represents a succession from fluvial gravel (riverbed) dynamics (layer 1) followed by the deposition of a fine-grained fluvial layer (layer 2), a hyperconcentrated flow deposit (layer 3) and long-term accumulation of irrigation sediments (layers 4, 5 and 7 through 11). While all layers except the fluvially deposited layers 1, 2 and 12 are characterised by CSR and LoI values and poor sorting, layer 4 (which pertains to Early/Middle Paracas through early Nasca using linear age interpolation, cf. Unkel 2006: 111) has the highest CSR (0.64) and almost the highest LoI (3.0). With the exception of layers 6 and 12, flood events did either not occur or are masked by agricultural activity. Layer 12 probably is the deposit of the 1999 flood event. The sand lenses in layers 5 and 8 could be the result of flood events, however, their very limited size lets an interpretation as results of irrigation channel flow appear more likely. Beside abundant plainware and fineware ceramic shards pertaining to the Middle Horizon and LIP, clear anthropogenic structures are present in this profile: a filled-in pit (7a) and a hearth with associated ceramic shards dating to the early Late Intermediate Period.

With the exception of one uncertain piece of ceramic from a filled-in pit, ceramic and radiocarbon dates are stratigraphically consistent. Sedimentation rates are 0.81–1.00 mm/a for the lower part of the profile (3,220–2,950 cal BP to 1,080–930 cal BP, layers 2 through 7) and 1.01–1.16 mm/a for the upper part of the profile (1,080–930 cal BP to AD 1999, layers 7 through 11).

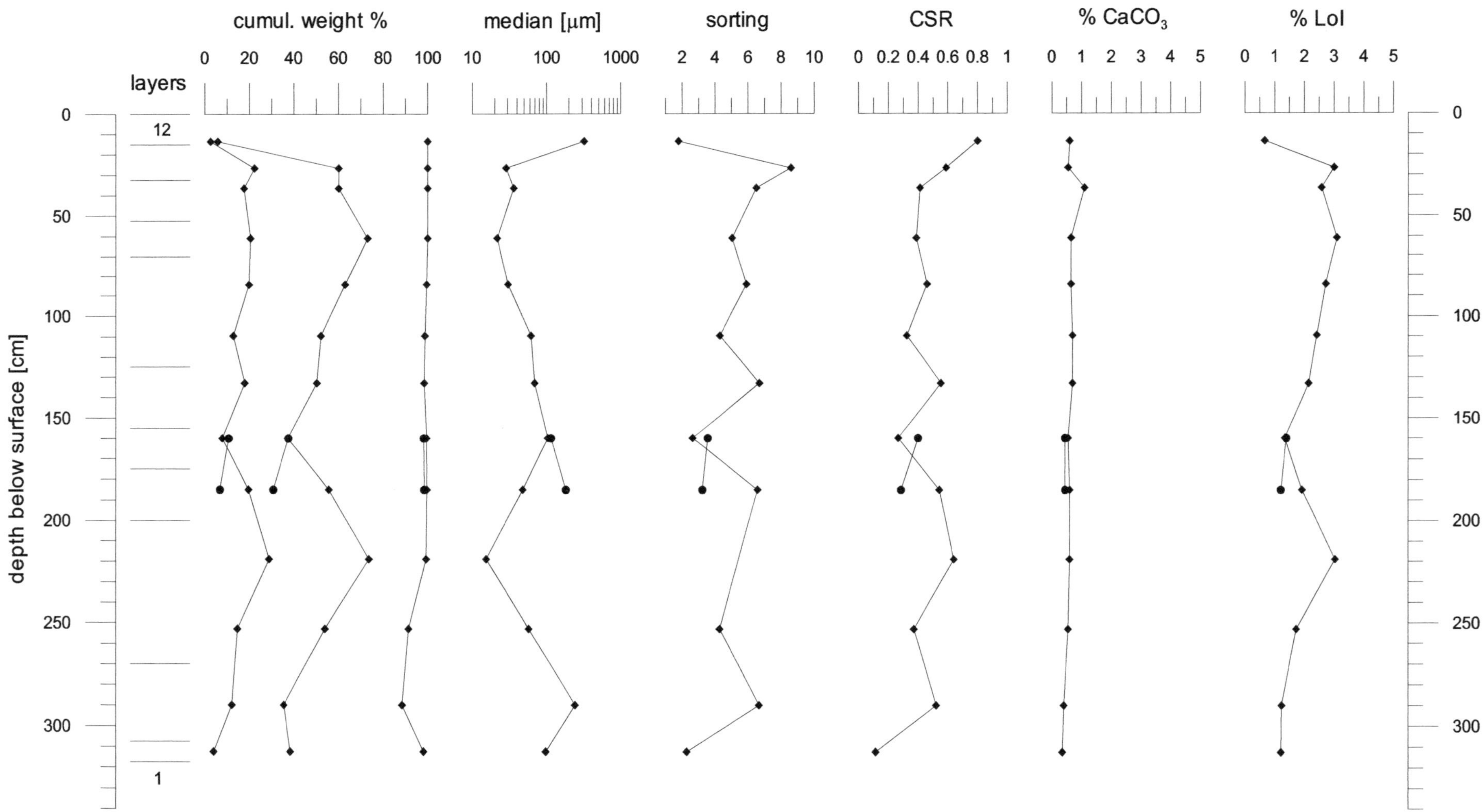

Fig. A-87: Laboratory results of samples from profile SF-2 (vertical scale 1: 30). Cumulative weight percent graph shows clay, silt and sand from left to right. Note that the lines connecting data points do not imply values etween data points but serve solely to aid visualisation.

Profile SF-3

Profile SF-3 (Figures A-88 and A-89) was a pit dug in a bush-covered fallow which had not been agriculturally used since probably 1999. It was at a distance of ca. 95 m from the current bed of the Rio Vizcas and between profiles SF-2 (80.8 m) SF-4 (37.5 m).

Layer 12 (the surface layer) is a 20–23 cm thick and somewhat cemented well-laminated and well-sorted medium to coarse sand. LoI is low (0.8%). It was probably deposited by the flood event of March 1999 when overbank flow from the Rio Vizcas inundated a large part of the central Palpa valley. Layer 11 is a silty sand with a very loose crumbly structure and some stones. The sharp and undulating upper surface of layer 11 is probably the result of erosional activity before the deposition of layer 12. Layer 10 is structureless sandy silt with numerous cohesive silt clods and a sharp, undulating upper boundary. Charcoal is present. Layer 9 is sandy silt with an undulating upper boundary which is pronounced by a sand lens. Numerous silt clods as well as some charcoal are present.

Layer 8 is structureless silty sand with a few stones (up to 12 cm) and a sharp horizontal upper boundary. A Late Intermediate Period (950–550 cal BP, see Unkel 2006: 111; ceramic identified by J. Palomino) ceramic shard was found at a depth of 0.85 m in layer 8. Layers 10 through 8 are characterised by LoI values between 1.9 and 3.3% and poor sorting and are interpreted as irrigation sediment. Layer 7 consists of structureless sandy silt / silty sand. Its upper boundary is marked by an uneven surface of well rounded stones (ca. 15 cm) which is present throughout the pit. The stone surface overlies the charred roots of a tree. At some points, vertical rows of similar-sized stones set into cone-shaped fine-grained material extend downward from this stone surface into layer 7. Within layer 7, some smaller stones as well as marine mollusc shell fragments are present. At a depth of 1.50 m, a paleosurface / usage horizon can be distinguished on the north-western side of the pit; on both the north-east and south-east sides, traces of intense heating are present as a reddened paleosurface. Layer 7 and in particular the stone surface and foundation-like structures are clearly of anthropogenic origin.

Layer 6 is structureless silty sand with a diffuse upper boundary, a few stones and charcoal. There are two pockets of this material extending down into layer 5. Layer 5 is well-bedded sand with a sharp, undulating upper boundary. A ca. 10 cm thick band of sand and fine gravel is present in this layer. The sand also fills a depression in the surface of layer 4. Layer 4 is cemented coarse sand with some fine and medium sand. Bedding is faint; the upper boundary is sharp and slightly sloping. Layer 3 is coarse gravel (ca. 60%) with a matrix of fine sand and a sharp horizontal upper boundary. Layer 2 is coarse imbricated gravel (ca. 80%) with a matrix of coarse sand and fine

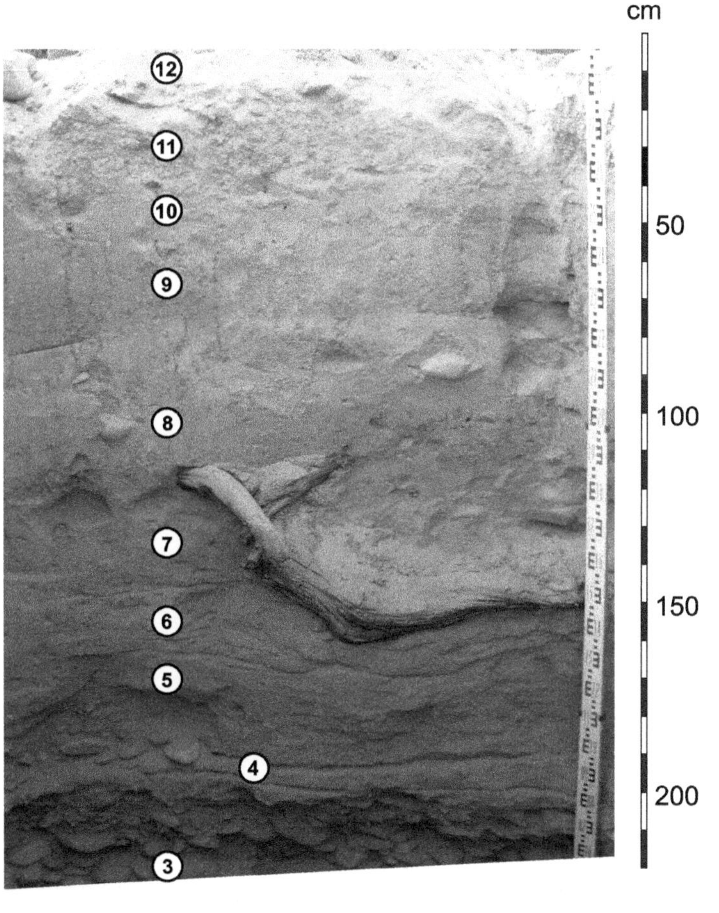

Fig. A-88: Photograph of profile SF-3 (vertical scale 1: 20).

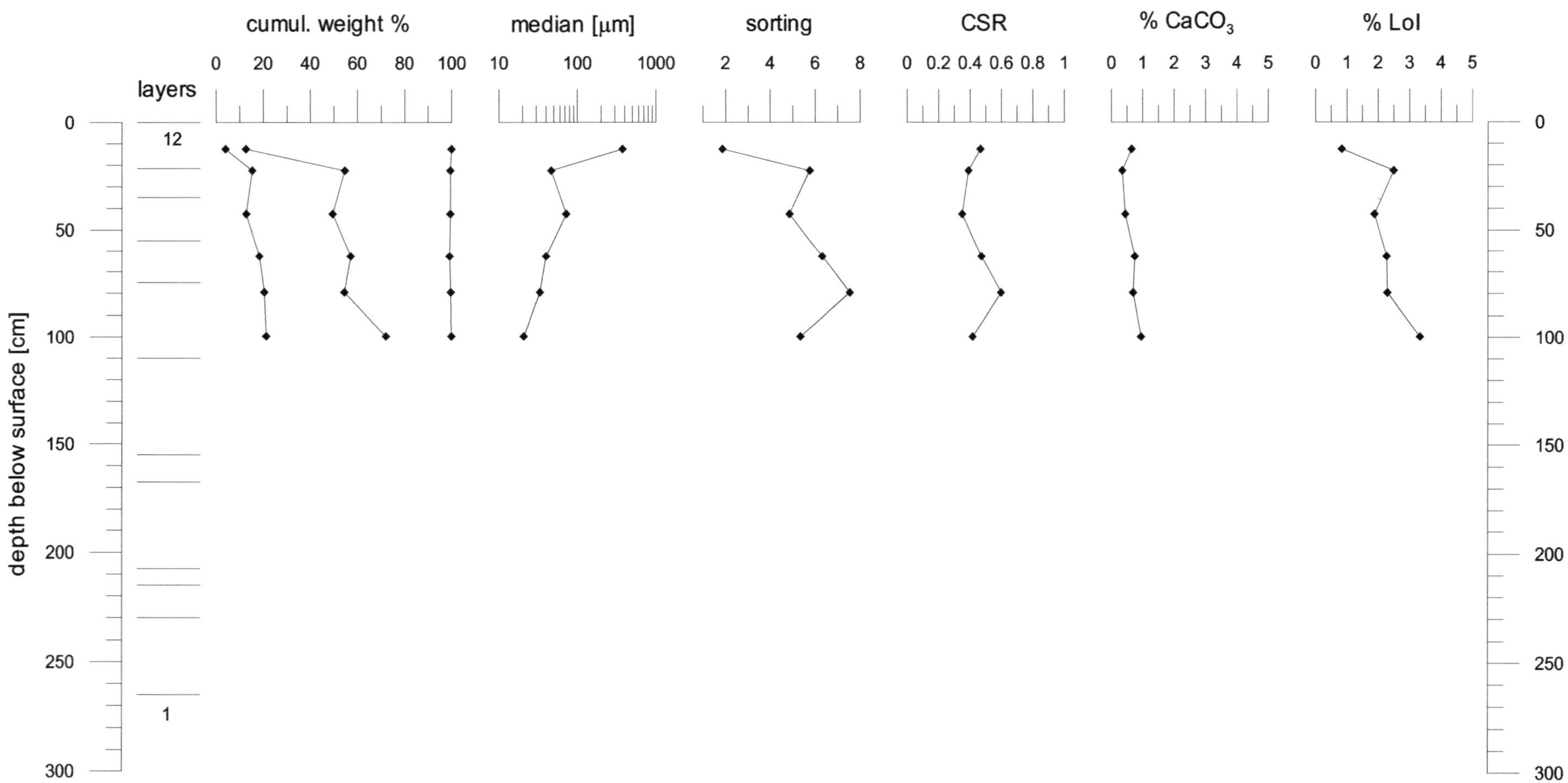

Fig. A-89: Laboratory results of samples from profile SF-3 (vertical scale 1: 30). Cumulative weight percent graph shows clay, silt and sand from left to right. Note that the lines connecting data points do not imply values etween data points but serve solely to aid visualisation.

gravel. The upper boundary is sharp and sloping. Marine mollusc shells are present. At depth between 2.10 and 2.50 m (layers 4 through 2), further Late Intermediate Period ceramic shards were found (950–550 cal BP, see Unkel 2006: 111; ceramic identified by J. Palomino). Layer 1 is coarse gravel to boulders (ca. 50%, up to 30 cm) in a upwards-fining matrix of coarse sand to fine gravel. Its upper boundary is sharp and slightly undulating. At the base of the profile (depth 3.00 m), a ca. 0.5 m^2 patch of fine-grained (silty sand) material is present within the gravel layer. The excavated material from the lower part of the pit also contained a Loro and possibly a Wari ceramic shard (Middle Horizon, 1,300–950 cal BP, cf. Unkel 2006: 111; ceramic identified by J. Palomino), however, layer assignment was not possible for these finds.

Profile SF-3 records a history of human impact at least since the Late Intermediate Period (950–550 cal BP, see Unkel 2006: 111) and possibly since the Middle Horizon (1,300–950 cal BP, cf. Unkel 2006: 111) as evidenced by the presence of ceramic shards, foundation-like structures (layer 7), marine mollusc shells, charcoal and strongly turbated sediment. Among other properties, poor sorting and high LoI values together with anthropogenic traces suggest agricultural use of the site. Fluvial impact is evidenced by river-bed gravel at the base and overbank flood sediment at the top of the profile.

Profile SF-4

Profile SF-4 (Figures A-90 and A-91) was dug into the right bank of a channel eroded by the Rio Vizcas ca. 1,000 m downstream of the Panamericana and 37.5 m from profile SF-3. Due to artificial embankments, the river now flows at a distance of ca. 60 m. The profile was cut 50 cm into the bank and downward into the fluvial gravels underlying the fine-grained bank material.

The surface layer 9 is laminated sand (alternating between fine and medium sand), the laminations are dipping slightly towards the Rio Vizcas. It is interpreted as overbank deposit related to the 1999 flood event. Layer 8 is structureless silty sand containing cohesive clods of silt (up to 2 cm). Its upper boundary is sharp and slightly undulating. At its base, it is mixed with material from the underlying layer 7. The observed turbation is interpreted as being related to agricultural activity. Layer 7 is laminated sand with a sharp horizontal upper boundary. There is a decrease in layer thickness away from the Rio Vizcas. At its base, the layer is characterised by a few stones. The layer is interpreted as an overbank gravel splay followed by fine-grained overbank sedimentation. Layer 6 is structureless sandy silt with some stones (ca. 2%, up to 15 cm) and a sharp and slightly undulating upper boundary. Layer 5 is a sequence of 5–7 cm silt with charcoal (top), 6-9 cm of silt (middle) and 5–8 cm of sandy silt (base). Its upper boundary is horizontal and slightly undulating.

Layer 4 is silty sand with 5% stones (up to 8 cm) and a slightly undulating upper boundary which is marked by a band of coarse sand to fine gravel. Sand content increases from top to base. Bones are present in this layer. Layer 3 consists of well-rounded and imbricated gravel to boulders (up to 30 cm) with a lens of sand and fine gravel. It has a diffuse horizontal upper boundary. Abundant ceramic shards (including a pot broken in situ) and marine mollusc shells are present. Layer 2 is silty sand with ca. 10% stones and a sharp, slightly undulating upper boundary. There are abundant ceramic shards. Layer 2 is interpreted as a Tritthorizont. Ca. 2 kg of plainware and several pieces of fineware ceramics were recovered from layers 3 and 2. Most of the fineware from layers 3 and 2 was classified by J. Palomino as pertaining to Loro (1,300–950 cal BP, cf. Unkel 2006: 111) and one piece from layer 3 as pertaining to Late Nasca (1,520–1,300 cal BP, cf. Unkel 2006: 111). A charcoal sample from a charcoal-rich section of layer 2 (depth 2.32 m) gave an age of 1,420–1,290 cal BP (Poz-16822, 1,510 ±40) which corresponds well with the found ceramic shards. Layer 1 is coarse gravel to boulders (up to 30 cm) with a diffuse upper boundary. The matrix (ca. 5%) consists of coarse sand to fine gravel; ca. 30% of the clasts are larger than 10 cm. Elongated clasts are imbricated, granites are not weathered. Layer 1 is interpreted as a high-energy (river-bed) fluvial sediment.

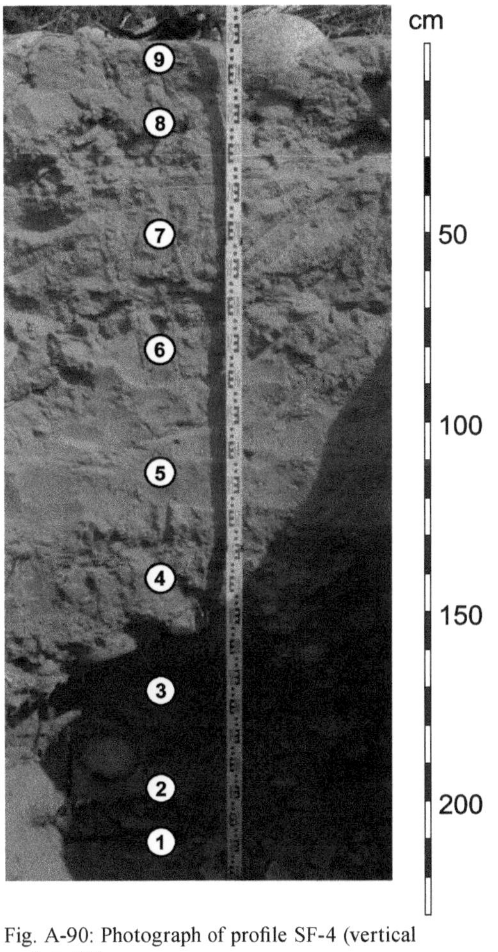

Fig. A-90: Photograph of profile SF-4 (vertical scale 1: 20).

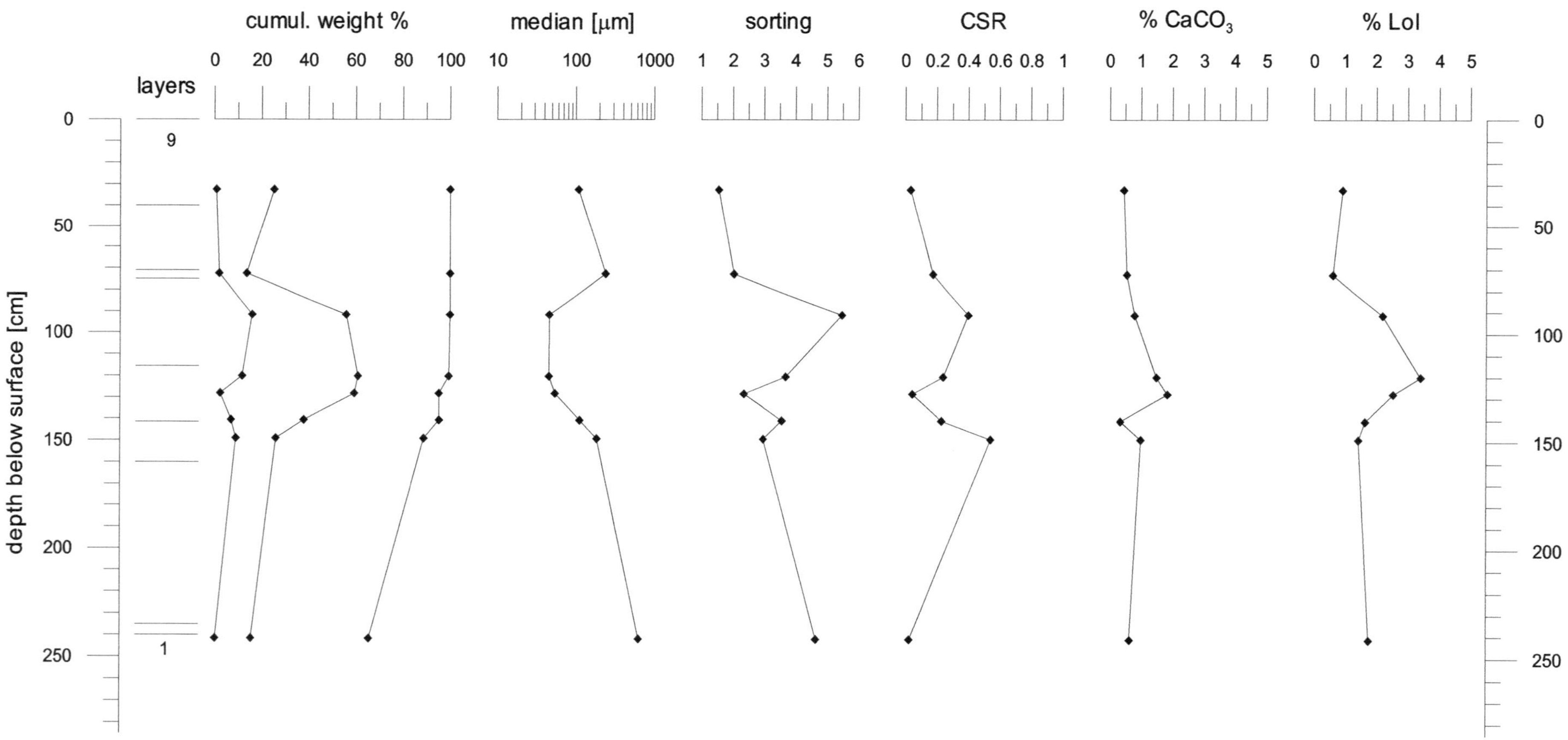

Fig. A-91: Laboratory results of samples from profile SF-4 (vertical scale 1: 30). Cumulative weight percent graph shows clay, silt and sand from left to right. Note that the lines connecting data points do not imply values etween data points but serve solely to aid visualisation.

Profile SF-4 records a sequence of fluvial activity and human impact. Very coarse river bed sediments were deposited at the site probably before 1,520–1,300 cal BP. The abundance of ceramic shards suggests that during Loro (1,300–950 cal BP, cf. UNKEL 2006: 111) and perhaps already during Late Nasca (1,520–1,300 cal BP, cf. UNKEL 2006: 111) times, there was a settlement near this site. The mixture of coarse gravel and well-preserved ceramic shards makes it likely that the site was situated at the edge of the river at that time. A more distal position of the river later allowed the deposition of poorly sorted, fine-grained sediment likely related to irrigation agriculture (layers 5 and 6) as indicated by strong turbation, increased LoI values and the presence of charcoal. The following layers (7 and 8) are interpreted as results of repeated fluvial overbank deposits (fine-grained as well as gravel) which were used for agriculture in the intervening periods. The top of the profile is marked by the 1999 flood sediment (well-sorted, laminated sand with low LoI).

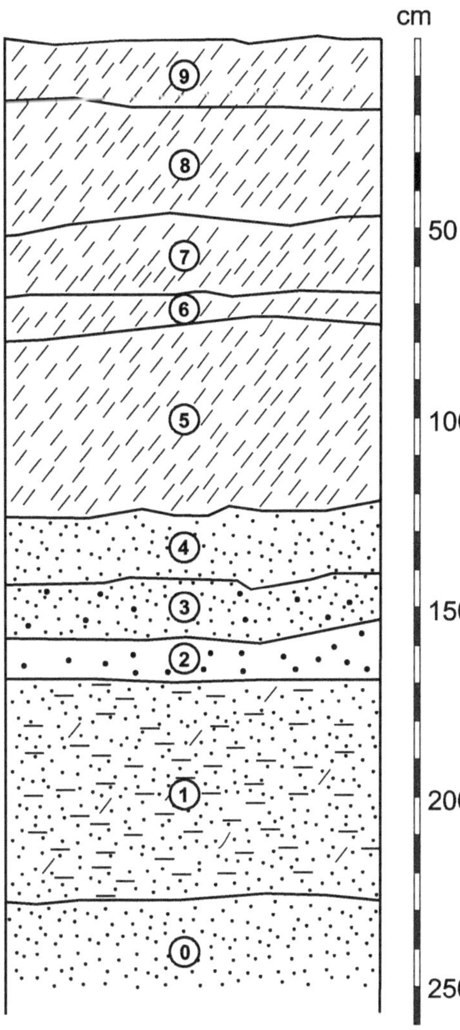

Fig. A-92: Sketch of profile SR (drawing: R. Mendler, vertical scale 1: 20).

Profile SR

Profile SR (Figures A-92 and A-93) was a pit on a asparagus field shortly (within a few days) after irrigation. The sketch shows the north-eastern side of the pit. The bottom 0.50 m were a narrow, hand-dug shaft; therefore, properties of layer boundaries could not be described below 2.00 m. Unfortunately, work had to be terminated before basal gravels were reached. Field determinations of grain sizes for this profile may err towards finer grain sizes since due to the high water content in this profile. There were no clasts larger 2 mm in this profile.

Layer 9 consists of loam. Layer 8 has a horizontal, diffuse upper boundary and is a loam with faint columnar structure. Layers 7 through 5 have horizontal, diffuse upper boundaries and consist of structureless, dark brown loam. Layer 5 contains diffuses charcoal and has a somewhat more crumbly structure than layers 7 and 6. It has a higher (2.1–3.1%) carbonate content than the rest of the profile (0.5–0.9%). Its base is strongly disturbed and contains numerous root molds. Layer 4 has a sharp and strongly undulating (pockets) but generally horizontal upper boundary and consists of bedded sand. Bedding is recognisable from colour changes between wide (approximately 5 cm) light and narrow (approximately 1 cm) dark bands. Layer 3 has an undulating and disturbed but generally horizontal upper boundary and consists of dark brown, fine to medium sand. The layer contains charcoal. Layer 2 has a generally horizontal and sharp, but disturbed upper boundary. It consists of grey, well-sorted medium sand with faint bedding. The disturbances in the top boundary may be plough traces. Oxidation stains are present. A charcoal sample from the base of the layer at a depth of 1.70 m gave a radiocarbon age of 470–310 cal BP (Poz-16824, 360 ±30 BP).

The broken-up upper boundaries of layers 4 through 2 are interpreted as indicators for tillage. Layer 1 has a sharp horizontal upper boundary and consists of dark brown sandy silt to loam. There is a high mica content. Layer 0 is a sand layer at least 20 cm strong. It's base was not reached.

Profile SR can be divided into two parts: While the top part (above 1.20 m, layers 9 through 5) is interpreted to be accumulated by irrigation, several unambiguous fluvial layers (4 through 2, 0) identified based on their grain size, sorting, CSR, LoI and bedding are present in the lower part. Layer 1 is interpreted as an irrigation sediment based on its grain size and sorting. Possible traces of tillage in layers 4 through 2 hint at agricultural use of fluvial deposits after flood events. Deposition of the upper 1.70 m occurred since 470–310 cal BP, i.e. during historic times. The high average sedimentation rate of 3.2–5.5 mm/a is interpreted to be due to both rapid aggradation by a small number of overbank flood events and intensive historic to modern irrigation with high water and sediment input.

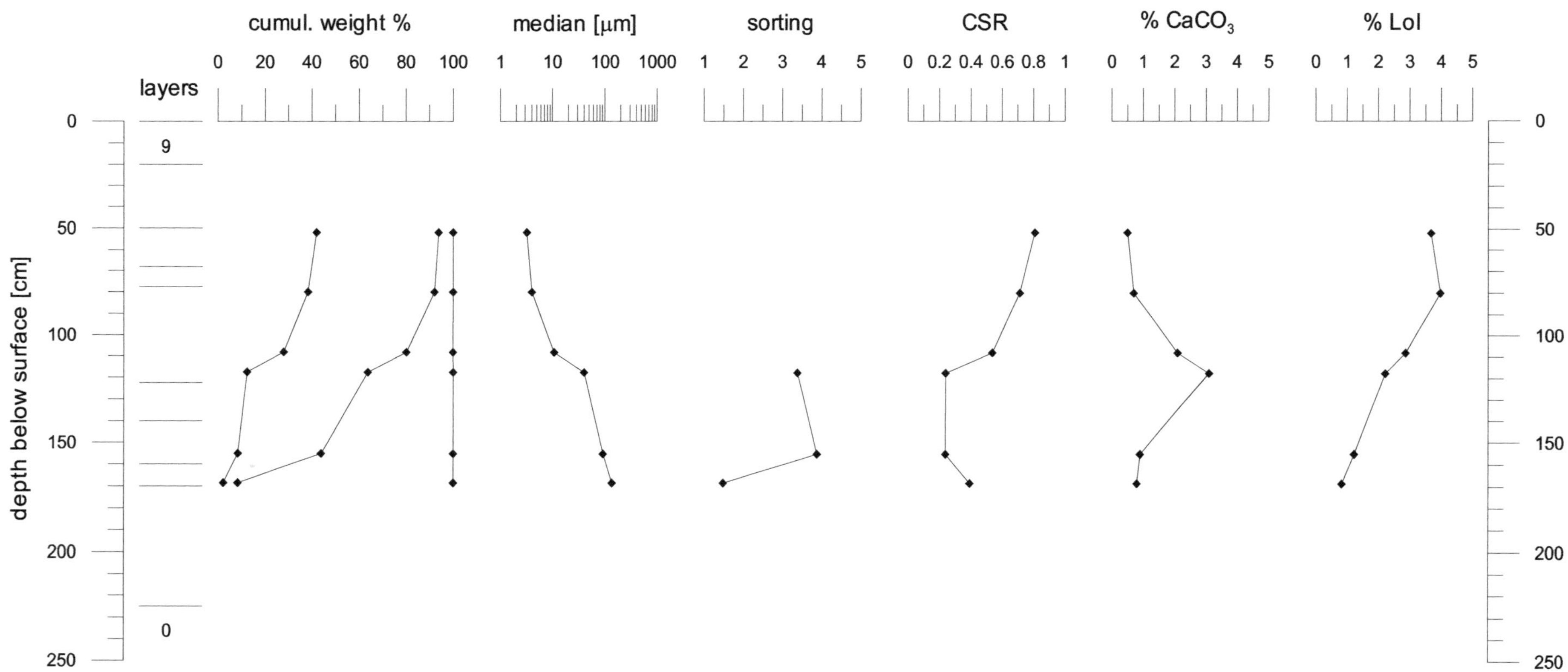

Fig. A-93: Laboratory results of samples from profile SR (vertical scale 1: 30). Cumulative weight percent graph shows clay, silt and sand from left to right. Note that the lines connecting data points do not imply values etween data points but serve solely to aid visualisation.

Profile RVU

Profile RVU (Figure A-94) was recorded at a more than 140-m-long and ca. 2 m high erosional bank exposure of the Rio Vizcas, the immediate vicinity of the bank being covered by trees and bushes. Aerial photographs from 1944 and 1970 show that this area had been used for agriculture at these times; however, fluvial structures / lineations visible in the aerial photographs suggest that this area had been eroded and re-filled at earlier times. Strong lateral erosion of the Rio Vizcas (re-)creating the exposure likely is related to the 1999 flood event. Due to modern high artificial gravel embankments, the river now flows at a distance of ca. 270 m, however, an irrigation channel flows along the base of the exposure. The following description of the exposure, lateral (along-bank) position refers to a base line running down-river from a point close to the up-river end of the exposed bank.

At 2.0 m, the stratigraphy is characterised by 1.20 of sandy silt with a few stones. Boundaries separating similar layers are discernible at depths of 0.50 cm and 1.20 m. A former near-surface situation marked by lateral roots of a small tree at a depth of 0.90 m indicates rapid infilling at this relatively low position.

Between 5.0 and 10.0 m lateral position, a depression filled with well-bedded, fine-grained sediment is present. From 10.0 m to 22.5m, an up to 40 cm thick, poorly sorted gravel layer is present at a depth between 25 to 70 cm. Maximum size of the mostly well-rounded clasts is 35 cm, elongated clasts are clearly imbricated. The upper boundary of the underlying fine-grained layer is erosional. The gravel layer is interpreted as overbank gravel splay related to a flood event.

At a lateral position of 13.0 m, a thin, charcoal-bearing layer is present at a depth of 0.65 m. At this position, at the base of the profile (below 1.20 m depth) there is a poorly-sorted layer containing abundant gravel-sized clasts in a fine-grained matrix. Granites in this layer are strongly weathered. The layers is interpreted as a hyperconcentrated flow deposit.

Between 24.0 and 28.5 m, two very large, well-rounded boulders (diameters 90 and 145 cm) enclose a depression or channel filled with bedded as well as structureless, poorly sorted material. At a depth of 0.85 and 1.45 m, bedded as well as structureless silt to sandy silt is present, the upper boundary of which is erosional. At a depth between 0.50 and 0.85 m, charcoal-bearing, very poorly sorted silt to gravel was deposited. The section between 24.0 and 28.5 m may be a filled in channel possibly related to irrigation. A basal hyperconcentrated flow deposit which can be traced from position 13.0 m is present at a depth of 1.45 m.

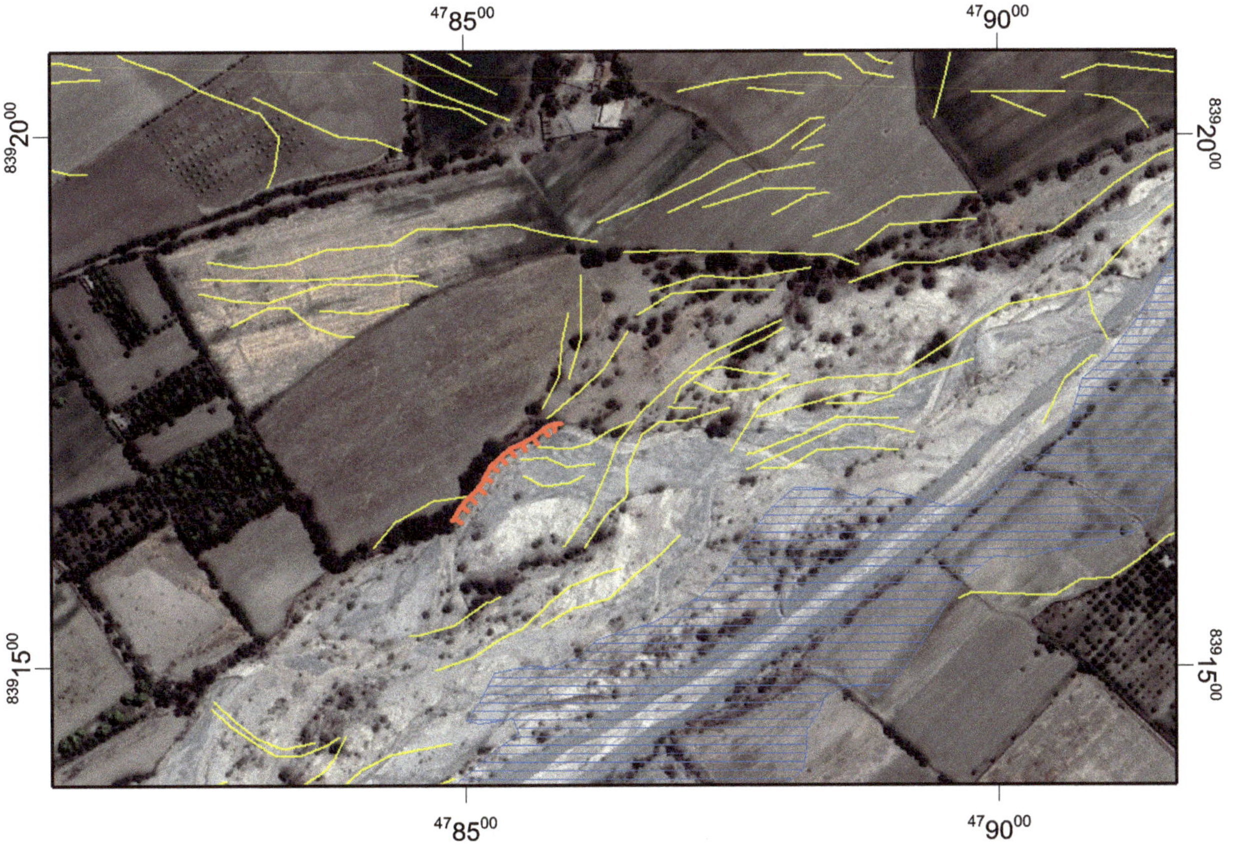

Fig. A-94: Satellite image (Quickbird, March 2005) of profile RVU (red line) with riverbed of the Rio Vizcas in 1944 (blue hatched) and fluvial lineaments (yellow lines) from the 1944 aerial photographs.

Between 28.5 and 31.0 m, a layer with ca. 50% rounded clasts (up to 30 cm) in a matrix of silty sand is present at a depth between 0.05 and 1.35 m. A layer of charcoal and ash at a depth between 1.30 and 1.50 m overlies structureless silty sand and (near the 28.5 m position) stones with a size of up to 6 cm in a matrix of silty sand.

At 32.5 m, 8 layers are present. Layer 1 (surface to 0.30 m depth) is a sandy silt with a few stones and a crumbly structure. Layer 2 (depth 0.30–0.80 m) is a structureless sandy silt with a few stones, some clods of cohesive silt and small sand lenses. The upper boundary of layer 2 is sharp and horizontal and may be due to sub-recent ploughing of layer 1. Layer 3 (depth 0.80–0.90 m) consists of bedded silt and fine sand. Bedding is partially broken-up. Layer 4 (depth 0.90–1.21 m) is structureless sandy silt with a few stones and some charcoal. The upper boundary of layer 4 is sharp and slightly undulating. Layer 5 (depth 1.21–1.29 m) is a gravel lens which extends from 32.5 to 34.5 m lateral position. To the left (down-valley), this layer continues as a gravel-rich layer of silty sand / sandy silt. Layer 6 (depth 1.29–1.35 m) consists of laminated silt. Layer 7 (depth 1.35–1.55 m) is structureless sandy silt with a few stones and charcoal.

At 37.5 m, 17 layers were recorded. Layer 1 (depth 0.00–0.05 m) is massive silt. Layer 2 (depth 0.05–0.08 m) is bedded silt and fine sand with a sharp horizontal upper boundary. Layer 3 (depth 0.08–0.20 m) is a massive sandy silt with a few stones (up to 2 cm) and a sharp horizontal upper boundary. Layer 4 (depth 0.20–0.40 m) is a sandy silt with crumbly to polyedric structure and a few stones. The undulating upper boundary is marked by a 1–3 cm thick gravel band (up to 5 cm). Layer 5 (depth 0.40–0.55/0.60 m) is structureless sandy silt with a few stones (up to 5 cm) and a sharp, horizontal upper boundary. Layer 6 (depth 0.55/0.60–0.90 m) is structureless sandy silt with a few stones (up to 1 cm); its sharp, undulating upper boundary is marked by a thin (0.5–2.0 cm) sand layer. Layer 7 (depth 0.90–0.97/1.02 m) consists of a well-bedded sequence of coarse sand, silt and fine sand. Its upper boundary is sharp and undulating. Layer 8 (depth 0.97/1.02–1.20 m) is structureless sandy silt with a few stones (up to 2 cm) and a sharp, undulating upper boundary. Layer 9 (depth 1.20–1.43/1.45 m) consists of gravel (up to 8 cm) in a matrix of silty sand. Gravel content decreases from 50% at the base to 5% at the top of the layer. The upper boundary of layer 9 is very diffuse; the lower boundary is characterised by thin (1–3 cm), interrupted beds of laminated silt. Layer 9 is interpreted as a flood-induced gravel splay into which fine-grained sediment was turbated, likely by agricultural activities. Layer 10 (depth 1.43/1.45–1.50/1.52 m) is structureless sandy silt with traces of bedding and some clods of cohesive silt. Coarse-grained material (> 2 mm) is lacking; the upper boundary is diffuse and undulating. Layer 10 is interpreted as somewhat turbated fine-grained (low-energy) overbank flood or still-water sediment resulting from a pond-like situation. Layer 11 (depth 1.50/1.52–1.65/1.68 m) is a sequence of structureless (5 cm thick) sandy silt overlain by 1–2 cm of laminated silt, 0.5 cm of charcoal-rich silt and 3 cm of laminated silt. The upper boundary is sharp and slightly undulating.

Layer 12 (depth 1.65/1.68–1.75/1.83 m) is structureless sandy silt with some coarse sand and fine gravel (5%, up to 1 cm). The sharp, undulating upper boundary is marked by a charcoal-rich band. Layer 13 (depth 1.75/1.83–1.85/1.95 m) is structureless silt to sandy silt, the sharp and undulating upper boundary of which is marked by a thin layer of ash and charcoal. Layer 14 (depth 1.85/1.95–2.25/2.30 m) is a sequence of nine fine-grained beds consisting of sandy silt, silt and fine sand. Within the respective bands, lamination is present. The upper boundary is sharp and undulating. Layer 14 is interpreted as repeated low-energy flow and still-water deposition possibly related to minor floods but (since repeated sedimentation occurred without intervening erosion or turbation) more likely caused by sedimentation in a pond-like situation fed by irrigation water. Layer 15 (depth 2.25/2.30–2.30/2.35 m) is laminated fine to medium sand with a sharp and slightly undulating upper boundary. Layer 16 (depth 2.30/2.35–2.70/2.75 m) is a sequence of silt and fine sand with each bed between few millimetres and a few centimetres thick. The upper boundary is sharp and slightly undulating. Layer 16 is clearly of low-energy fluvial origin (overbank sedimentation). Layer 17 (below 2.70/2.75 m) is coarse gravel to boulders (well-rounded, up to 100 cm) almost without fine-grained matrix (ca. 1%). The upper boundary of layer 17 is sharp and undulating.

At 53.5 m, 13 layers were recorded. Layer 1 (depth 0.00–0.20 m) consists of beds of silt and fine to coarse sand with lamination within the beds. Layer 2 (depth 0.20–0.23 m) is a thin bed of medium to coarse sand and silt with a horizontal upper boundary. Layer 3 (depth 0.23–0.33 m) is sandy silt with a crumbly structure and a slightly undulating upper boundary. Layer 4 (depth 0.33–0.45 m) is sandy silt with a crumbly structure and a horizontal upper boundary. Layer 5 (depth 0.45–0.60 m) is structureless sandy silt with a horizontal upper boundary. Ceramic shards (not classifiable) and charcoal are present. Layer 6 (depth 0.60–0.70 m) is structureless sandy silt with an undulating upper boundary. Layer 7 (depth 0.70–0.75 m) is a well-laminated but partially broken-up band of silt which is present between positions 53.0 m and 60.0 m. Layer 8 (depth 0.75–0.95 m) is structureless sandy silt with some stones, charcoal and ceramic shards (not classifiable). Layer 9 (depth 0.95–1.00 m) is well-bedded coarse sand. This layers is partially broken up and continues from position 50.0 m to 60.0 m. Layer 10 (depth 1.00–1.45 m) is structureless silt with some stones. Layer 11 (depth 1.45–1.62 m) is a gravel layer (with clasts of up to 15 cm) beginning at position 50.0 m. Layer 12 (depth 1.62–1.65 m) is a charcoal-rich layer consisting of silt and coarse sand with a sharp upper boundary. Layer

13 (below 1.65 m) is poorly sorted and bedded coarse sand with some gravel-size clasts and a horizontal upper boundary marked by a thin (1 cm) veneer of sandy silt.

At 57.0 m, 19 layers were recorded. Layers 1 through 8 are similar to layers 1 through 8 at position 53.5 m. At 57.0 m, the lower boundary of layer 8 is at 1.10 m. Layer 9 (depth 1.10–1.15 m) is a fluvial layer of silt and sand similar to layer 9 at position 53.5 m. Layer 10 (depth 1.15–1.50 m) is structureless silt with some stones and abundant charcoal and can be traced to layer 10 at position 53.5 m. Layer 11 (depth 1.50–1.79 m) is a gravel-rich (ca. 30%, up to 18 cm) layer of loose sandy silt. larger clasts are not imbricated, but bedding within the layer is visible through the presence of several charcoal-rich bands. The upper boundary of layer 11 is generally diffuse but clearly marked by a charcoal layer at some places. Layer 11 is interpreted as showing repeated human impact by burning and deposition of very poorly sorted material. Layer 12 (depth 1.79–1.89 m) consists of bedded silt with a sharp, erosional upper boundary which in places shows sings of fire impact (reddening) from the then-surface which indicates that at least the lower charcoal-rich material of the overlying layer 11 is due to local fire. Layer 13 (depth 1.89–2.15/2.25 m) is sandy silt with some stones and a strongly turbated, 5–10 cm thick charcoal layer. The upper boundary is horizontal in the along-bank direction and dips towards the river. Layer 14 (depth 2.15/2.25–2.20/2.30 m) is well-sorted fine to medium sand with an undulating upper boundary which is marked by a thin layer of charcoal and ash. Layer 15 (depth 2.20–2.25 m) is a well-sorted lens of fine sand with faint bedding which is present between positions 55.5 and 56.5 m. Layer 16 (depth 2.25–2.47 m) is sandy silt with a crumbly structure and a sharp, undulating upper boundary. It contains some stones (up to 5 cm) and charcoal. Layer 17 (depth 2.47–2.60 m) is crumbly sandy silt with a thin silt band at its upper boundary. Layer 18 (depth 2.60–2.95 m) is gravel with a matrix of silty sand and a sharp, horizontal upper boundary. Layer 1 (below 2.95 m) is gravel with a sandy matrix.

At 59.0 m, a triangular piece of rusted iron (ca. 1.5 cm diameter and 5 cm long) was found at a depth of 1.07 m (i.e. at the base of layer 8 as recorded at positions 53.5 and 57.0 m).

At 66.0 m, 8 layers were recorded. Layer 1 (depth 0.00–0.06 m) is structureless silty sand. Layer 2 (depth 0.06–0.10 m) is well-bedded sand. Layer 3 (depth 0.10–0.55 m) is sandy silt with some stones (up to 7 cm) and a sharp upper boundary. Within layer 3, several similar layers are separated by thin veneers of sand. Layer 4 (depth 0.55–0.80 m) is structureless, poorly sorted silty sand with fine gravel and some larger clasts. Charcoal is present. The upper boundary is marked by a discontinuous band of sand. Layer 5 (depth 0.80–0.95 m) is a band of gravel with a fine-grained matrix. At some points, the layer is only represented by a few single stones. Layer 5 is interpreted as an overbank gravel splay subject to later turbation. Layer 6 (depth 0.95–1.45 m) is structureless silt with sand to fine gravel. Some unclassifiable ceramic shards are present. Layer 7 (depth 1.45–1.60 m) is structureless sand to fine gravel (up to 5 mm). Some cohesive clods of silt are present, however, the matrix contains only ver little silt. Layer 7 is interpreted as a hyperconcentrated flow deposit with only a short transport path, possibly related to upstream bank collapse during a flood event. Layer 8 (below 1.60 m) is structureless sandy silt with some angular clasts (up to 6 cm) and a sharp, undulating upper boundary. Carbonate precipitation is present in the form of small white dots. Layer 8 is also interpreted as a hyperconcentrated flow deposit.

At 75.0 m, 9 layers were recorded. Layer 1 (depth 0.00–0.30 m) is sandy silt with a crumbly structure. It is interpreted as a subrecent ploughing horizon. Layer 2 (depth 0.30–0.60 m) is structureless sandy silt with some stones, cohesive clods of silt and charcoal. Layer 3 (depth 0.60–1.05/1.20 m) is poorly sorted silty sand with ca. 30% stones (up to 20 cm) which fills a local depression and also extends laterally. Layer 4 (depth 1.05/1.20–1.60/1.70 m) is poorly sorted silt to coarse sand with some fine gravel and an undulating upper boundary. The base of layer 4 is at places marked by a thin band of fine sand. Layer 5 (depth 1.60/1.70–2.60 m) is structureless silty sand with carbonate precipitation (white dots) and some charcoal. Its upper boundary is sharp and undulating (erosional). Layer 6 (depth 2.60–2.90 m) is structureless silt with some stones (10%). and a diffuse upper boundary. Stone content decreases from base to top of the layer. Carbonat precipitation is less than in layer 5. Layer 7 (depth 2.90–3.03 m) is imbricated gravel (up to 10 cm) in a matrix of medium sand. Layer 8 (depth 3.03–3.15 m) is structureless sandy silt with oxidation stains and a slightly undulating upper boundary. Layer 9 (below 3.15 m) is coarse gravel to boulders with a matrix of sand to fine gravel. Layer 9 is interpreted as river-bed sediment.

Between 75.0 m and 85.0 m, the upper gravel-rich layer (layer 5 at position 66.0 m) peters out in a series of lenses. From 85.0 to 135 m (and extending down-river over the recorded length of the exposure), a continuous layer of coarse gravel to boulders with a surface at a depth between 0.10 and 0.95 m is present. Maximum layer thickness (ca. 1.50 m) and grain size (ca. 40 cm) were recorded between positions 98.0 and 110.0 m. Bedding and imbrication are clearly visible. This gravel layer is interpreted to be the result of a very strong erosional event which had cut a new river channel in the underlying sediments and then deposited river-bed material in this channel.

At 133 m (Figure A-95), 95 to 105 cm of structureless and cemented silty sand (base not exposed) with a very sharp, erosional upper boundary are overlain by 80 to 110 cm of well-bedded medium to coarse gravel with occasional larger grain sizes and imbricated clasts. A radiocarbon age of 5,500–4,950 cal BP (Poz-16826, 4,630 ±70 BP) from 10 cm below the gravel base at 128 m (i.e. 5 m to the north-east) indicates a mid-Holocene deposition of the silty sand and a maximum age for the erosional event followed by gravel deposition. A ca. 60 cm deep pit filled with a mix of silty sand and rounded stones is cut into this gravel layer. The fill also contains ceramic shards (mostly Loro, some may be Loro or Nasca) and charcoal. The radiocarbon age of 1,260–1,060 cal BP (Poz-16827, 1,275 ±30 BP) from this fill (depth 0.70 m) corresponds to the Loro (Middle Horizon) age of the ceramics (1,300–950 cal BP, cf. UNKEL 2006: 111). The top 30–40 cm of the profile are sandy silt with a few stones.

Profile RVU records a complex (albeit incomplete) pattern of erosion and deposition in the proximity of the Rio Vizcas since the mid-Holocene (5,500–4,950 cal BP). During the mid-Holocene, a structureless and homogenous fine-grained sediment (silty sand) with a minimum thickness of 95 cm was deposited, the origin of which is unknown. None of the possible interpretations (turbated fluvial overbank / low energy flow deposit, irrigation sediment or hyperconcentrated flow deposit) can be securely rejected or supported on the base of available data, although the presence of charcoal weakly suggests a possible early human influence. The mid-Holocene was followed by repeated erosion and deposition as evidenced by erosional surfaces as well as several coarse gravel sheets and fine-grained sediments. The earliest definite human impact appears in the form of Loro ceramic shards and coeval (1,260–1,060 cal BP) charcoal, giving a minimum age for the last gravel sheet deposition at this site. Repeated fluvial as well as anthropogenic influence of probably mainly historic age is evidenced throughout a large part of the profile by well-bedded or laminated fluvial sediments on one hand and strongly turbated and charcoal-rich layers locally with ceramic shards on the other hand. Bedded fluvial overbank deposits likely related to the 1999 erosional event are present along several parts of the exposure.

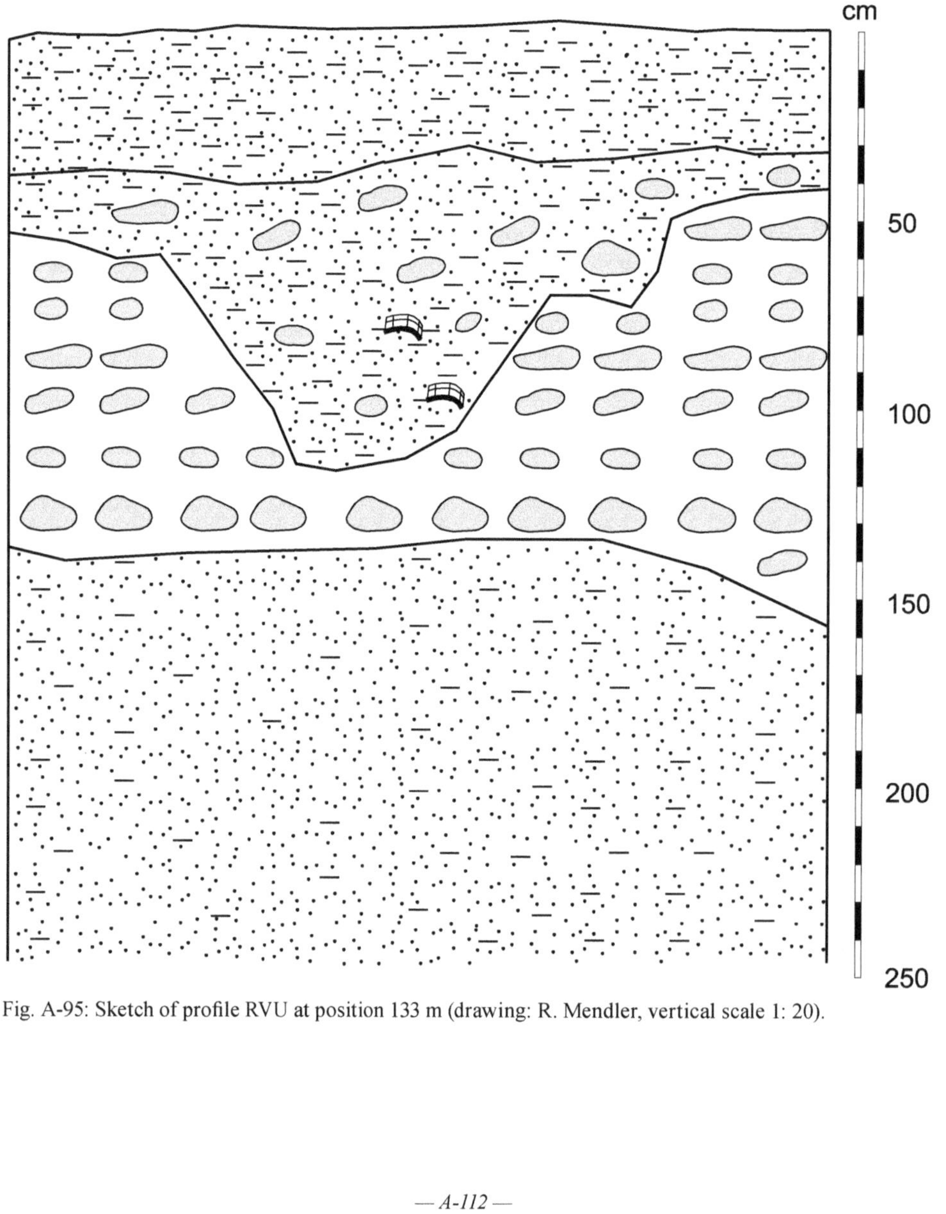

Fig. A-95: Sketch of profile RVU at position 133 m (drawing: R. Mendler, vertical scale 1: 20).

Profiles south-east of the Rio Vizcas

Tab. A-8: Positions of profiles south-east of the Rio Vizcas.

Profile	Depth [m]	Northing [m]	Easting [m]	Altitude [m]
LLPT	4.25	8,389,375.97	476,943.89	281.58
BF-2	6.80	8,390,028.88	477,152.55	286.11
BJ	10.00	8,391,829.42	480,104.64	342.08
FV	3.30	8,392,120.77	479,850.77	338.74
RV	3.40	8,392,127.21	479,578.16	335.59
LU	2.90	8,389,636.41	476,900.59	282.22

Profile LLPT

Profile LLPT (Figures A-96, A-97 and A-98) was recorded in the side of a large pit near the village of Llipata where material for adobe production is removed. For the purpose of recording the profile, the depth was locally increased to 4.35 m. Basal gravels were reached at 4.00 m below the surface.

Layer 11 through 9 are structureless, porous silty sand with a few (ca. 1%) small pebbles. Layer 10 is capped by a thin (ca. 1 cm) veneer of medium to coarse sand. The top of layer 9 is marked by up to 5 cm thick charcoal-enriched and oxidised material which can be traced throughout the pit and which indicates large-scale burning. A radiocarbon date from this layer (depth 1.00 m) indicates an age younger than 260 cal BP (Poz-15240, 120 ±30 BP). Layer 8 is a structureless silty sand which shows traces of turbation in the form of cohesive clods of silt and clay. Its upper boundary is marked by a colour change; however, grain-size and measured chemical parameters remain unchanged. Layer 7 is a silty clay with a west-sloping upper and lower boundary. It is characterised by increased values of Cu, Fe, Mn, K and Mg. Layer 6 is a structureless silty sand with a sharp and slightly undulating upper boundary. Layer 5 is a lens of fine sand and coarse silt. Layer 4 is a structureless silty sand with some stones (up to 7 cm, both rounded and angular) and disperse charcoal. The upper boundary of layer 4 is diffuse. In grain size and chemical parameters it is similar to layers 8 through 10 with the exception of slightly increased carbonate content. Top (255 cm) and basal (330 cm) radiocarbon samples yielded ages of 1,830–1,610 cal BP (Poz-15239, 1,860 ±30 BP) and 2,720–2,360 cal BP (Poz-8027, 2,525 ±30 BP), respectively. This layer was interpreted in the field as a fossil soil. Layer 3 is poorly sorted medium sand to fine pebbles with slight grading from finer grain sizes at the bottom to coarser grain sizes at the top. Layer 2 is well-sorted medium sand without recognisable bedding and with brown oxidation stains. Its lower boundary slopes towards the west. Layer 1 is well-sorted and bedded fine to medium gravel which is sloping to the west. It is characterised by oxidation stains and lacks a fine-grained matrix.

Layer 0 is poorly sorted, coarse, well-rounded to sub-rounded gravel. It shows imbrication and contains a matrix of sand to fine gravel.

Profile LLPT is characterised by basal fluvial gravels (layer 0) overlain by a gravel-sand sequence (layers 1 and 2) interpreted as a gravel/sand bar complex which in turn is overlain by an upwards-coarsening layer (layer 3) which can be interpreted as an overbank splay sheet. A conspicuous change in sedimentation separates the lower (layers 0 through 3) from the upper (layers 4 through 11) portion of the profile. The deposition of 3.30 metres of silty sand is only interrupted by thin lenses or layers of silt (layer 5), silty clay (layer 7) and sand (top of layer 10).

The entire upper sequence is interpreted as irragric anthrosol with the inception of irrigation agriculture by early to middle Paracas times (2,360–2,720 cal BP, cf. Unkel 2006: 111). Quasi-continuous sedimentation lasted at least until early to middle Nasca times (1,610–1,830 cal. BP, cf. Unkel 2006: 111) after which three undated interruptions (as shown by silt / silty clay deposition and layer boundaries) occurred until late historic times (260 cal BP to present) without a strong change in average sedimentation rates (0.68–1.41 mm/a for layer 4 and 0.82–1.15 mm/a for layers 5 through 9). The anomalously high sedimentation rate for the top 1.00 m (≥3.85 mm/a) in late historic times may be due to very high irrigation water input for modern high yield-high demand plants, an intentional sedimentation in the adobe pits or mechanised earth movement.

Profile BF-2

Profile BF-2 (Figures A-99 and A-100) was recorded in a 7.30 m deep well close to the Rio Vizcas. The well had a diameter of ca. 3 m and had been dug in 2004. The profile was recorded in 2005. Since the open water in the well had to be protected from dirt, only a narrow (ca. 10–20 cm) vertical strip could be cleaned. Therefore, indistinct sedimentary structures or layer boundaries may have been missed. In March 2006, groundwater rose to within 2 m of the surface and caused the walls to collapse.

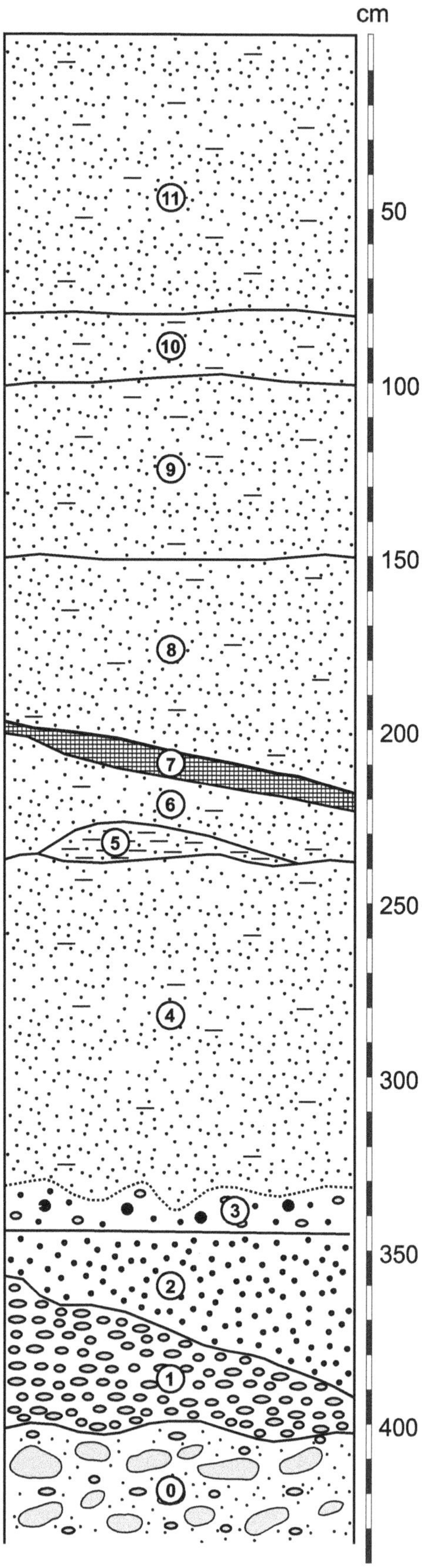

Fig. A-96: Sketch of profile LLPT (drawing: R. Mendler, vertical scale 1: 20).

Layer 1 is a sandy silt with columnar structure. Layer 2 is a thin (ca. 5 cm) layer of poorly sorted silty sand which laterally grades into a charcoal-rich pocket on the north side of the well. A radiocarbon sample from this pocket (at a depth of 1.25 m below the surface) gave an age of ≤270 cal BP (Poz-15287, 135 ±30 BP). Layer 3 consists of structureless sandy silt with a few pebbles (up to 3 cm). A radiocarbon sample from a depth of 2.70 m gave an age of ≤260 cal BP (Poz-15288, 95 ±30 BP). The very high sedimentation rates resulting from these ages (≥4.6 mm/a and ≥10.4 mm/a, respectively) during late historic times may be due to very high irrigation water input or mechanised earth movement.

Layer 4 is a thin layer of well-rounded gravel (ca. 4 cm) which is interpreted as an overbank gravel splay deposit. Layer 5 is a structureless layer of fine to medium sand with some silt. Layer 6 is a thin blackish layer which is mainly present on the south-east wall. In the field it was thought to contain charcoal due to its colour, however, in the laboratory this colour turned out to be of mineral origin. Layer 7 consists of structureless and slightly cemented fine sand with some silt; bedding is not present. Layer 8 consists of well-sorted medium sand without discernible bedding. Layer 9 is coarse sand to medium gravel, bedding is present.

A radiocarbon sample from a depth of 4.98 m dates the top of this layer to 6,750–6,490 cal BP (Poz-13548, 5,880 ±50 BP). Layer 10 consists of well rounded, very coarse (up to 80 cm length) gravel to boulders in a matrix of coarse sand. The large clasts are strongly weathered and the material is cemented.

The bottom 1.60 m of profile BF-2 expose strongly weathered and cemented material (layer 10) reminiscent of the road cuts through the bajadas (e.g. the Cresta Sacramento). This material is probably of pre-Holocene age and likely corresponds to the early period of bajada deposition. It is overlain by a sequence (layers 9 through 4) of well-sorted material with low LoI values which is interpreted as result of fluvial channel and overbank deposition during the mid-Holocene. The top 3.57 m of the profile (layers 3 through 1) are characterised by increasingly poorer sorting and higher LoI values as well as the inclusion of dispersed small stones in a matrix of

Fig. A-97 (following page): Laboratory results of samples from profile LLPT (vertical scale 1: 30). Cumulative weight percent graph shows clay, silt and sand from left to right. Note that the lines connecting data points do not imply values between data points but serve solely to aid visualisation.

Fig. A-98 (page A-116): AAS measurement results of samples from profile LLPT (vertical scale 1: 30). Note that the lines connecting data points do not imply values between data points but serve solely to aid visualisation.

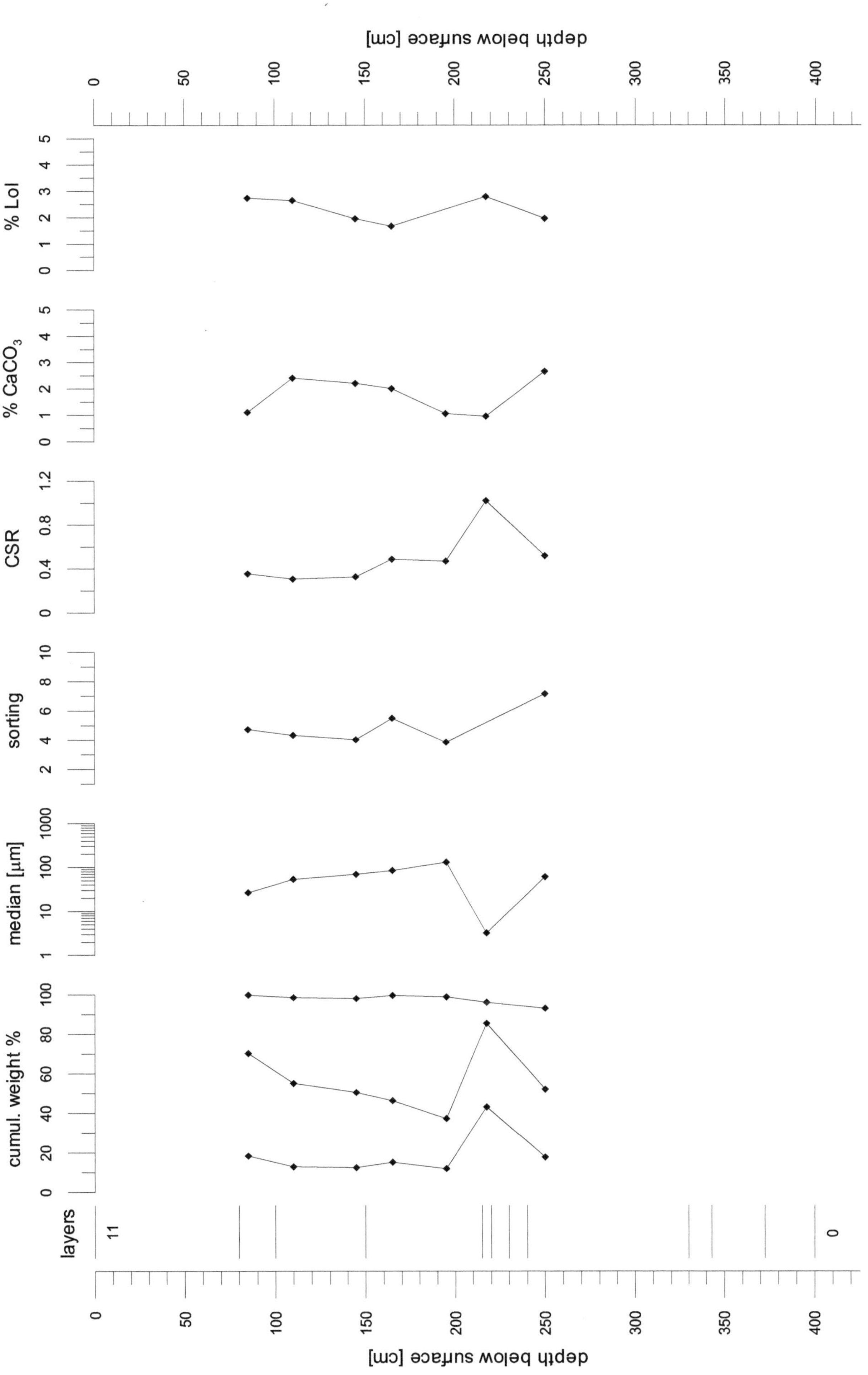
depth below surface [cm]
% LoI
% $CaCO_3$
CSR
sorting
median [µm]
cumul. weight %
layers
11
0

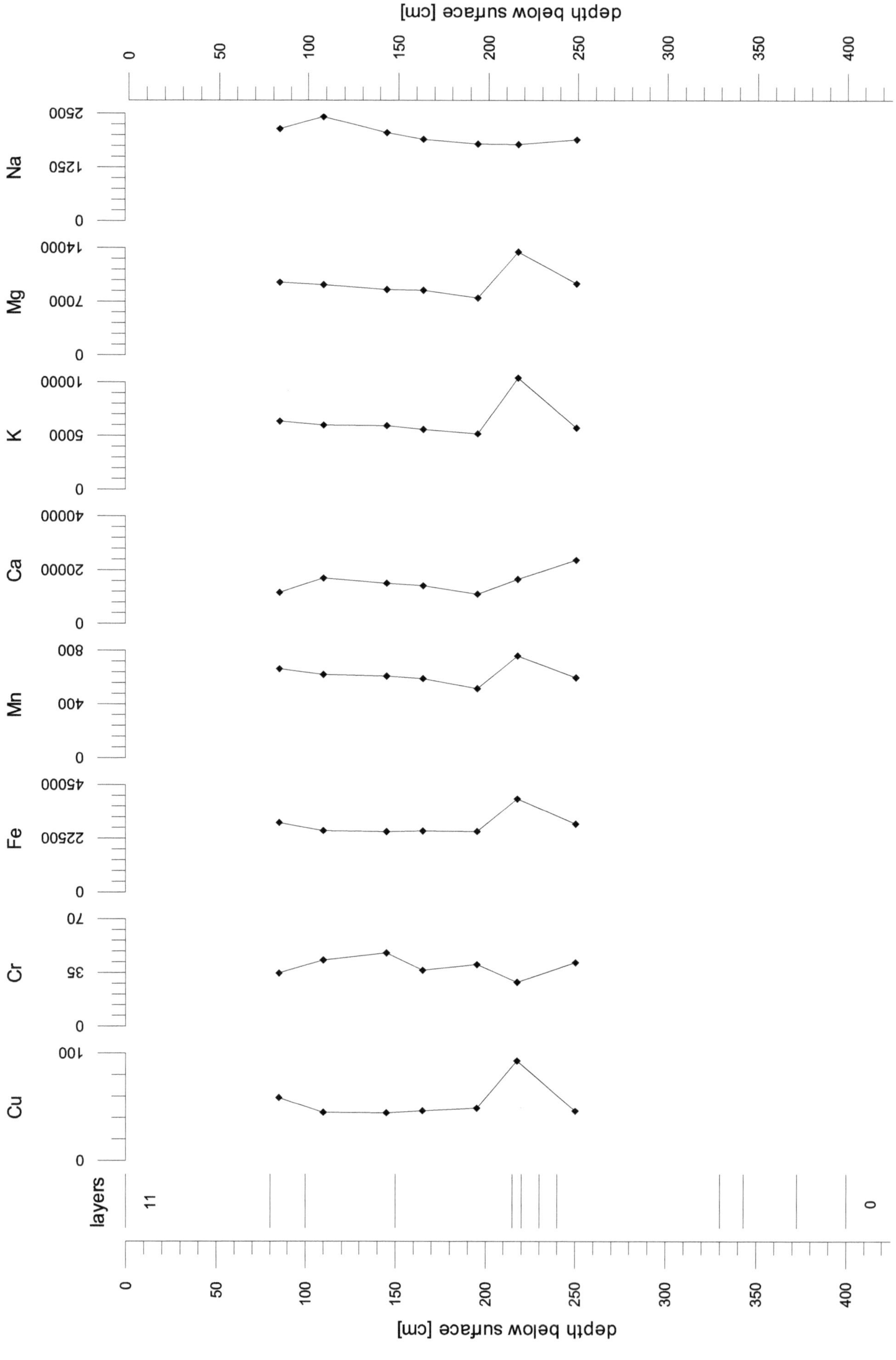
depth below surface [cm]
Na
Mg
K
Ca
Mn
Fe
Cr
Cu
layers
11
0

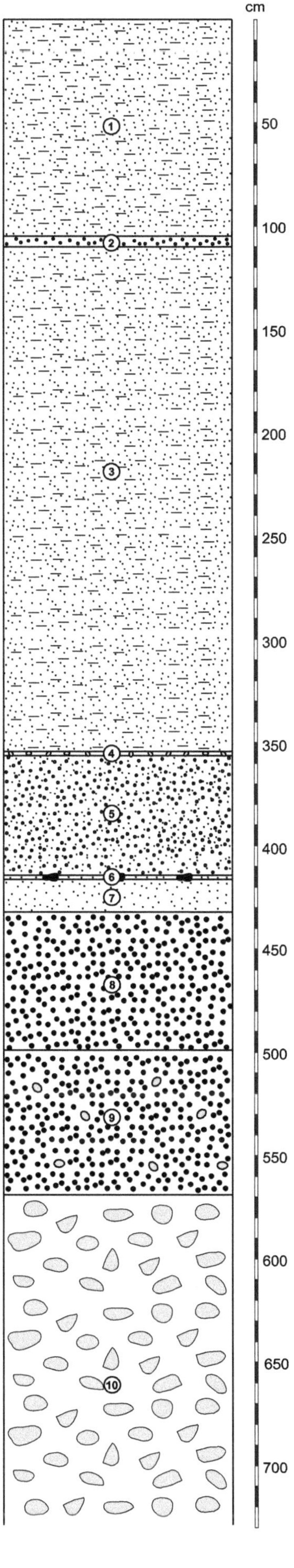

Fig. A-99: Sketch of profile BF-2 (drawing: R. Mendler, vertical scale 1: 30).

silty sand/sandy silt of a structureless to columnar structure. Therefore, these layers are interpreted as irragric anthrosol.

Profile BJ

Profile BJ (Figures A-101, A-102, A-103 and A-104) was recorded in a more than 10 m deep well excavated in 2005. While the lower part of the profile is characterised by roughly horizontal layers, the upper 2.15 m show a very complex structure.

The recent (until 2004) A_P horizon (layer 1) consists of sandy silt. In the south-western sector of the well, layer 1 is underlain by layer 5, a 50 cm thick sequence of bedded silt, sand and fine gravel with bone fragments and ceramic shards. Laterally, this sequence is separated by an almost symmetric dam-like upward extension of layer 8 from a structureless, poorly sorted silty sand without bedding. Into this, a small channel had been incised and filled repeatedly with sandy silt and well-bedded fine to coarse sand with cross-bedding (layers 2 through 4). This channel had also incised through layer 6 and into layer 8. Layer 6 is a bedded layer of silt and sand and can be followed onto the dam-like structure of layer 8. Layer 8 consists of structureless silty sand and has a diffuse upper boundary. In the south-east and north-west, the top of layer 8 forms dam-like structures ca. 20 cm high.

Layer 9 consists of structureless silty sand; the contact to layer 8 is diffuse. Layer 7 is a refilled hole cutting through layers 8 and 9. The material is poorly-sorted silty sand with some rounded stones. At the base of layer 7, a high concentration of ceramic shards, bone fragments, glass shards and rounded stones (up to 15 cm). Layer 10 is a layer of charcoal-enriched silty sand. A radiocarbon assay from layer 10 (1.42 m below surface) gives an age of 470–300 cal BP (Poz-13544, 355 ±30 BP). Layer 11 consists of structureless silty sand and has a diffuse upper boundary; it is not present in the south and south-west of the well where it laps onto layer 12. In the north-eastern part, where the lower boundary of layer 11 is deepest, a number of large (ca. 30 cm) rounded clasts are present at the base of the layer. LoI values from layer 11 are the lowest in the top 2.50 m of the profile.

Layer 12 consists of structureless loam with a conspicuous columnar structure. Some (< 1%) disperse stones (up to 8 cm) and root molds are present. The colour ranges from brownish grey at the base to blackish grey at the top; this corresponds with a slight increase in both clay content and LoI from base to top of the layer. The upper bound-

Fig. A-100 (following page): Laboratory results of samples from profile BF-2 (vertical scale 1: 50). Cumulative weight percent graph shows clay, silt and sand from left to right. Note that the lines connecting data points do not imply values between data points but serve solely to aid visualisation.

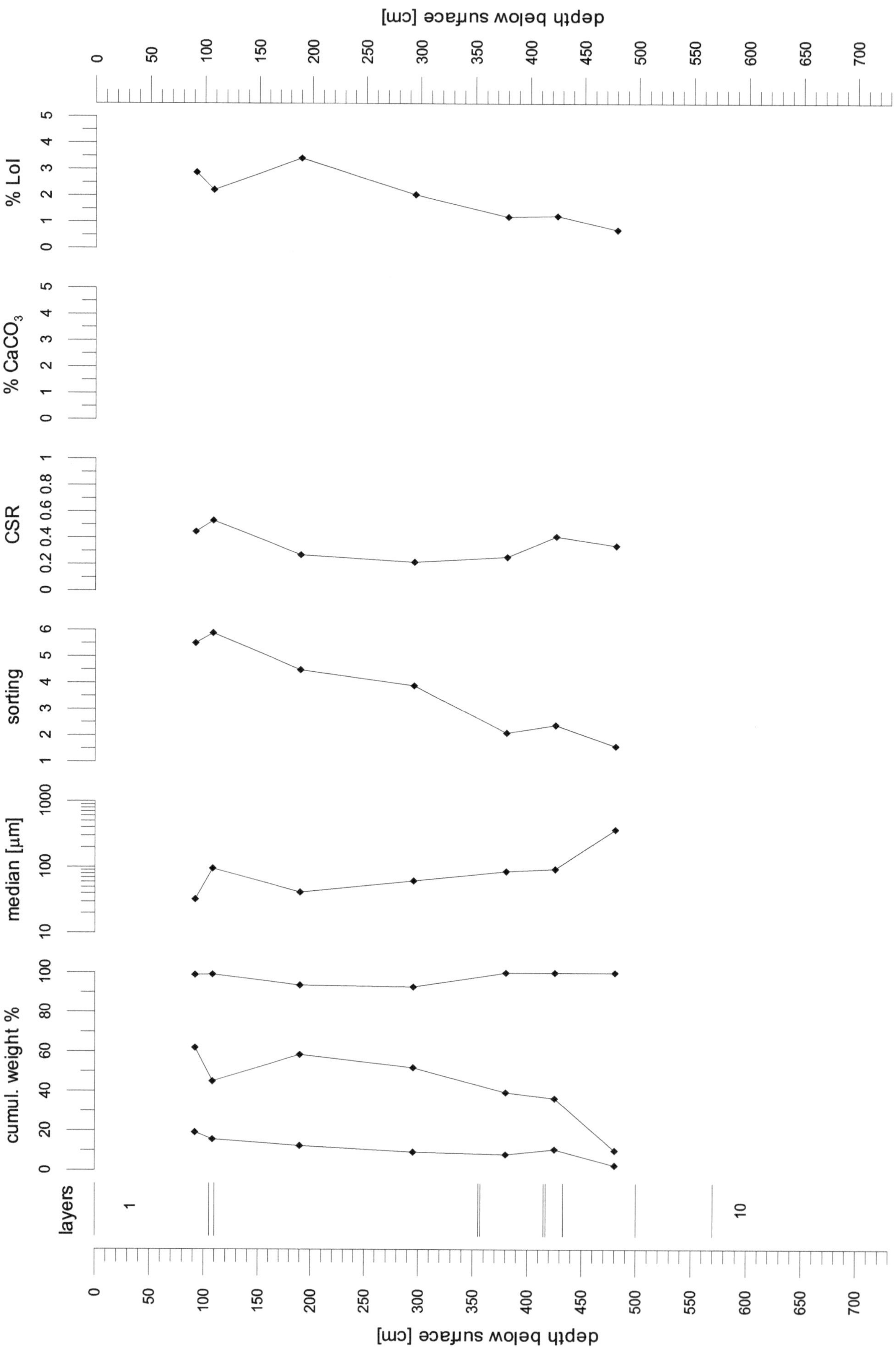
depth below surface [cm]
% LoI
% CaCO3
CSR
sorting
median [µm]
cumul. weight %
layers
1
10
depth below surface [cm]

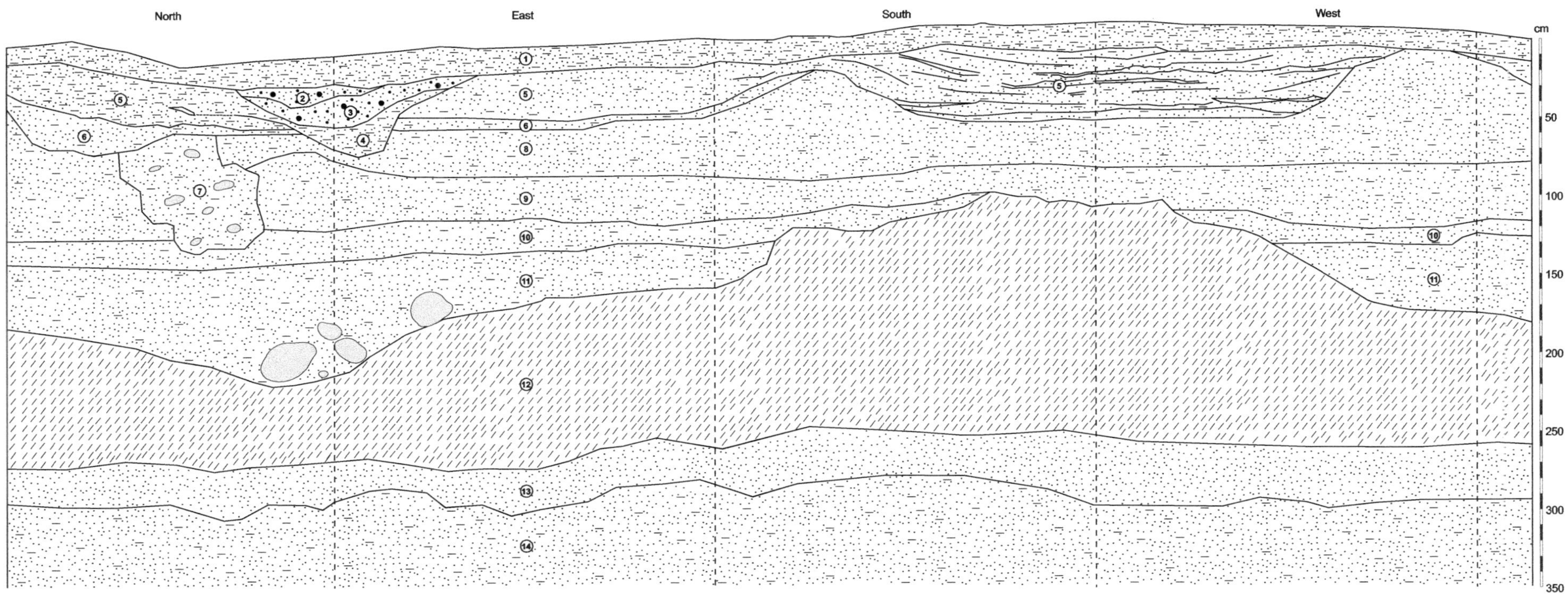

Fig. A-101: Sketch of the upper portion of profile BJ (drawing: R. Mendler, vertical scale 1: 40).

ary of layer 12 is highest in the south and south-west and lowest in the north-east. The sharp, sloping upper boundary of layer 12 together with the large stones in the base of the overlying layer 11 suggest an erosional truncation of the upper 1.20 m in the north-eastern part. Layer 12 was interpreted in the field an a fossil soil; the high LoI values comparable to AP horizons and the presence of disperse stones and charcoal together with the poorly sorted fine- grained material support the interpretation as irragric anthrosol. Two radiocarbon dates from the boundary between layers 12 and 11 in the north-eastern sector (2.12 m below surface) and from the top of the underlying layer 13 (2.75 m below surface) bracket the age of layer 12 between 1,610–1,410 cal BP (Poz-15232, 1,690 ±30 BP) and 2,720–2,360 cal BP (Poz-13293, 2,525 ±30 BP).

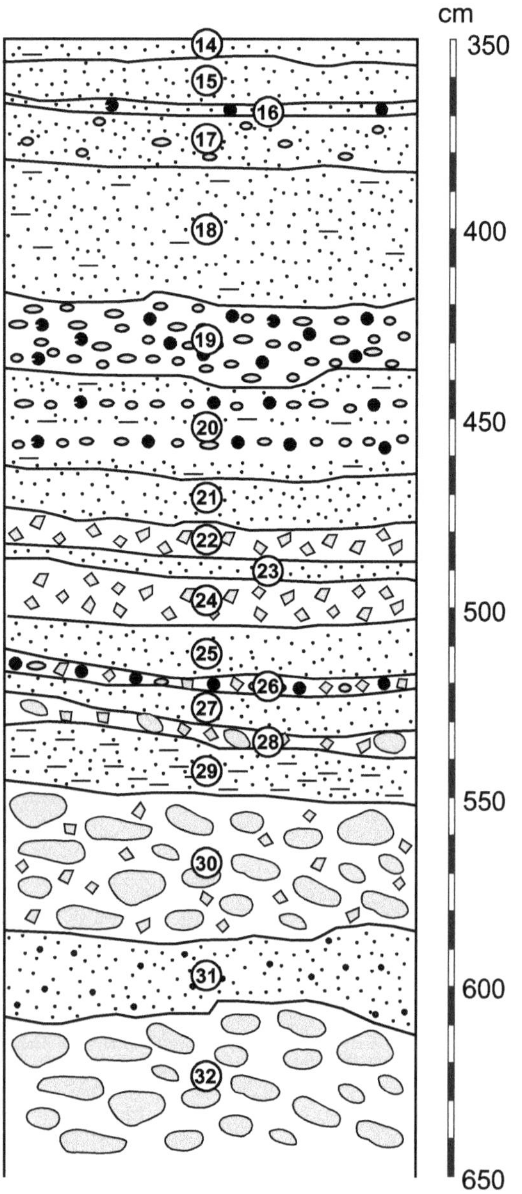

Fig. A-102: Sketch of the lower portion of profile BJ (drawing: R. Mendler, vertical scale 1: 20).

Layer 13 is structureless sand with some disperse stones (up to 6 cm) and some pockets with fine gravel. Layer 14 has an indistinct, undulating upper boundary and consists of structureless silty sand with some disperse gravel and with oxidation stains. Layer 15 has a horizontal and slightly diffuse upper boundary and consists of sand with some fine gravel in the top; oxidation stains are present. Layer 16 has a horizontal and sharp but broken-up upper boundary and consists of fine to coarse sand with discernible bedding.

Layer 17 has a diffuse horizontal upper boundary and consists of sand and fine gravel; oxidation stains are present. Layers 13 through 18 have low LoI values; layers 13 to 17 are well-sorted. Sorting is poorer in layer 18 which consists of structureless silty sand with embedded rounded and angular clasts (up to 1 cm). The upper boundary of layer 18 is sharp and horizontal. Layer 19 is coarse to angular gravel (up to 6 cm) in a matrix of coarse sand. Layer 20 is a sequence of fine gravel and coarse sand interbedded with thin sheets of fine sand to silt. Layer 21 is structureless fine sand with oxidation stains. Layer 22 is structureless sandy angular fine gravel without rounded clasts. Layer 23 is structureless, silty fine to coarse sand with some fine gravel. Layer 24 is angular fine gravel with oxidation stains. Layer 25 consists of structureless fine sand with some thin silt sheets and some small stones.

Layer 26 is coarse sand to angular fine gravel with some rounded clasts and distinct oxidation stains. This layer is thickest in the north-western sector (towards the Rio Vizcas) and is not present in the south-eastern sector (towards the Pampa San Ignacio). Layer 27 is well-sorted fine sand with a fining-upwards tendency. Layer 28 is gravel (5 to 10 cm) in a matrix of angular fine gravel. This layer is thickest in the south-east (towards the Cresta San Ignacio). Layer 29 is sandy silt with a gravel lens. Layer 30 is sub-rounded coarse gravel (up to 7 cm) in a matrix of angular fine gravel; a lens of angular fine gravel is also present. The upper boundary of layer 30 is slightly sloping north-west (towards the Rio Vizcas). Layer 31 is fine to medium sand. Layer 32 is rounded coarse gravel to boulders reminiscent of the bajada material at Cresta Sacramento in terms of strongly weathered and cemented clasts.

Fig. A-103 (following page): Laboratory results of samples from the north-eastern side of profile BJ (vertical scale 1: 30). Cumulative weight percent graph shows clay, silt and sand from left to right. Note that the lines connecting data points do not imply values between data points but serve solely to aid visualisation.

Fig. A-104 (page A-122): Laboratory results of samples from fossil irragric anthrosol on the south-western side of profile BJ (vertical scale 1: 30). Cumulative weight percent graph shows clay, silt and sand from left to right. Note that the lines connecting data points do not imply values between data points but serve solely to aid visualisation.

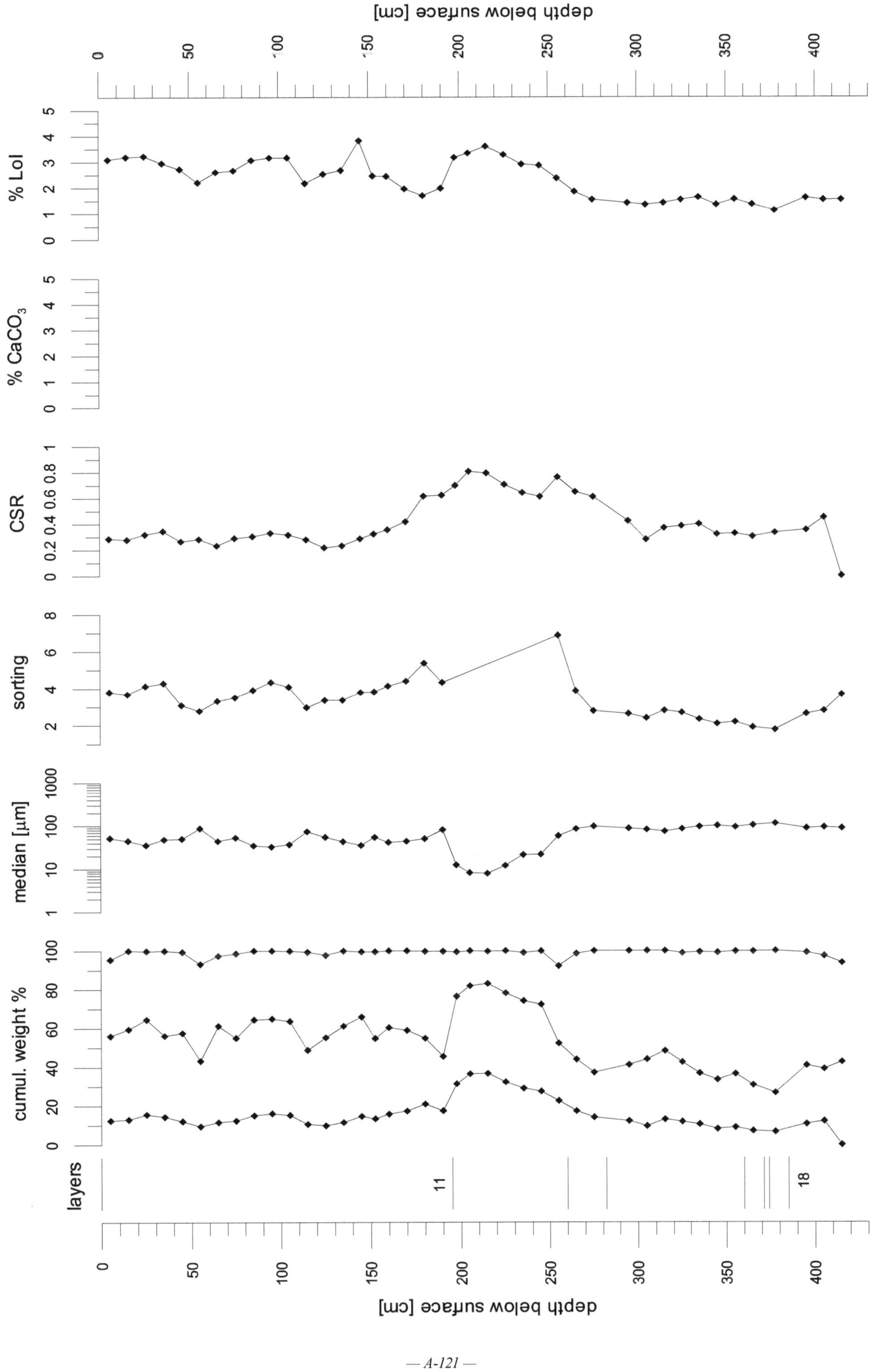
depth below surface [cm]
% LoI
% CaCO3
CSR
sorting
median [µm]
cumul. weight %
layers
11
18
depth below surface [cm]

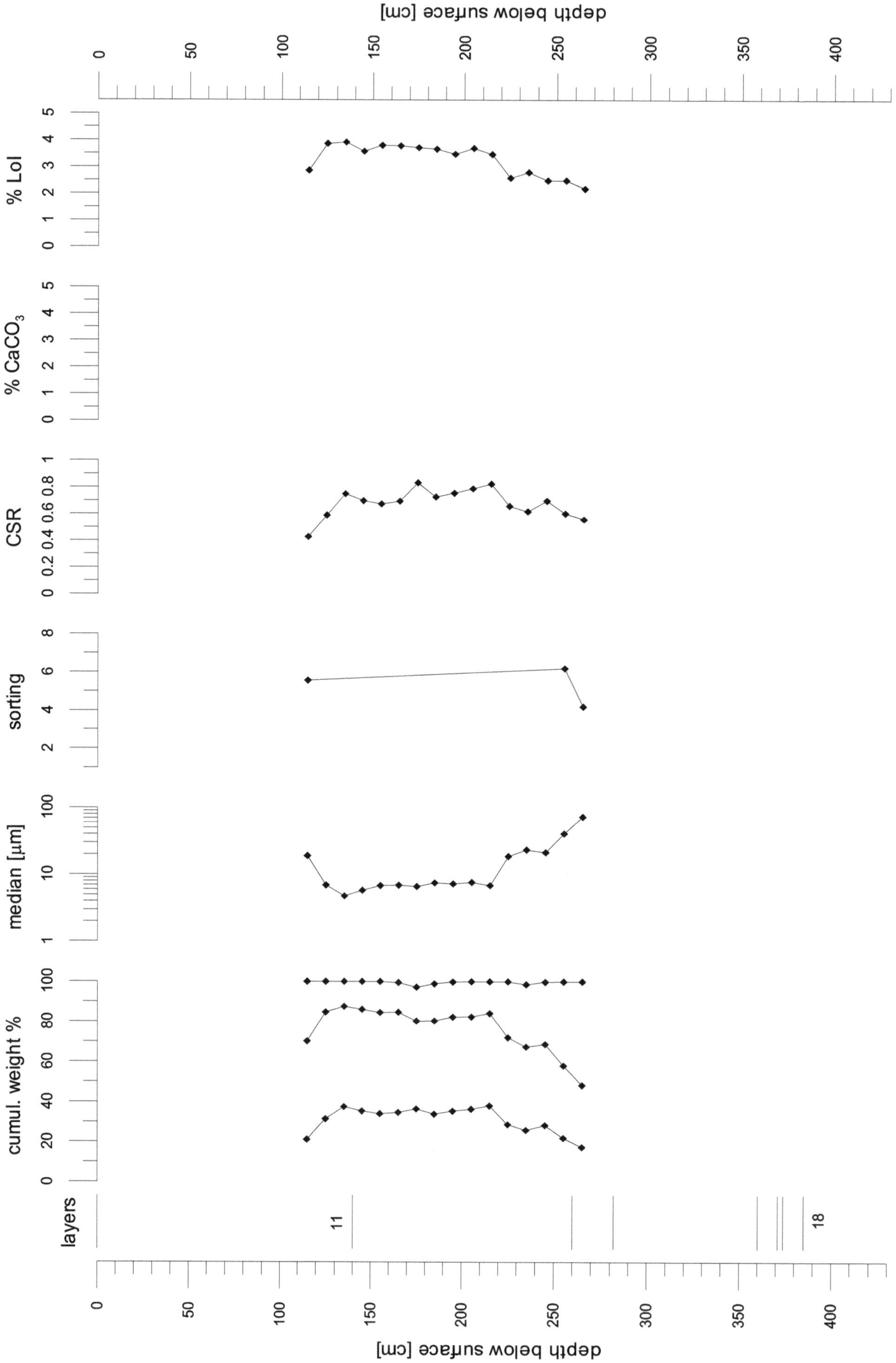
depth below surface [cm]
% LoI
% CaCO$_3$
CSR
sorting
median [µm]
cumul. weight %
layers
11
18
depth below surface [cm]

At the base of profile BJ, 3.90 m of strongly weathered and cemented gravel (layer 32) of probably pre-Holocene age, likely coeval with early bajada deposition. It is overlain by a sedimentary sequence of fluvial overbank deposition (layers 31, 29, 27, 25, 23, 21, 20 and 17–15) and hyper-concentrated flow / debris flow deposition (layers 30, 28, 26, 24, 22, 19 and 18) originating from a nearby quebrada in the flank of the Cresta San Ignacio. Layers 14 and 13 appear to represent the beginning human impact at the site as indicated by the appearance of charcoal. Given the still low LoI values and the proximity of a quebrada mouth, the turbation of gravel with silty sand may be due to a hyperconcentrated flow event. Layer 12 is interpreted as irragric anthrosol with a quasi-continuous (seasonal) sedimentation from shortly after 2,720–2,360 cal BP until at least 1,610–1,410 cal BP with a mean sedimentation rate between 0.46 and 0.80 mm/a. The upper section of profile BJ records a complex sequence of erosion, deposition and human impact. At an unknown time between 1,610–1,410 cal BP and 470–300 cal BP, deposition or irrigation sediment was terminated and followed by erosion of at least 0.70 m in the northern and north-eastern sector, perhaps related to the current position of the profile close (ca. 5 m) to a terrace edge. This was followed by the deposition of either colluvial (perhaps layer 11 as judged from anomalously low LoI values) or irrigation (probably layers 10 through 8) sediment. A 90 cm deep pit (layer 7, probably a grave) was dug into this sequence in historic times (after 470–300 cal BP).

Apparently, this grave was later looted since human bones fragments as well as glass and ceramic shards were found in the fill of the pit as well as in the layer which must have represented the concurrent surface. A low dam then separated a low-energy fluvial or still-water from a turbated sector into which shallow channels had been incised. This is interpreted as being related to the current position near a lower terrace/field edge with a shallow pond situation, a field edge and a collector channel for excessive irrigation water. The top 15 cm (layer 1) represent the A_p horizon of an irragric anthrosol.

Profile FV

Profile FV (Figures A-105 and A-106) was a pit on a field between the Rio Vizcas and the Panamericana. Sketch and description refer to the NE wall of the pit. At the time of investigation, the field was bare of vegetation. Unfortunately , work had to be terminated before basal gravels were reached.

Layer 1 consists of sandy silt and has a crumbly structure. It is the A_p horizon of the field. Layer 2 has a sharp, horizontal upper boundary and consists of structureless sandy silt. A columnar structure is weekly developed. Layer 3 has a horizontal, diffuse upper boundary, consists of hard-packed loam with polyedric structure. Layer 4 has a horizontal, diffuse upper boundary and consists of structureless, silty sand with a coarse polyedric structure. While similar to layer 2 in structure, layer 4 is characterised by higher clay content and a higher CSR. Layer 5 has a horizontal upper boundary and consist of hard-packed loam with polyedric structure. A charcoal sample from a depth of 0.80 m was radiocarbon dated at 1,710–1,540 cal BP (Poz-15231, 1,765 ±30 BP). Layer 6 has an undulating upper boundary and consists of loam with coarse polyedric structure.

All layers from 1 through 6 contain 1–2% gravel-sized stones up to 3 cm long and consist of loam, silty sand or sandy silt. They significantly differ in their structure (crumbly, columnar, polyedric). These properties contrast clearly from the layers 7 through 17 where stone content is fluctuating from layer to layer and crumbly, columnar or polyedric structure are not present. Layer 7 has a horizontal, diffuse upper boundary and consists of silty sand. Stone content is higher than in layers 1 through 6 (10%), clasts are angular to sub-angular with long axes up to 3 cm. Some of the clasts are strongly weathered granite. Layer 8 has a horizontal, diffuse upper boundary and consists of structureless, silty sand with 2–5% stones (up to 2 cm). Layer 9 consists of very loose, bedded sand and fine gravel (up to 0.5 cm) and is therefore of fluvial origin. The upper boundary is cut by clearly delimited cones of cohesive, silty sand extending downward from layer 8. At their wider (top) end, these cones have diameters of 10 to 20 cm; the points of some of them reach the base of layer 9. In the centre of the profile, one of the cones extends with a rounded, block-like appearance from layer 8 through layer 9 and into layer 10. A charcoal sample from this structure at a depth of 1.65 m was radiocarbon dated at 2,800–2,150 cal BP (Poz-13545, 2,510 ±110 BP). Since the sides of the cones are rather steep and the material of layer 9 is very loose but its bedding undisturbed, the possibility of incision and later infilling can be excluded. Furthermore, lateral excavation by ca. 30 cm revealed that they are indeed cone-shaped structures and not narrow channels. A plausible explanation for these structures is the (at least partial) deposition of layer 8 on top of layer 9 and the creation of the cones as planting holes dug through the then-surface layer 8 and into layer 9.

Layer 10 has a sharp, horizontal upper boundary and consists of structureless silty sand. Porosity (possibly root molds) is high. Layer 11 has a sharp but broken-up upper boundary and consists of silty sand. There are 2 pockets of lighter-coloured fine sand extending downward from the upper boundary through the layer. Layer 12 has a horizontal, diffuse upper boundary and consists of structureless, silty fine sand with 0.5% stones. There is a high concentration of fine pores. Layer 13 also has a horizontal, diffuse upper boundary and consists of sand with 0.5% stones (up to 2 cm). There are numerous small white dots (carbonate) and diffuse oxidation stains. The upper boundary of layer 14 is sloping towards the north-west and diffuse.

The layer consists of silty sand with 5% stones (up to 2 cm). Visible pores are present. The upper boundary of layer 15 is undulating and diffuse; the layer consists of sand with 30% stones (long axes up to 8 cm). The stones are angular to subangular; the matrix is structureless and very hard. Based on these properties, the layer is interpreted to be a hyperconcentrated flow deposit. Layer 16 has a horizontal, sharp upper boundary and consists of silty sand with 1% stones (up to 0.5 cm). Within the layer, there is a vertical change of colour from yellowish brown to dark brown. Layer 17 has a horizontal, undulating upper boundary and consists of sand with 20% stones (up to 3 cm). It is interpreted as a hyperconcentrated flow deposit. The base of this layer was not reached.

Profile FV represents a transition from natural dynamics (hyperconcentrated flow deposits, layers 15 and 17) to anthropogenic influence (irrigation agriculture, layers 1 through 6). In the upper 2.70 m (layers 1 through 14, une-

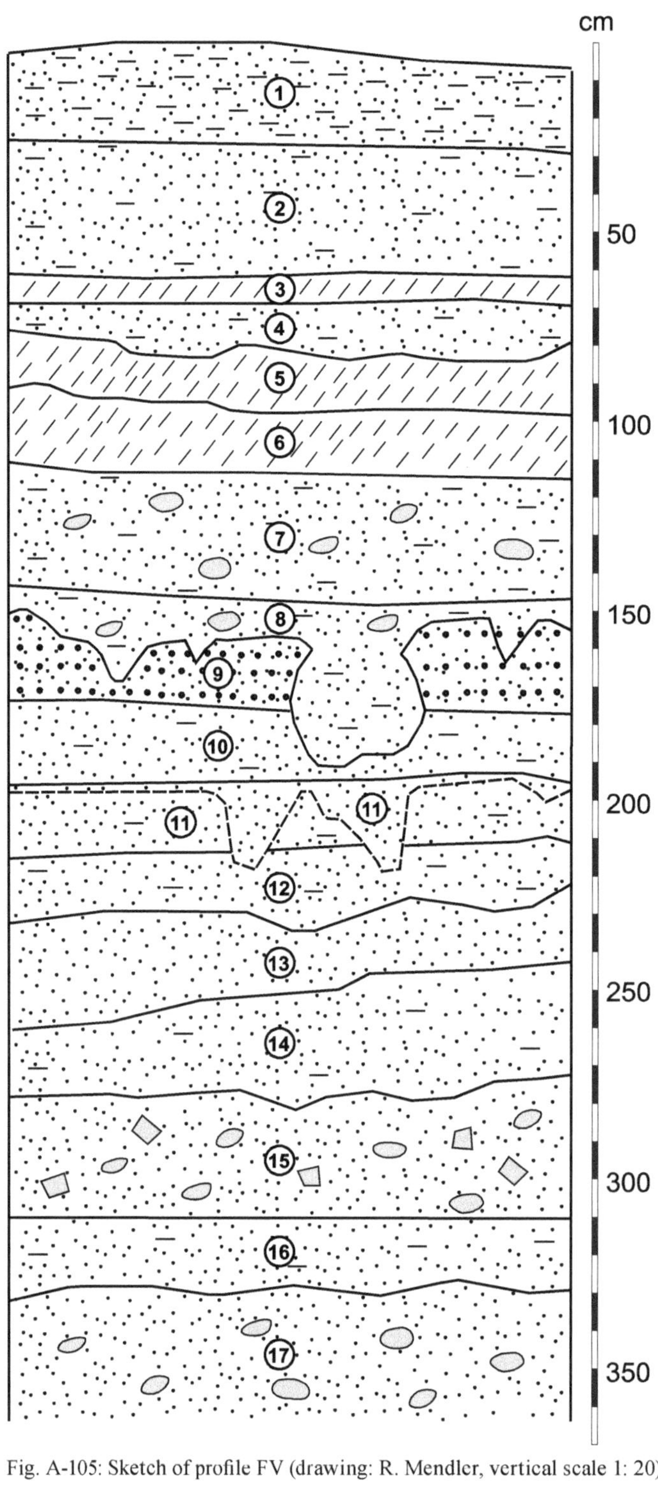

Fig. A-105: Sketch of profile FV (drawing: R. Mendler, vertical scale 1: 20).

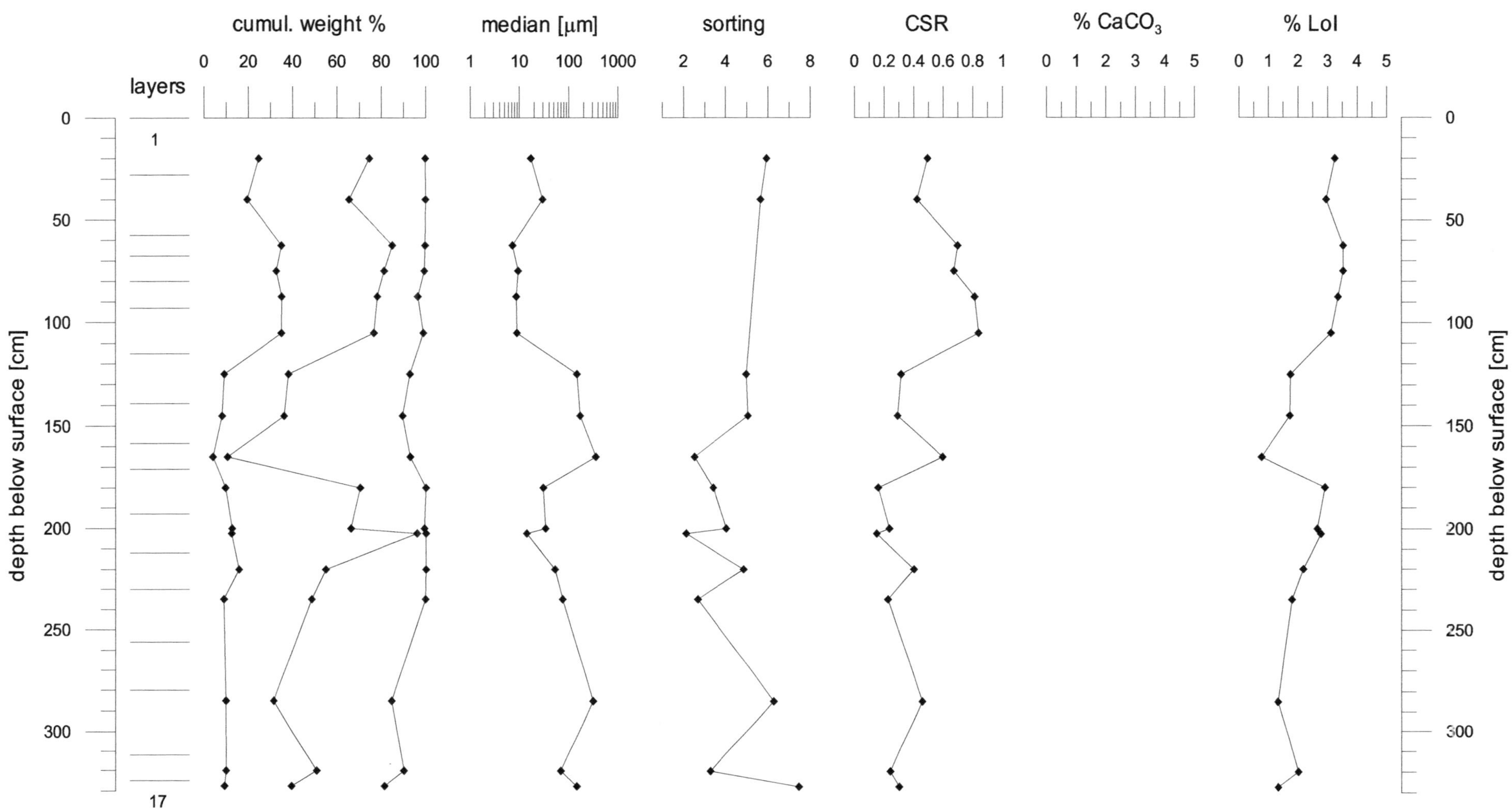

Fig. A-106: Laboratory results of samples from profile FV (vertical scale 1: 30). Cumulative weight percent graph shows clay, silt and sand from left to right. Note that the lines connecting data points do not imply values between data points but serve solely to aid visualisation.

quivocal fluvial influence is only present as layer 9. Structures interpreted as planting holes (extending from layer 8 into layer 9) are evidence for agricultural activity at this site by 2,800–2,150 cal BP after a fluvial deposition event which left bedded sand to fine gravel (layer 9). The planting structures as well as low CSR and LoI values and high sand content of layers 7 and 8 suggest that a significant part of the fluvial sediment was turbated into the overlying irrigation sediment. Between 2,800–2,350 cal BP and 1,710–1,540 cal BP, irrigation sediment (layers 5 through 8) was deposited at a rate of 0.68–1.93 mm/a. The sedimentation rate for the top layers (1 through 4) is 0.45–0.52 mm/a.

Profile RV

Profile RV (Figure A-107) was recorded on a north-west-facing exposure at the mouth of a gully on the left bank of the Rio Vizcas. The river (which in this stretch is today confined by artificial embankments and does therefore not reach this site) had obviously been actively eroding its bank here at most a few years before this profile was recorded, likely during the 1999 flood event. However, comparison with an aerial image from 1944 shows that river bank erosion at this point has been less than 10 m during the past 60 years. An irrigation overflow/drainage canal had cut a gully extending down to the former river-bed level and a few metres backwards into the adjoining plantation. The exposure was at the mouth of this gully. A low (20–30 cm) artificial bank along the edge of the river-bank apparently serves to hold back irrigation water in the plantation to improve infiltration. Intended or unintended, it thereby also facilitates fine-grained suspended sediment deposition from irrigation water.

Layer 1, the recent top soil, is grey, silty sand. Layer 2 has a horizontal but indistinct upper boundary and consists of structureless grey sand with silt clods. It has a faint polyedric structure. Based on the surface features described above, it is interpreted as part of the low artificial bank. Layer 3 has a horizontal and partially broken-up upper boundary and consists of structureless grey sand. At the lower boundary, there is a thin silty layer. Layer 4 has a sharp, slightly undulating but generally horizontal upper boundary. It consists of bedded grey sand. In this layer, a synthetic textile fragment was found, placing this layer firmly in the (probably late) 20th century. Layer 5 has a sharp, slightly undulating upper boundary and consists of light grey, well-laminated silt. Layer 6 has a sharp, north-east-sloping and slightly undulating upper boundary. It consists of grey loam with a somewhat crumbly structure and a few disperse clods of clay. Layer 7 has a slightly undulating, diffuse upper boundary. It is a light grey, gently nort-east-sloping band of silt.

Layer 8 has a slightly undulating and somewhat diffuse upper boundary. It consists of grey loam with a crumbly structure and disperse clods of clay (more than in layer 6). Layers 6 through 8 appear to represent the edge of an artificial field-edge bank like the one found on the surface but positioned slightly more towards the river. As sedimentation from irrigation raised the field or plantation surface behind the bank, the bank itself had to be raised. Layer 5 is interpreted to be an irrigation sediment which was not disturbed by land use due to its position at the edge of the field or plantation. Layer 9 has a broken-up upper boundary and consists of light grey, well-laminated silt. The laminations are undulating, indicating deposition in slowly flowing rather than standing water. Their preserva-

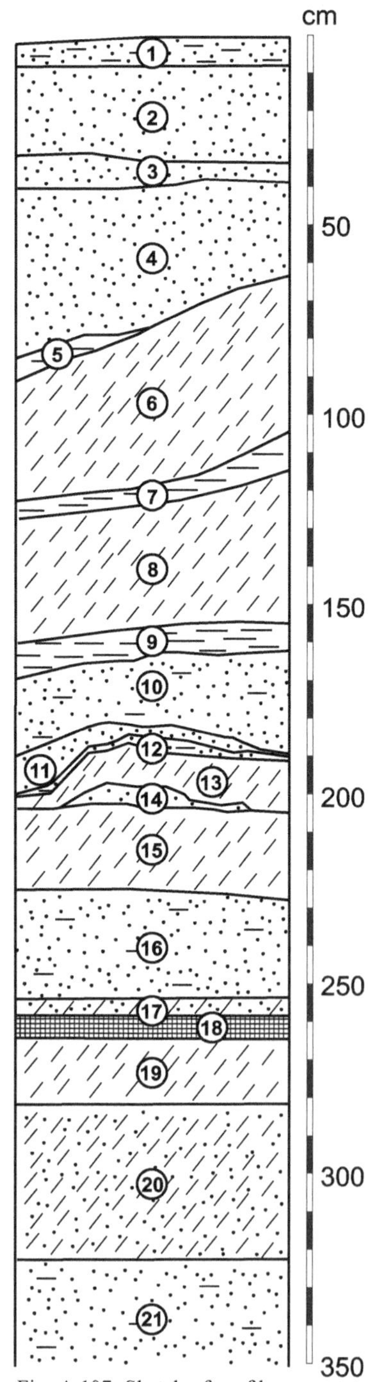

Fig. A-107: Sketch of profile RV (drawing: R. Mendler, vertical scale 1: 20).

tion has to be attributed to lacking tillage. Layer 10 has an undulating upper boundary and consists of grey silty sand with a crumbly structure. At the base of the layer there is a broken-up clay layer and clay clods. Layer 11 has a very diffuse upper boundary and consists of grey sand with indistinct bedding and oxidation stains. Layer 12 has a strongly undulating and broken-up upper boundary and consists of structureless, grey, silty fine sand with oxidation stains. Layer 13 has an undulating and indistinct upper boundary and consists of reddish brown to black loam. Structure is crumbly with vertical cracks; oxidation stains are present. Layer 14 has a sharp, slightly undulating upper boundary and consists of structureless, grey fine sand with oxidation stains.

Layers 11 through 14 appear to represent a low artificial bank comparable to but smaller than the one represented by layers 6 through 8. Layer 15 has a sharp upper boundary defined by a light silt layer (2 mm) and a dark clay layer (3 mm). The layer consists of reddish brown to black loam with oxidation stains and a crumbly structure with vertical cracks. Layer 16 has a diffuse, horizontal upper boundary and consists of reddish brown silty sand with vertical grey marks (root molds). Silt content diminishes towards the base of the layer (i.e., the grain size has a upwards-fining tendency). Oxidation stains are present. Since significant grain size grading is not expected in sediments deposited by quasi-continuous irrigation, this layer is interpreted as an overbank deposit of the Rio Vizcas. Layer 17 has a sharp but partially broken-up upper boundary and consists of yellowish brown sandy loam with yellow and dark stains.

Layer 18 has a sharp horizontal upper boundary and consists of blackish brown, laminated clay with oxidation stains. Layer 19 has an uneven horizontal upper boundary and consists of dark brown, structureless loam. It contains a flat, round stone of approximately 20 cm diameter. Layer 20 has a sharp horizontal upper boundary and consists of structureless, dark brown, sandy loam with clay clods. Layer 21 has an indistinct upper boundary and consists of silty sand. To the north-east (outside of the profile recorded in the sketch) there is a lens of sand up to 20 cm strong in the top of the layer. A radiocarbon date from a charcoal from layer 21 at a depth of 3.34 m gives an age of 490–310 cal BP (Poz-13542, 375 ±35 BP).

Profile RV records a succession of fluvial (layer 16, sand lens in top of layer 21) and irrigation (layers 3 through 5, 9 and 10, 15, 17 through 21) sediments as well as several phases of direct anthropogenic influence as seen as heaped-up material of low field-edge banks (layers 1 and 2, 6 through 8, 11 through 14). The laminated silt layers 5 and 9 are interpreted as low-energy deposits as would be expected e.g. from slow irrigation water sheet flow. Oxidation stains are present below 1.78 m and indicate the influence of a strongly fluctuating ground water level. The entire sequence was deposited since 490–310 cal BP; the average sedimentation rate was 6.13–10.8 mm/a.

Profile LU

Profile LU (Figure A-108) was recorded on a ca. 3 m high erosional bank exposure of the Rio Vizcas. The near-vertical bank was ca. 50 m wide, however, much of it was not accessible. Furthermore, the surface (in particular of layer 5) was strongly affected by earth wasps so that sedimentary features may have been missed.

Layer 0 consists of structureless silty sand with clods of cohesive silt. Layer 8 is a silty sand with a sharp horizontal upper boundary. Close to the upper boundary, faint traces of bedding are discernible. Layer 7 is a structureless sandy silt with some stones (up to 5 cm) and ceramic shards (not identifiable). The upper boundary is undulating and sharp. Layer 6 consists of silt with some small stones (up to 2 cm) as well as Late Intermediate Period ceramic shards (1,400–1,000 cal BP, cf. Unkel 2006: 111). It has a sharp horizontal upper boundary and shows traces of bedding near its top. Layer 5 is sandy silt with some stones (up to 4 cm). Layer 4 is a silty sand with an undulating upper boundary and contains stones as well as ceramic shards. It represents the filling of a shallow channel cut into layer 3. Layer 3 consists of cemented sand and fine gravel with faint bedding and has a sharp, undulating upper boundary. Layer 2 consists of cemented, structureless, well-sorted coarse sand with some silt; the upper boundary is diffuse and undulating. Layer 1 is structureless silty sand with a sharp horizontal upper boundary. It contains some rounded and angular clasts and some charcoal. Laterally, it grades into a charcoal-rich layer in the northern part of the exposure.

Profile LU represents both anthropogenic (layers 1 and 4 through 9) and fluvial (layers 2 and 3) influence. Besides Late Intermediate Period ceramic shards in layer 6 of the recorded profile, Nasca and Wari shards were found laterally in bank exposure at depths between 1.00 and 2.00 m. While a reworked or colluvial nature of some of the sediments (in particular the shallow channel fill of layer 4) can not be excluded, layers 5 through 8 are interpreted as irrigation sediment. The bedding present in the tops of layers 6 and 8 is interpreted as low energy / still water sedimentation related to irrigation because of the fine (silt, silty sand) grain size atypical for proximal flood deposits. Assuming primary deposition of the Late Intermediate Period shards, average sedimentation rate for layers 6 through 8 is 0.69–1.28 mm/a.

Fig. A-108: Photograph of profile LU (vertical scale ca. 1: 20).

Fig. A-109: Legend for profile sketches in Appendix A (drawing: R. Mendler).

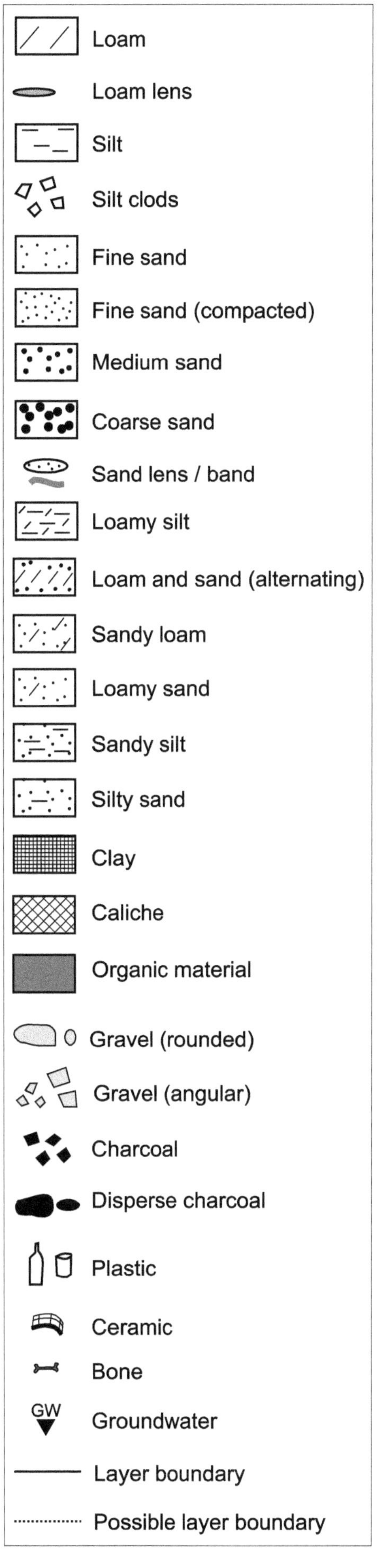

Appendix B – additional graphs and tables

Appendix B contains the full list of radiocarbon dates aquired in the course of this project (Table A-9) as well as sedimentation rates reconstructed from these calibrated radiocarbon dates and from diagnostic archaeological finds (Table A-10).

Magnetic susceptibility profiles for profiles LM-1, LM-2, LM-5 and LM-6 as well as SF-1, SF-2, SF-3 and SF-4 are shown in Figure A-110 and Figure A-111, respectively.

Tab. A-9: Radiocarbon dating results. ^{14}C measurements were performed by Poznan Radiocarbon Laboratory. Calibration was performed using OxCal Version 3.10 (Bronk Ramsey 1995, 2001) and southern hemisphere correcion shcal04.14c (McCormac et al. 2004) (continued on page A-132).

Sample-ID	^{14}C BP	Location	Depth below surface [cm]	cal BP (2σ)
Poz-5259	28 ±530	LM-1	70	450 – 150
Poz-5260	525 ±30	LM-1	160	545 – 495
Poz-5262	385 ±30	LM-1	200	490 – 320
Poz-5263	195 ±30	LM-2	68	≤300
Poz-5264	85 ±30	LM-4	200	≤260
Poz-5266	105 ±30	LM-3	98	≤260
Poz-5270	115 ±30	RPE (4.9 m)	150	≤260
Poz-5271	100 ±30	RPE (5.1 m)	45	≤260
Poz-8006	395 ±25	Jau-1	100	500 – 320
Poz-8007	1,545 ±25	Jau-1	320	1,490 – 1,300
Poz-8008	545 ±25	Jau-1	230	550 – 505
Poz-8010	925 ±25	LM-1	293	910 – 730
Poz-8011	520 ±30	LM-2	122	545 – 495
Poz-8012	450 ±25	LM-2	97	520 – 330
Poz-8013	335 ±25	LM-2	83	450 – 300
Poz-8014	255 ±25	LM-5	175 – 180	320 – 150
Poz-8015	205 ±25	LM-5	72	≤300
Poz-8016	210 ±25	LM-5/6	105	≤300
Poz-8020	830 ±30	YN-1	205	750 – 665
Poz-8021	145 ±25	RGB-1	145	≤270
Poz-8022	370 ±25	RGB-3	78	480 – 310
Poz-8023	905 ±30	RGB-3	155	910 – 680
Poz-8024	1,330 ±25	RG-FL	168	1,290 – 1,140
Poz-8026	1,705 ±30	RPW	258	1,690 – 1,410
Poz-8027	2,525 ±30	LLPT	330	2,720 – 2,360
Poz-8047	725 ±35	LM-2	200	680 – 560
Poz-8049	110 ±30	LM-6	112	≤260
Poz-13292	22,650 ±140	RGE	layer 5	
Poz-13293	2,525 ±30	BJ	275	2,720 – 2,360
Poz-13491	22,610 ±230	RGE	layer 4	
Poz-13540	2,050 ±40	RPF-2	150	2,070 – 1,820
Poz-13541	2,040 ±35	RGE	near surface	2,040 – 1,820
Poz-13542	375 ±35	RV	334	490 – 310
Poz-13544	355 ±30	BJ	142	470 – 300
Poz-13545	2,510 ±110	FV	165	2,800 – 2,150

Tab. A-9 (continued from page A-131)

Sample-ID	^{14}C BP	Location	Depth below surface [cm]	cal BP (2σ)
Poz-13546	380 ±30	DO	164	490 – 320
Poz-13547	1,125 ±35	FL	180	1,060 – 920
Poz-13548	5,880 ±50	BF-2	498	6,750 – 6,490
Poz-13606	22,060 ±120	RGE	layer 4	
Poz-13607	1,885 ±30	R-1	251	1,870 – 1,630
Poz-13608	1,780 ±30	R-2	170	1,720 – 1,540
Poz-13609	2,460 ±30	R-2	228	2,700 – 2,340
Poz-14006	1,595 ±30	R-1	210	1,530 – 1,350
Poz-15230	1,905 ±30	RG-BG	126	1,880 – 1,710
Poz-15231	1,765 ±30	FV	80	1,710 – 1,540
Poz-15232	1,690 ±30	BJ	215	1,610 – 1,410
Poz-15234	815 ±35	RG-Mo	90	745 – 655
Poz-15235	880 ±40	DO	239	900 – 670
Poz-15236	385 ±30	R-1	140	490 – 320
Poz-15238	415 ±30	FL	104	510 – 320
Poz-15239	1,860 ±30	LLPT	255	1,830 – 1,610
Poz-15240	120 ±30	LLPT	100	≤260
Poz-15287	135 ±30	BF-2	125	≤270
Poz-15288	95 ±30	BF-2	270	≤260
Poz-16814	1,730 ±35	Jau-Co	155	1,700 – 1,420
Poz-16815	255 ±35	SF-1	200	440 – 140
Poz-16816	2,890 ±35	RP-06	140	3,080 – 2,840
Poz-16817	5,320 ±40	RP-06	285	6,190 – 5,920
Poz-16818	2,980 ±35	SF-2	320	3,220 – 2,950
Poz-16820	1,165 ±30	SF-2	134	1,080 – 930
Poz-16821	5,410 ±40	SF-1	350	6,280 – 5,990
Poz-16822	1,510 ±40	SF-4	232	1,420 – 1,290
Poz-16824	360 ±30	SR	170	470 – 310
Poz-16825	2,150 ±35	R-3	247	2,300 – 1,950
Poz-16826	4,630 ±70	RVU (128 m)	170	5,500 – 4,950
Poz-16827	1,275 ±30	RVU (133 m)	70	1,260 – 1,060
Poz-16828	130 ±30	ES	158	≤270
Poz-16829	460 ±30	AR	137	530 – 330
Poz-16830	21,800 ±150	RGE	158	

Table A-10: Sedimentation rates for profiles onvestigated in this thesis. Calculations are based on radiocarbon dates listed in Table A-9 except for two ages based on diagnostic archaeological finds (*). (continued on page A-134)

Profile	Bottom date			Top date			Sedimentation rate		Facies
	depth [m]	age range [cal BP]	mean age [cal BP]	depth [m]	age range [cal BP]	mean age [cal BP]	minimum [mm/a]	mean [mm/a]	
LM-2	2.00	680 – 560	620	0.00	-55	-55	2.72	2.96	F3/F4
LM-3	0.98	260 – -55	103	0.00	-55	-55	3.11		F3/F4 + F1/F2
LM-4	2.00	260 – -55	103	0.00	-55	-55	6.35		F3/F4
LM-5	0.72	300 – -55	123	0.00	-55	-55	2.03		F1/F2 + F3/F4
LM-5	1.78	320 – 150	235	0.72	300 – -55	123	2.83		F3/F4
LM-6	1.12	260 – -55	103	0.00	-55	-55	3.56		F1/F2 + F3/F4
LM-5/6	1.05	300 – -55	123	0.00	-55	-55	2.96		F1/F2 + F3/F4
RGB-1	1.45	270 – -55	108	0.00	-55	-55	4.46		F3/F4 + F1/F2
RGB-3	0.78	480 – 310	395	0.00	-55	-55	1.46	1.73	F3/F4 + F1/F2
RGB-3	1.55	910 – 680	795	0.78	480 – 310	395	1.28	1.93	F3/F4
RG-Mo	0.90	745 – 655	700	0.00	-55	-55	1.13	1.19	F1/F2
RG-BG	1.25	1,880 – 1,710	1,795	0.00	-55	-55	0.65	0.68	F1/F2
RPF-2	0.90	1,620 – 1,520	1,570	0.00	-55	-55	0.54	0.55	F1/F2
RPF-2	1.50	2,070 – 1,820	1,945	0.90	1,620 – 1,520	1,570	1.09	1.60	F2
RPW	2.58	1,690 – 1,410	1,550	0.00	-55	-55	0.93	1.02	F1/F2
YN-1	2.05	750 – 665	708	0.00	-55	-55	2.55	2.69	F3/F4 + F1/F2
R-2	1.70	1,720 – 1,540*	1630	0.25	-49	-49	0.82	0.86	F2
R-2	2.28	2,700 – 2,340	2520	1.70	1,720 – 1,540*	1,630	0.50	0.65	F2
R-3	1.70	950 – 555	753	0.40	-49	-49	1.30	1.62	F2
R-3	2.47	2,300 – 1,950	2125	1.70	950 – 555	753	0.44	0.56	F2
ES	1.58	270 – -55	108	0.00	-55	-55	4.86		F3/F4 F1/F2
DO	1.64	490 – 320	405	0.00	-55	-55	3.01	3.57	F3/F4 + F1/F2
DO	2.39	900 – 670	785	1.64	490 – 320	405	1.29	1.97	F3/F4
Jau-1	1.00	500 – 320	410	0.00	-55	-55	1.80	2.15	F3/F4 + F1/F2
Jau-1	2.30	550 – 505	528	1.00	500 – 320	410	5.65		F3/F4
Jau-1	3.20	1,490 – 1,300	1,395	2.30	550 – 505	528	0.91	1.04	F3/F4 + F2
FL	1.04	510 – 320	415	0.00	-55	-55	1.84	2.21	F3/F4 + F1/F2
FL	1.80	1,060 – 920	990	1.04	510 – 320	415	1.03	1.32	F3/F4
RP-06	1.10	2,300 – 2,150*	2,225	0.00	-55	-55	0.47	0.48	F1/F2
RP-06	1.40	3,080 – 2,840	2,960	1.10	2,300 – 2,150*	2,225	0.32	0.41	F2
RP-06	2.85	6,190 – 5,920	6,055	1.40	3,080 – 2,840	2,960	0.43	0.47	F8 + F3/F4
SF-1	3.50	6,280 – 5,990	6,135	0.00	-55	-55	0.55	0.57	F1/F2
R-1	1.40	490 – 320	405	0.00	-55	-55	2.57	3.04	F1/F2
R-1	2.10	1,530 – 1,350	1,440	1.40	490 – 320	405	0.58	0.68	F2
R-1	2.51	1,870 – 1,630	1,750	2.10	1,530 – 1,350	1,440	0.79	1.32	F2
SF-2	1.34	1,080 – 930	1,005	0.20	-49	-49	1.01	1.08	F1/F2

Tab. A-10 (continued from page A-133).

Profile	Bottom date			Top date			Sedimentation rate		Facies
	depth [m]	age range [cal BP]	mean age [cal BP]	depth [m]	age range [cal BP]	mean age [cal BP]	minimum [mm/a]	mean [mm/a]	
SF-2	3.20	3,220 – 2,950	3,085	1.34	1,080 – 930	1,005	0.81	0.89	F2
SR	1.70	470 – 310	390	0.00	-55	-55	3.24	3.82	F3/F4 + F1/F2
LLPT	1.00	260 – -55	103	0.00	-55	-55	3.17		F1/F2?
LLPT	2.55	1,830 – 1,610	1,720	1.00	260 – -55	103	0.82	0.96	F2
LLPT	3.30	2,720 – 2,360	2,540	2.55	1,830 – 1,610	1,720	0.68	0.91	F2
BF-2	1.25	270 – -55	108	0.00	-55	-55	3.85		F1/F2
BF-2	2.70	260 – -55	102.5	0.00	-55	-55	8.57		F1/F2
BF-2	4.98	6,750 – 6,490	6,620	2.70	260 – -55	102.5	0.34	0.35	F3/F4 + F2
BJ	2.12	1,610 – 1,410	1,510	0.00	-55	-55	1.27	1.35	F1/F2
BJ	2.75	2,720 – 2,360	2,540	2.12	1,610 – 1,410	1,510	0.48	0.61	F2
FV	0.80	1,710 – 1,540	1,625	0.00	-55	-55	0.45	0.48	F1/F2
FV	1.65	2,800 – 2,150	2,475	0.80	1,710 – 1,540	1,625	0.67	1.00	F2
RV	3.34	490 – 310	400	0.00	-55	-55	6.13	7.34	F3/F4 + F1/F2
LU	1.00	1,400 – 1,000	1,200	0.00	-55	-55	0.69	0.80	F1/F2

Fig. A-109 (following page): Magnetic susceptibility graphs for profiles LM-1, LM-2, LM-5 and LM-6 (10^{-5} SI units; measured in the field using a ZH Instruments SM 30 Magnetic Susceptibility Meter; average values for three measurepents per data point). Vertical scale 1:30. Note the very high values in layer 18 of profile LM-2. Note that the lines connecting data points do not imply values between data points but serve solely to aid visualisation.

Fig. A-110 (page A-136): Magnetic susceptibility graphs for profiles SF-1, SF-2, SF-3 and SF-4 (10^{-5} SI units; measured in the field using a ZH Instruments SM 30 Magnetic Susceptibility Meter; average values for three measurepents per data point). Vertical scale 1:30. Note that the lines connecting data points do not imply values between data points but serve solely to aid visualisation.

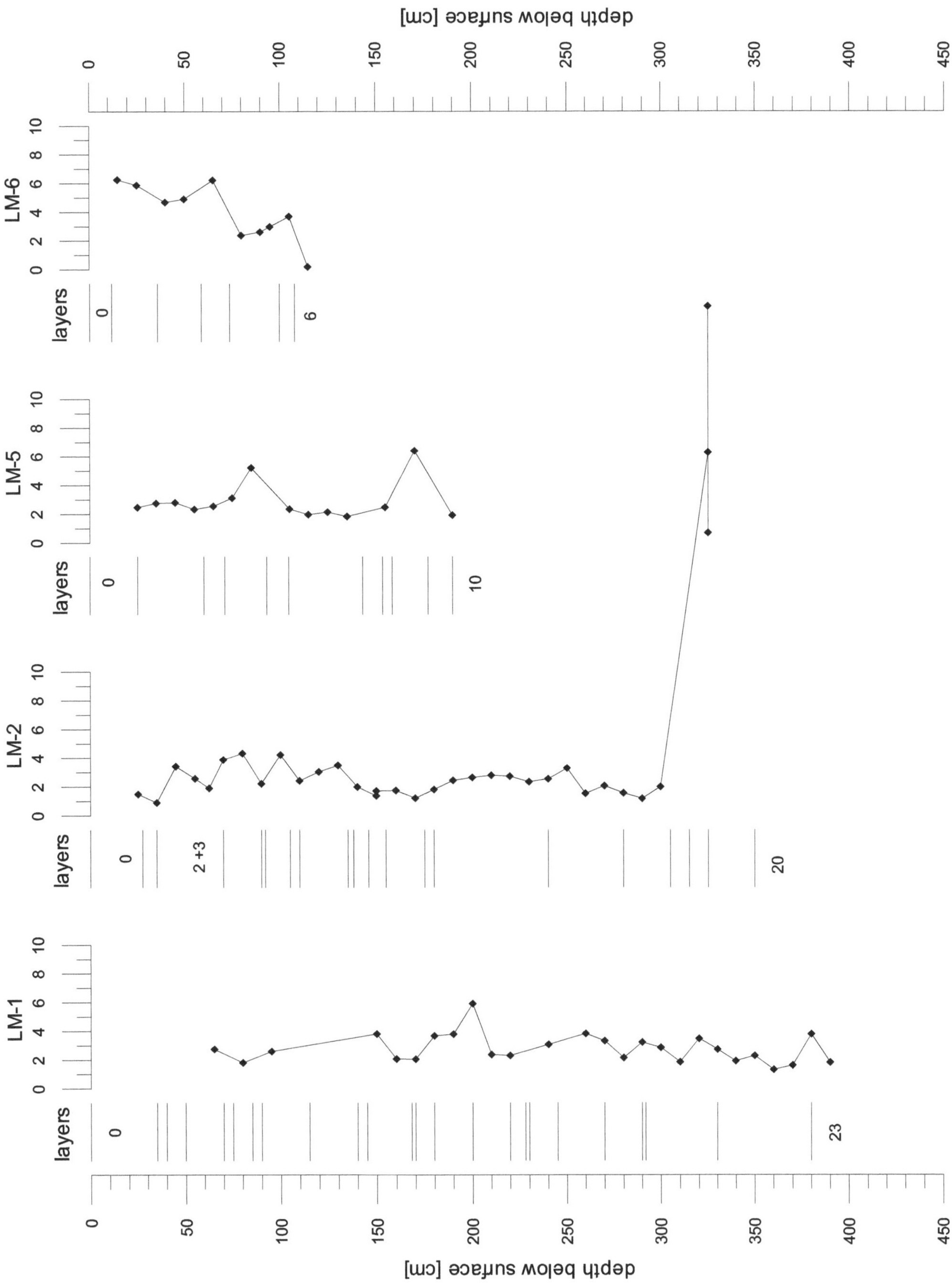
depth below surface [cm]
LM-1
LM-2
LM-5
LM-6
layers

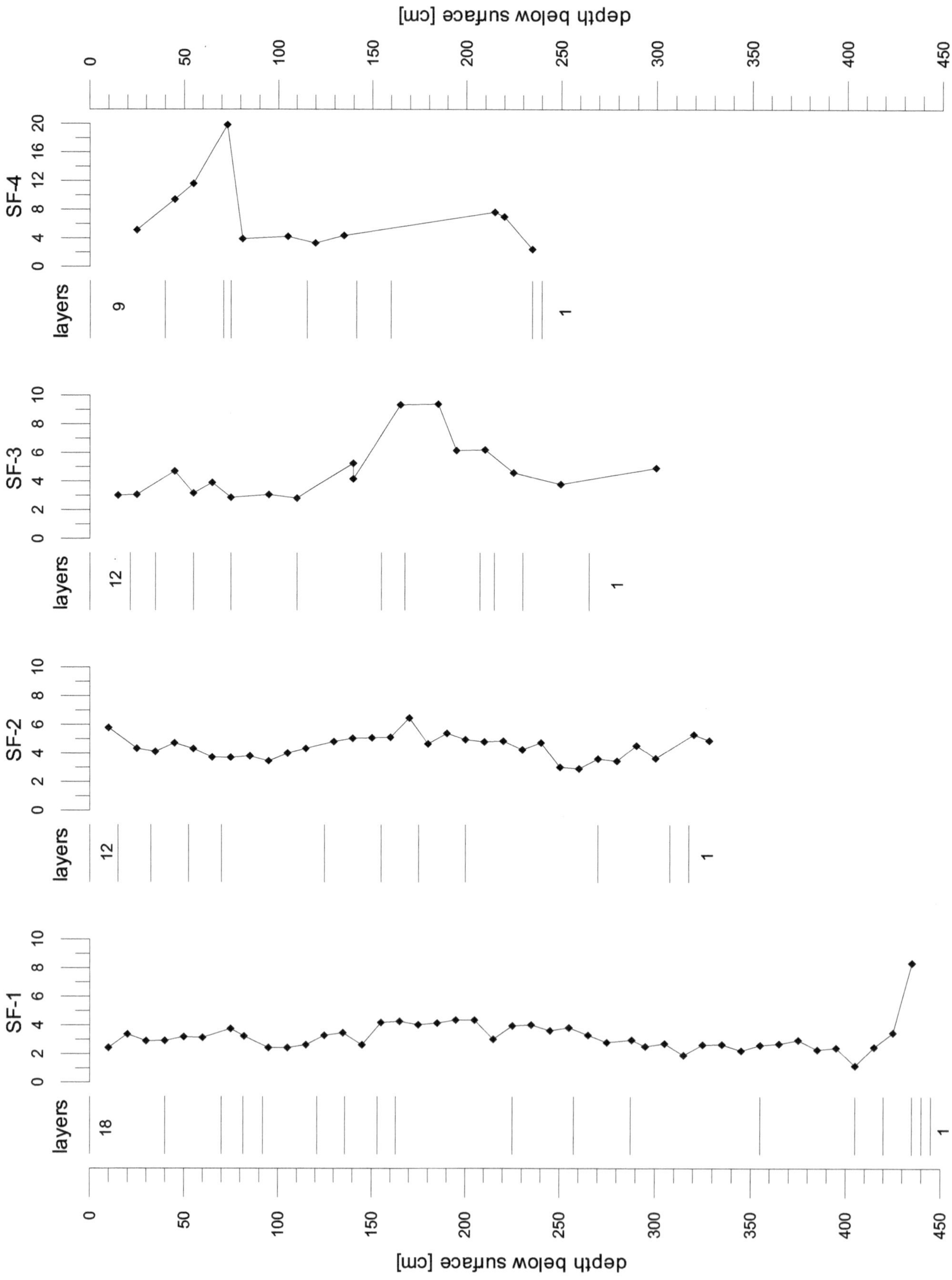
depth below surface [cm]
SF-1
SF-2
SF-3
SF-4
layers
18
12
12
9
1

www.ingramcontent.com/pod-product-compliance
Lightning Source LLC
LaVergne TN
LVHW072251300926
849160LV00018B/1643